Solubilization and Dispersion of Carbon Nanotubes

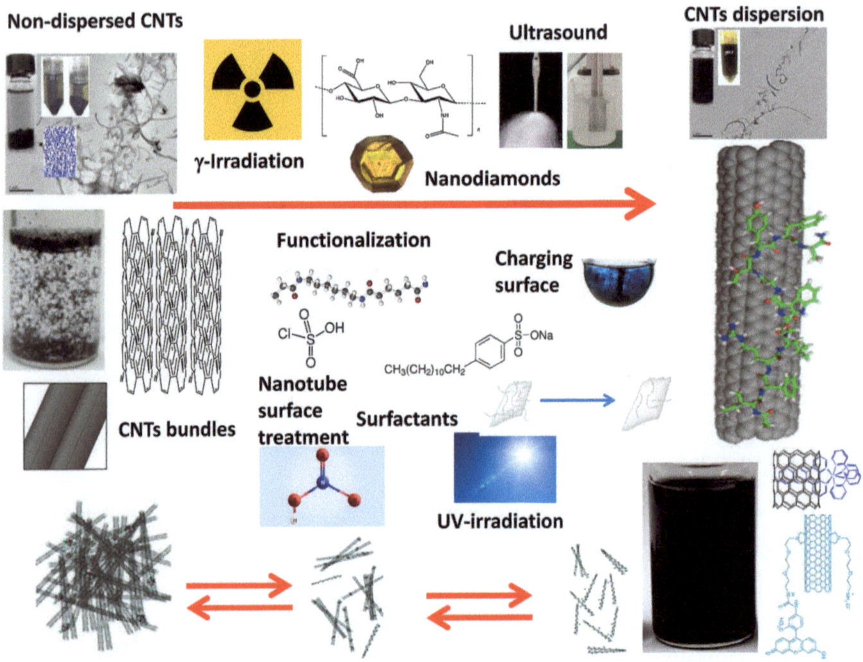

Non-dispersed CNTs
γ-Irradiation
Ultrasound
CNTs dispersion
Nanodiamonds
Functionalization
Charging
surface
CNTs bundles
Nanotube
surface
treatment
Surfactants
UV-irradiation

Oxana Vasilievna Kharissova
Boris Ildusovich Kharisov

Solubilization and Dispersion of Carbon Nanotubes

Springer

Oxana Vasilievna Kharissova
Universidad Autónoma de Nuevo León
Monterrey, Mexico

Boris Ildusovich Kharisov
Universidad Autónoma de Nuevo León
Monterrey, Mexico

ISBN 978-3-319-62949-0 ISBN 978-3-319-62950-6 (eBook)
DOI 10.1007/978-3-319-62950-6

Library of Congress Control Number: 2017950873

Printed on acid-free paper

This Springer imprint is published by Springer Nature
The registered company is Springer International Publishing AG
The registered company address is: Gewerbestrasse 11, 6330 Cham, Switzerland

Preface

Beginning with the first evidence of the tubular nature of nanosized carbon filaments, which dates as far back as 1952, and their further "rediscoveries", including their high-impact report in 1991, the field of carbon nanotubes (CNTs) continues to be one of the hottest topics in nanoscience and nanotechnology for already more than 25 years. Since their discovery, the development of their functionalization methods has led to the creation of a host of the carbon nanotube-functionalized derivatives and composites possessing interesting (although sometimes unusual) and useful properties. It is well-known that commercially fabricated single-wall and multi-wall carbon nanotubes (SWCNTs and MWCNTs, respectively) are practically insoluble in water and organic solvents; on the other hand, in order to be utilized in many distinct areas, from cement and polymer reinforcement to medical applications, CNTs should be well-dispersed in water or solvents without re-precipitation for extended periods of time. This is the underlying reason why their dispersion methods are being developed very intensively during the last decade years, applying a series of contemporary physical, chemical, and biological techniques, as well as their combinations.

The multidisciplinary study of carbon nanotubes is currently an enormous field of scientific and applied research. CNT consumption increases annually, novel CNT applications continually appear, and the demand for their production grows. Most CNT use requires well-suspended phases, so the importance of CNT dispersion methods in solvents is obvious, primarily in water, since water is most preferable among solvents relevant to potentially more existing and possible CNT applications (in particular for medical and biological purposes). In this monograph, we attempt to catalogue all known and available physical, chemical and biological methods and techniques for CNT dispersion in the liquid phase, describe their mechanisms of solubilization, and showcase important applications of dispersed CNTs.

Monterrey, Mexico

Oxana Vasilievna Kharissova
Boris Ildusovich Kharisov

Contents

Authors

Oxana V. Kharissova (born in 1969 in Ukraine, former USSR, has lived in Mexico from 1995, and naturalized in Mexico in 2004) is currently a professor and researcher at the *Universidad Autónoma de Nuevo León* (UANL). She obtained her M.Sc. in crystallography in 1994 from *Moscow State University*, Russia, and Ph.D. in materials from the *Universidad Autónoma de Nuevo León*, Mexico. She has memberships in the *National System of Researchers* (SNI, Level II), *Materials Research Society*, and *Mexican Academy of Sciences*. She is the co-author of five books, 12 book chapters, and 80 articles and holds eight patents. Her specialties include materials, nanotechnology (carbon nanotubes, graphene, nanostructurized metals, fullerenes), microwave irradiation and crystallography, and nanotechnology-based methods for petroleum treatment. Dr. Kharissova holds the awards *Flama, Vida y Mujer* (2017) and *Tecnos* (2004). She is an expert of the *National Council on Science and Technology* of Mexico (CONACYT). Dr. Kharissova has three children.

Boris I. Kharisov (born in 1964 in Russia, has lived in Mexico from 1994, and naturalized in Mexico in 2003) is currently a professor and researcher at the *Universidad Autónoma de Nuevo León* (UANL). He took part in the liquidation of the consequences of the Chernobyl accident, working in the contaminated zone in 1987. He earned his M.Sc. in radiochemistry in 1986 and Ph.D. in inorganic chemistry in 1993 from *Moscow State University*, Russia, and Dr. hab. in physical chemistry in 2006 from *Rostov State University*, Russia. His specialties include materials chemistry, coordination and inorganic chemistry, phthalocyanines, ultrasound, nanotechnology, chemical treatment of petroleum, and environmental remediation. He has memberships in the *Mexican Academy of Sciences* and *National System of Researchers* (SNI, Level II). He is the co-author of nine books, 157 articles, and eleven book chapters and holds eight patents. He is a co-editor of three invited special issues of international journals. He is a member of the editorial board of four journals. Dr. Kharisov has three children. His biography was published in *Who's Who in the World*, *Outstanding People of the Twentieth Century*, and so on.

Chapter 1
Introduction

The carbon nanotubes (CNTs), one of the best novel nanostructures [1] and classic objects in nanotechnology, form bundle-like structures with very complex morphologies with a high number of Van der Waals interactions, causing extremely poor solubility in water or organic solvents. Due to their exceptional combination of mechanical, thermal, chemical, and electronic properties, single-walled (SWNTs or SWCNTs) and multiwalled carbon nanotubes (MWNTs or MWCNTs) are considered as unique materials, with very promising future applications, especially in the field of nanotechnology, nanoelectronics, and composite materials. Additionally, CNTs are becoming highly attractive molecules for applications in medicinal chemistry. At present, potential biological and medical applications [2] of CNTs have been little explored, in particular for drug delivery purposes [3]. The main difficulty to integrate such materials into biological systems derives from their lack of solubility in physiological solutions. Functionalization of CNTs with the assistance of biological molecules remarkably improves the solubility of nanotubes in aqueous or organic environment and, thus, facilitates the development of novel biotechnology, biomedicine, and bioengineering. Many of these applications require an increased "solubility" of CNTs in solvents, first of all in water, especially for biological applications. This could be reached by their functionalization, which is a very actively discussed topic in contemporary literature because the planned modification of CNT properties is believed to open the road toward real nanotechnology applications [4]. It is difficult to prepare an aqueous dispersion of CNTs stable for months; their insolubility has been a limitation for the practical applications of this unique material. Proper dispersion of CNT materials is important to retaining the electronic properties of nanotubes. The redissoluble functional compound/CNT composites are needed for post-processing because CNT dispersions usually easily aggregate and therefore make additional processing very difficult.

"True water-soluble nanotube solutions" are those solutions that entropically favor individualized nanotubes [5], where the reaggregation of CNTs in a solvent is less favored, on a thermodynamic basis, than their continued solvated state [6]. In some embodiments, the extent of functionalization is dependent upon a number of

O.V. Kharissova, B.I. Kharisov, *Solubilization and Dispersion of Carbon Nanotubes*, DOI 10.1007/978-3-319-62950-6_1

factors, e.g., the reactivity of the CNTs, the reactivity of the functionalizing agent, steric factors, etc. In some such embodiments, as a result of such dependencies, the extent of functionalization can be in the range of from at least about 1 functional group per every 1000 CNT carbons to at most about 1 functional group per every 2 CNT carbons. Analyzing the published scientific research articles, it is evident that there is an important issue or misunderstanding between the terms "dispersion" and "solubilization" [7]. Many researchers use the terms interchangeably, particularly when stating the interaction of CNTs with liquids, which causes confusions. The fundamental question when dealing with carbon nanotubes in liquids, particularly in water, is: *are they dissolved or dispersed*? It was suggested [8] using the term "dispersion" rather than "solution." Also, one of the most problematic issues affecting the CNT research community is the lack of standards and uniform characterization methods for CNT dispersion. A 2005 NASA-NIST workshop aimed to address this issue and made a clear distinction between "nanodispersion" of individual CNTs and "macrodispersion" of CNT bundles [9]. Unfortunately, this distinction has yet to percolate through the CNT dispersion literature.

The history of CNT solubilization is not long. One of the first fundamental reviews [10] in the field of CNT dissolution and, more recently, a book [11], several chapters [12–16], and reviews [17–27] have been published, which are devoted to particular aspects of CNT dispersion and stability in liquid media. In particular, some reviews were dedicated to the strategic approaches toward the solubilization of CNTs using chemical and physical modifications [28–30]; environmental, toxicological, and pharmacological studies related with use of CNTs [31, 32]; the main methods for the modification of CNTs with polymers [33, 34]; applications of functionalized CNTs [35, 36], in particular as biosensors [37]; and discussions on the possibility of the existence of SWCNTs in organic solvents in the form of clusters [38]. In addition, certain attention is paid to CNT nanofluids [39–43].

In this book, we try to discuss all available methods leading to CNT dispersion in water and organic solvents and present a state of the art for this very important problem of nanoscience and nanotechnology. The book does not deal with CNTs in polymers (except application of polymers for CNT solubilization in solvents) and other solid materials, although the term "CNT dispersion" is frequently used for those processes.

References

1. A.A. Rashad, R. Noaman, S.A. Mohammed, E. Yousif, Synthesis of carbon nanotube: A review. J. Nanosci. Technol **2**(3), 155–162 (2016)
2. N. Bhandare, A. Narayana, Applications of nanotechnology in cancer: A literature review of imaging and treatment. Nuclear medicine & radiation therapy. J. Nucl. Med. Radiat. Ther **5**, 4 (2014.) 9 pp
3. A.C. Tripathi, S.A. Saraf, S.K. Saraf, Carbon nanotropes: A contemporary paradigm in drug delivery. Materials **8**, 3068–3100 (2015)

4. H. Kuzmany, A. Kukovecz, F. Simon, M. Holzweber, C. Kramberger, T. Pichler, Functionalization of carbon nanotubes. Synth. Met. **141**(1), 113–122 (2004)
5. F. Liang, E.W. Billups. Water-soluble single-wall carbon nanotubes as a platform technology for biomedical applications. US20070110658, 2007
6. J.M. Tour, J.L. Hudson, C. Dyke, J.J. Stephenson Functionalization of carbon nanotubes in acidic media. WO05113434, 2005
7. T. Premkumar, R. Mezzenga, K.E. Geckeler, Carbon nanotubes in the liquid phase: Addressing the issue of dispersion. Small **8**(9), 1299–1313 (2012)
8. K.E. Geckeler, T. Premkumar, Carbon nanotubes: Are they dispersed or dissolved in liquids? Nanoscale Res. Lett. **6**(1), X1–X3 (2011)
9. M.J. Green, Analysis and measurement of carbon nanotube dispersions: Nanodispersion versus macrodispersion. Polym. Int. **59**(10), 1319–1322 (2010)
10. J. Hilding, E.A. Grulke, Z.G. Zhang, F. Lockwood, Dispersion of carbon nanotubes in liquids. J. Dispers. Sci. Technol. **24**(1), 1–41 (2003)
11. C. Backes. Noncovalent Functionalization of Carbon Nanotubes: Fundamental Aspects of Dispersion and Separation in Water. Springer Theses, (2016) pp. 220
12. M. Wiesner, J.-Y. Bottero, *Environmental Nanotechnology* (McGraw-Hill Professional, Blacklick, 2007), p. 540
13. K. Gonsalves, C. Halberstadt, C.T. Laurencin, L. Nair, *Biomedical Nanostructures* (Wiley, New York, 2007), p. 507
14. S.-K. Choi, *Synthetic Multivalent Molecules: Concepts and Biomedical Applications* (Wiley-Interscience, Hoboken, 2004), p. 418
15. S. Reich, C. Thomsen, J. Maultzsch, *Carbon Nanotubes: Basic Concepts and Physical Properties* (Wiley-VCH, Weinheim, 2004), p. 224
16. A. Jorio, G. Dresselhaus, M.S. Dresselhaus, *Carbon Nanotubes: Advanced Topics in the Synthesis, Structure, Properties and Applications* (Springer, Berlin/Heidelberg, 2008), p. 720
17. Y. Maeda, M. Yamada, T. Hasegawa, T. Akasaka, J. Lu, S. Nagase, Interaction of single-walled carbon nanotubes with amine. Nano **7**(1), art. no. 1130001 (2012)
18. Y.Y. Huang, E.M. Terentjev, Dispersion of carbon nanotubes: Mixing, sonication, stabilization, and composite properties. Polymer **4**, 275–295 (2012)
19. J. Labille, J. Brant, Stability of nanoparticles in water. Nanomedicine **5**(6), 985–998 (2010)
20. A. Di Crescenzo, V. Ettorre, A. Fontana, Non-covalent and reversible functionalization of carbon nanotubes. Beilstein J. Nanotechnol. **5**, 1675–1690 (2014)
21. X. Xin, G. Xu, H. Li, Dispersion and property manipulation of carbon nanotubes by self-assemibles of amphiphilic molecules, in *Physical and Chemical Properties of Carbon Nanotubes*, (INTECH, Rijeka, Croatia, 2013), pp. 255–273
22. M. Sanchez-Dominguez, C. Rodriguez-Abreu (eds.), *Nanocolloids: A Meeting Point for Scientists and Technologists*, 1st edn. (Elsevier, Amsterdam, The Netherlands, 2016), p. 536
23. G. Babatunde Olowojoba, P. Fraunhofer, *Assessment of Dispersion Evolution of Carbon Nanotubes in Shear-Mixed Epoxy Suspensions by Interfacial Polarization Measurement* (Fraunhofer Verlag, Stuttgart, 2013), p. 128
24. S. Won Kim et al., Surface modifications for the effective dispersion of carbon nanotubes in solvents and polymers. Carbon **50**, 3–33 (2012)
25. H. Li, Q. Li. Selective separation of single-walled carbon nanotubes in solution hongbo. in *Electronic Properties of Carbon Nanotubes*, ed. by J. M. Marulanda, (INTECH, Rijeka, 2011), pp. 69–91, ISBN: 978-953-307-499-3, Available from: http://www.intechopen.com/books/electronic-properties-of-carbon-nanotubes/selectiveseparation-of-single-walled-carbon-nanotubes-in-solution
26. S. Prakash Yadav, S. Singh, Carbon nanotube dispersion in nematic liquid crystals: An overview. Prog. Mater. Sci. **80**, 38–76 (2016)
27. J. Njuguna, O. Arda Vanli, R. Liang, A review of spectral methods for dispersion characterization of carbon nanotubes in aqueous suspensions. J. Spectrosc. **2015**, 11 (2015.) Article ID 463156

28. M. Hiroto, N. Naotoshi, Soluble carbon nanotubes and their applications. J. Nanosci. Nanotechnol. **6**(1), 16–27 (2006)
29. D. Tasis, N. Tagmatarchis, V. Georgakilas, M. Prato, Soluble carbon nanotubes. Chemistry **9** (17), 4000–4008 (2003)
30. N. Nakashima, T. Fujigaya, Fundamentals and applications of soluble carbon nanotubes. Chem. Lett. **36**(6), 692 (2007)
31. L. Lacerda, A. Bianco, M. Prato, K. Kostarelos, Carbon nanotubes as nanomedicines: From toxicology to pharmacology. Adv. Drug Deliv. Rev. **58**(14), 1460–1470 (2006)
32. A. Helland, P. Wick, A. Koehler, K. Schmid, C. Som, Reviewing the environmental and human health knowledge base of carbon nanotubes. Environ. Health Perspect. **115**(8), 1125–1131 (2007)
33. P. Liu, Modifications of carbon nanotubes with polymers. Eur. Polym. J. **41**(11), 2693–2703 (2005)
34. R. Atif, F. Inam, Reasons and remedies for the agglomeration of multilayered graphene and carbon nanotubes in polymers. Beilstein J. Nanotechnol. **7**, 1174–1196 (2016)
35. N. Nakashima, Soluble carbon nanotubes. Int. J. Nanosci. **4**, 119–137 (2005)
36. H. Murakami, N. Nakashima, Soluble carbon nanotubes and their applications. J. Nanosci. Nanotechnol. **6**, 16–27 (2006)
37. Y. Yun, Z. Dong, V. Shanov, W.R. Heineman, H.B. Halsall, A. Bhattacharya, L. Conforti, M.J. Schulz, Nanotube electrodes and biosensors. Nano Today **2**(6), 30–37 (2007)
38. F. Torrens, G. Castellano, Effect of packing on the cluster nature of C nanotubes: An information entropy analysis. Microelectron. J. **38**(12), 1109–1122 (2007)
39. M. Jama, T. Singh, S.M. Gamaleldin, M. Koc, A. Samara, R.J. Isaifan, M.A. Atieh, Critical review on nanofluids: Preparation, characterization, and applications. J. Nanomater. **2016**, 22 (2016.) Article ID 6717624
40. M.S. Patil, J.-H. Seo, S.-K. Kang, M.-Y. Lee, Review on synthesis, thermo-physical property, and heat transfer mechanism of nanofluids. Energies **9**, 840 (2016.) 17 pp
41. C. Kleinstreuer, Z. Xu, Mathematical modeling and computer simulations of nanofluid flow with applications to cooling and lubrication. Fluids **1**, 16 (2016.) 33 pp
42. S.S.J. Aravinda, S. Ramaprabhu, Graphene–multiwalled carbon nanotube-based nanofluids for improved heat dissipation. RSC Adv. **3**, 4199–4206 (2013)
43. S. Delfani, M. Karami, M.A.A. Akhavan Bahabadi, Experimental investigation on performance comparison of nanofluidbased direct absorption and flat plate solar collectors. Int. J. Nano Dimens. **7**(1), 85–96 (2016)

Chapter 2
Physical Methods

2.1 Ultrasonication

Ultrasonication is often used to disperse nanoparticles in aqueous solution. However, a good dispersion of nanoparticles in aqueous solution is not always achieved, due to the fact that incoming ultrasonic waves in liquid are usually reflected and damped at the gas/liquid interface. For the case of carbon nanotubes, this is a classic debundling method: the MWCNTs can be effectively ultrasonically dispersed in the water solution [1]. Under the ultrasound action, the *cavitation* process produces the strong shear force, leading to the exfoliation of carbon nanotube bundles, bubble formation, and collapse, providing homogeneity of nanosuspension [2].

The *role of sonication energy* on SWCNT and MWCNT (with distinct lengths) dispersions in the presence of sodium dodecylbenzene sulfonate (SDBS, see in sections below) was evaluated [3]. It was revealed, in particular, that the concentration of dispersed carbon nanotubes in the SDBS solution depends on the *sonication energy*, but *not* output power of the sonicator alone or the sonication time. More quantity of CNTs is dissolved upon addition of more SDBS; smaller or shorter nanotubes lead to less nanotube dispersion. It was also established that the optimal energy depends on CNT diameter and it was independent on amounts of CNTs or SDBS, surface functional groups, and CNT length. In another report [4], the authors distinguish between *stable cavitation*, leading to chemical modification of the CNT surface, and *inertial cavitation*, favoring CNT length reduction and exfoliation. The second type of cavitation (which is depending on surfactant concentration) was found to be responsible for effective CNT dispersion. However, the CNTs can be damaged upon ultrasonication; [5] this fact should be taken into account.

A considerable number of reports described in chapters below, dedicated to the use of surfactants and functionalizing agents for CNT dispersion/solubilization, imply simultaneous application of ultrasound of *distinct power*, weak (Fig. 2.1) or elevated (Fig. 2.2). A union of both ultrasonic sources is also known (Fig. 2.3).

O.V. Kharissova, B.I. Kharisov, *Solubilization and Dispersion of Carbon Nanotubes*, DOI 10.1007/978-3-319-62950-6_2

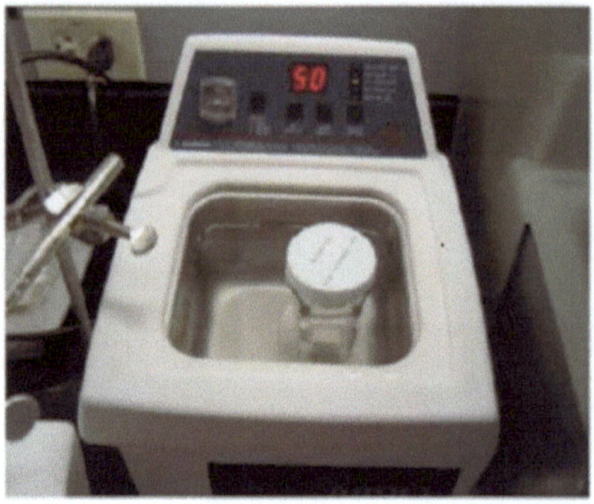

Fig. 2.1 Water bath sonicator

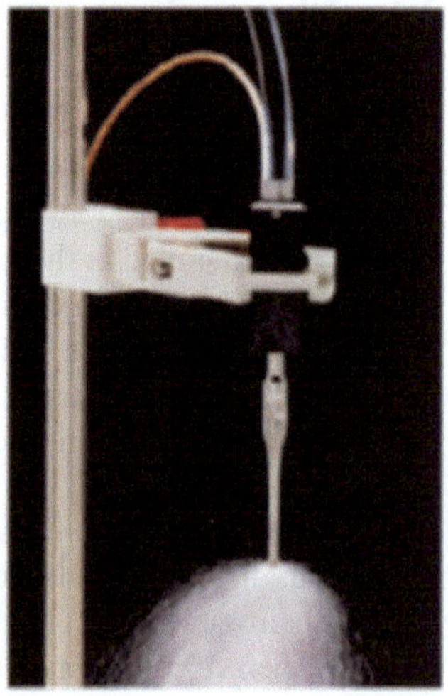

Fig. 2.2 Probe/horn sonicator

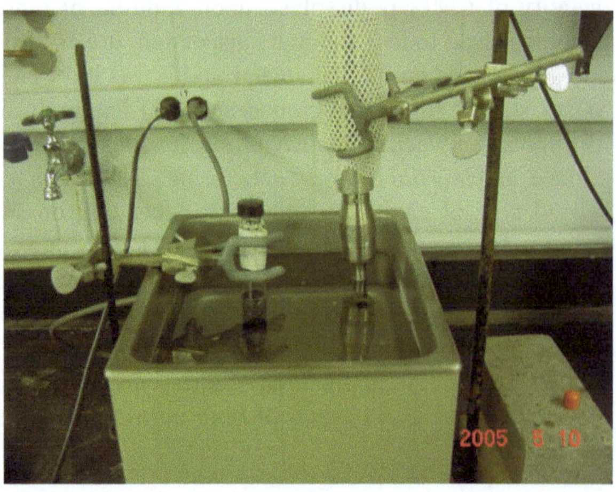

Fig. 2.3 Typical double sonication system used in dispersing SWCNTs. The system consists of an ultrasonic bath and an ultrasonic probe on the right [8] (Reproduced with permission from *Scientific Research*)

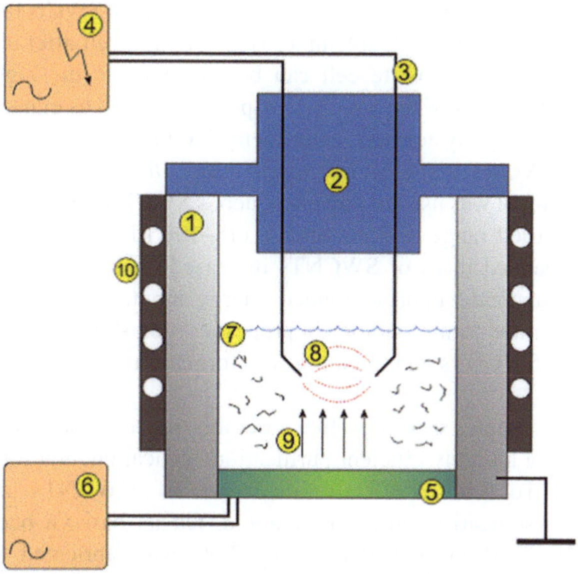

Fig. 2.4 Diagram of chamber for efficient CNT dispersion, (*1*) the chamber made of stainless steel, (*2*) the cover made of polytetrafluoroethylene, (*3*) electrodes, (*4*) the power supply, (*5*) the ultrasonic/megasonic transducer, (*6*) the control/driver unit for transducer, (*7*) the fluid with carbon nanotube, (*8*) the electric field, (*9*) the acoustic wave from transducer, (*10*) the heater/heat sink (Reproduced with permission from Proceedings of the *34th International Microelectronics and Packaging IMAPSCPMT Poland Conference*, 2010)

Depending on work purpose, the ultrasonic dispersion of carbon nanotubes can be used *in combination* with other techniques, for example, spraying. Thus, thin films of CNTs were obtained by sonication-assisted suspension in organic solvents (N-methyl-2-pyrrolidone or N-cyclohexyl-2-pyrrolidone), spraying onto a heated substrate using an ultrasonic spray nozzle, and fast solvent evaporation, leading to a thin film of randomly oriented nanotubes [6]. Another combination, designed to improve dispersion of CNTs in solvents, represented a method uniting dielectrophoresis and megasonification [7]. This full system is shown in Fig. 2.4. Combined alternating electric field together with mechanical wave from ultrasonic transducer led to faster and better dispersion of threads formulated from CNTs in the presence of strong electric field. Both electric field and ultrasonic transducer had controllable frequency and power. Both dispersion and agglomeration of CNTs can be observed, while the parameters of electric field were changed. Additionally, different fluids required different frequencies for optimal dispersion ratio.

Sonicators with different modes for CNT dispersion:

Several studies to *increase CNT dispersibility* without application of surfactants have been carried out, sometimes leading to useful CNT applications. Thus, a more efficient cooling cell for probe-type ultrasonication was reported [9]. As compared with a conventional cylindrical cell, the concentration of the SWCNTs solubilized in water was found to be almost double in a rosette cooling cell after ultracentrifugation. The efficiency of a rosette cell can be attributed to the higher efficiency in circulation and cooling of the SWCNT dispersion as well as enhancement of the cavitation process. A systematic calorimetry-based technique was developed to standardize SWCNT dispersion protocol using *ultra-dismembrator* [10]. The standardized protocol was used to prepare aqueous SWCNT suspensions sonicating SWCNTs in a varied range of input energy (18–100 kJ) in water. It was demonstrated that suspended mass of SWCNTs increased up to 18 kJ of energy input with no further increase upon continued energy input. The significant changes observed were limited to morphological properties, i.e., shorter length, debundled, and sharp-edged SWCNTs and fractal cluster formation (lower Df) with increased input energy.

In addition, a simple, surfactant-free assembly process was used to MWCNT buckypapers using a highly efficient purification, sonication, and filtration process (Fig. 2.5) [11]. To achieve effective dispersion of MWCNTs into ethanol, a minimum 5-min sonication time was required. On its basis, a buckypaper with pore size of 41 ± 10 nm and porosity of 72.9% was fabricated with a 10-min sonication. The as-prepared buckypaper was used as a membrane for humic acid (HA) removal from water. The buckypaper prepared from purified CNTs exhibited excellent removal of HA (>93%) and a long lifetime for filtration.

The *modification of the SWCNT surface* in dispersions upon distinct sonication parameters was monitored by several techniques, in particular by Raman spectroscopy [12]. It was shown that standard conditions in water (1–6-h sonication, power 3–80 W) do not result in covalent modifications. In case of addition of

Fig. 2.5 CNT dispersions in ethanol after (**a**) 1 min, (**b**) 3 min, and (**c**) 5 min sonication. A typical buckypaper produced from 50 mg of CNTs and a 10-min sonication time (**d**), and SEM images of the cross-section (**e**) and the surface (**f**) of the corresponding buckypaper. A typical TEM image of the purified CNTs (*inset* of **f**) (Reproduced with permission from *Elsevier Science*)

single-stranded DNA (ssDNA) to this dispersion, it can be noncovalently absorbed on SWCNT surface without covalent linkage. Less than 10% of defect sites can appear upon higher sonication power or longer sonications. Studying the length and diameter of SWCNTs in sonication conditions (in the presence of 1 w/v.% sodium deoxycholate) by AFM [13], the authors considered the sonication power to be most important for SWCNT dispersion. A shortening of SWCNT length was observed at long sonication time. Also, the ultracentrifugation was shown to be useful to eliminate large-diameter bundles from SWCNT dispersion to get a homogeneous system.

The monitoring of SWCNT surface was also carried out by SEM and UV-vis-NIR spectroscopy [14], resulting that the maximum concentration of dispersed SWCNTs corresponds to the maximum UV-vis-NIR absorbance of the solution. To disperse higher SWCNTs concentrations, longer time is required for sonication. Optimal conditions are as follows: 1.5 wt.% of surfactant and 0.4 mg/mL of SWCNTs dispersed. In a related report [15], these data are slightly different: 9–10 mg/mL for SDBS and 8–9 mg/mL for sodium deoxycholate lead to 0.25 mg/mL of SWCNT dispersion (0.7–2.5 nm of SWCNT diameter). The SEM analysis also showed the presence of surfactant particles on the SWCNT surface. In case of organic solvents, 4 h of treatment are sufficient to get best dispersions in 1,2-DCE or DMF.

Concluding, the ultrasonic treatment is often used for dispersion of carbon nanotubes [16, 17] (as well as other nanomaterials [18]) alone or, more frequently, as an additional assistance to chemical methods described in chapters below, and resulting CNT nanocomposites can be applied, in particular, for modification of glassy carbon electrodes [19], cement reinforcement [20–24], polymer [25, 26], and protein [27] composites or for creation of pharmaceuticals [28] and for other medical applications [29], among many others. Efficiency of ultrasonic dispersion can depend on many factors, in particular on the surfactants [30] or catalysts [31], applied for CNTs synthesis, and, therefore, properties of thus formed CNTs.

2.2 Plasma Treatment of CNTs

Distinct types of plasma techniques have been applied to improve CNT dispersibility, frequently without surfactants, giving different quantitative results. Thus, carbon nanotube microcapsules were prepared [32] by oil in water (O/W) Pickering emulsions (Figs. 2.6 and 2.7) without any surfactant used. CNTs were treated with oxygen plasma (radio frequency of 13.56 MHz) at a power of 100 W and a pressure of 200 mTorr for several different periods. The oxygen plasma treatment introduced several hydrophilic groups on carbon nanotubes resulting in the improved aqueous dispersion. The plasma-treated carbon nanotubes were self-assembled at the interface between water and oil phases. In similar conditions (Fig. 2.8), being pretreated in pure ethanol by a supersonic homogenizer and then plasma treated, the MWCNTs in the presence of citric acid resulted in many

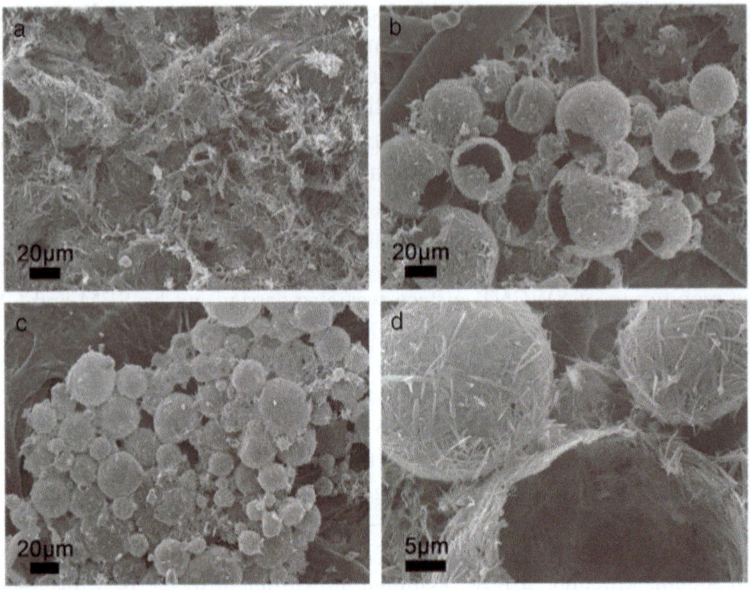

Fig. 2.6 SEM images of CNT microcapsules prepared with various different contents of CNTs in O/W Pickering emulsion (water, 10 ml; cyclohexane, 3 ml): (**a**) 0.0025 g, (**b**) 0.005 g, (**c**) 0.01 g, and (**d**) magnified image of sample (**b**) (Reproduced with permission from *Elsevier Science*)

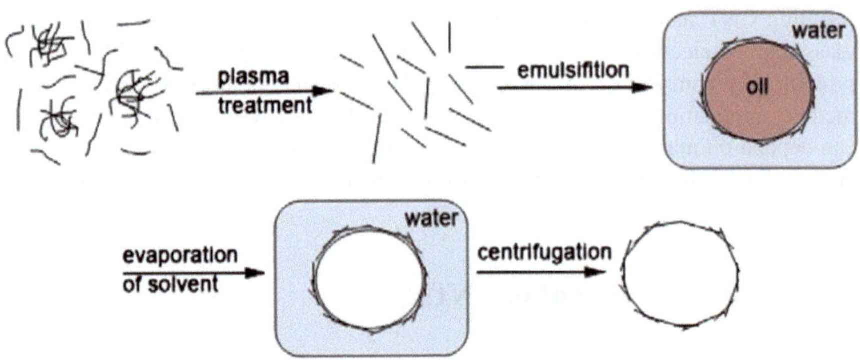

Fig. 2.7 Schematic illustration of the preparation of CNT hollow microspheres by O/W Pickering emulsions (Reproduced with permission from *Elsevier Science*)

-COOH functional groups, attached onto the MWCNT surfaces leading to a stable aqueous dispersion [33].

CNTs synthesized by direct thermal CVD growth from stainless steel mesh were surface functionalized via an $Ar/O_2/C_2H_6$ capacitively coupled RF plasma discharge [34]. The open and rigid network of CNTs allowed for a high degree of the CNT surface to be treated. As a result, the aqueous nanofluids that were produced by removing the CNTs from the substrate were found to remain stable

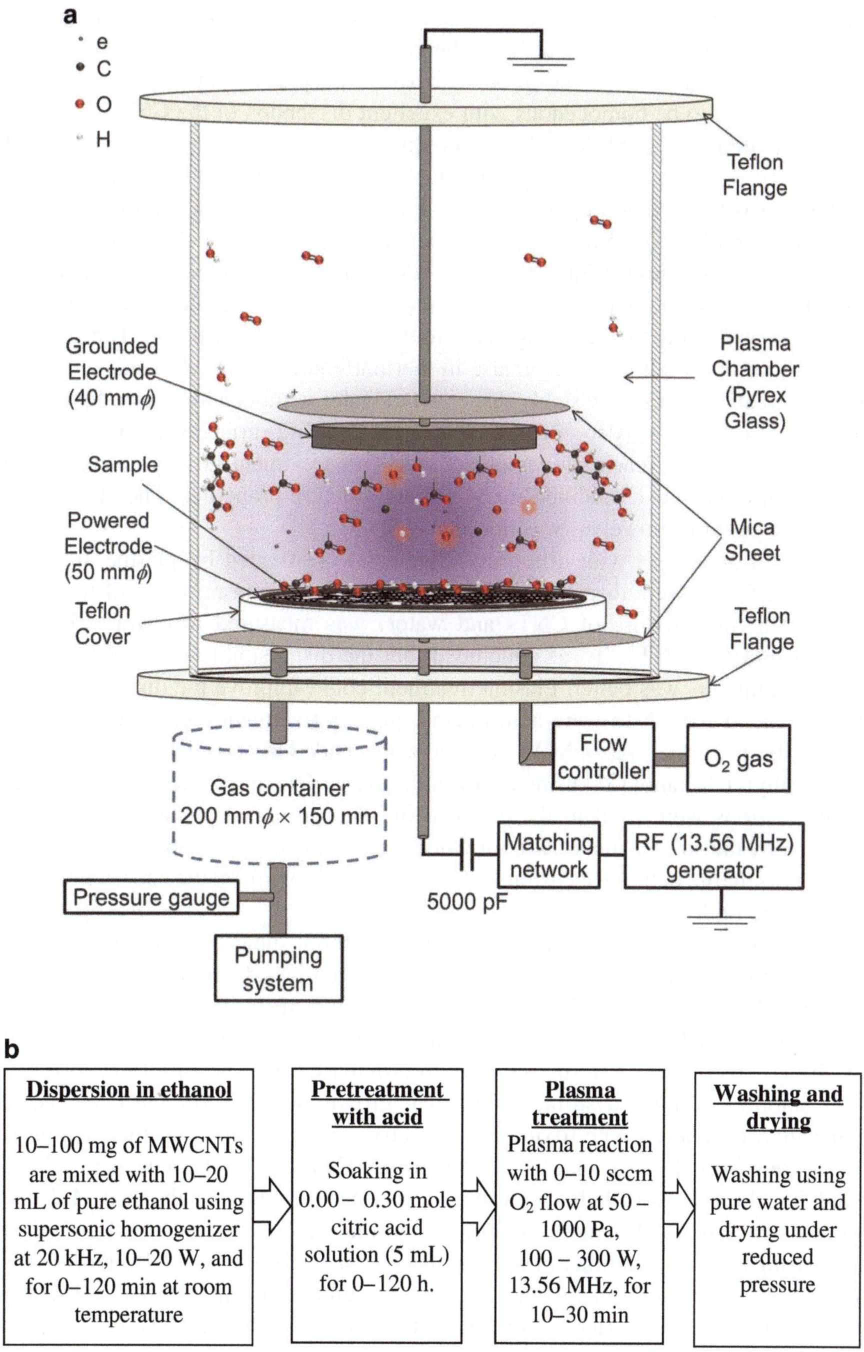

Fig. 2.8 (**a**) Schematic diagram of the plasma reactor. (**b**) Flow chart of the functionalization process (Reproduced with permission from *Hindawi Publishing Corporation*)

for extended periods of time (greater than 6 months). It was revealed that oxygen comprised approximately 21 at.% of the functionalized CNT surface. Poly(vinyl alcohol)/CNT nanocomposites produced using the plasma functionalized CNTs were observed to be homogenous with excellent dispersion of the CNTs.

In a related work [35], MWCNTs were plasma treated using glow discharges of argon, oxygen, and methane/oxygen mixtures and then dispersed into a base fluid of water. It was found that proper plasma treatments of CNTs using nanoscale plasma coatings significantly improved the dispersion and stabilized the suspension of CNTs in the base fluid. With 0.01 vol.% addition of plasma-treated CNTs, a 25% initial increase in thermal conductivity was achieved, and a stabilized 20% increase was observed with the resulting nanofluids after 5 days' settling. The authors pointed out that such a large increase in thermal conductivity was achieved when plasma-treated CNTs were stably dispersed in water without adding any dispersing agents or surfactants. Also, microwave excited Ar/H_2O surface-wave plasma was used [36] trying to reach similar objectives. This treatment greatly enhanced the content of oxygen and modified surface microstructure properties. The integrity of nanotube patterns, however, was not damaged.

In addition, CNTs with different diameters were treated by plasma and acid [37]. The CNTs were dispersed in water, and the light transmittance of the CNT dispersoid (the mixture of CNTs and water) was measured to characterize the dispersibility of CNTs. It was concluded that the dispersibility of the CNTs with bigger diameters was better. Plasma treatment could improve the dispersibility of CNTs, and if the CNTs were acid treated, plasma treatment could improve their dispersibility further. Also, MWCNTs were suspended in octylalcoholes 2-octanol and 2-ethyl-1-hexanol [38] using a barrier discharge. The needle-to-plane electrode configurations were used in the experiment. The grounded plane electrode was covered by the solid insulator with or without an isolated metallic plate. Successive barrier discharge in the suspension was obtained by applying to the needle electrode a 60 Hz ac high voltage.

In a related report [39], plasma-induced (method: dielectric barrier discharge plasma in water vapor-saturated air at 70 °C) oxygen-based functional groups were confirmed on the MWCNT surfaces and improved their dispersion in aqueous solutions. Indeed, oxygen plasma treatment leads to the appearance of -CO and -COO groups on the CNT surface [40], improving CNT dispersibility in solvents. The stability of formed surface functional groups on CNT surface depends, in particular, on plasma type (oxidative, reductive, and neutral) and other treatment conditions [41]. In case of carbon nanohorns (CNHs), closely related to CNTs, studying the effects of helium plasma, it was unexpectedly established [42] that the existence of C–O is more important than that of the OH functional group for increasing their dispersibility in water and decreasing the size of CNH agglomerates in water. Interesting conclusions were made comparing this plasma and ozone treatment dielectric barrier discharge at atmospheric pressure. Thus, (1) oxygen-containing functional groups were generated by O_3 use; (2) hydroxyl functional groups on the surface of CNHs were formed by He plasma treatment; (3) sizes of CNHs in water were smaller after ozone treatment, and, respectively, plasma-treated

CNHs were bigger; (4) the dispersed CNHs after ozone treatment were stable for 3 months, and plasma-treated CNHs were considerably less stable, leading to the flocculation in some minutes.

2.3 Irradiation Techniques and Labeling

Radiation methods, described in a recent monograph [43], have been applied for CNT solubilization too. It is well known that strong sources of irradiation can produce defects and imperfections not only in biological molecules but also in inorganic materials. Carbon nanotubes are not an exception of this rule: additional defects cause higher-scale formation of attached -COOH or other functional groups. The most common structural defects of CNTs, caused by neutron and gamma irradiation, are dangling bonds, topological defects, atomic vacancies, Stone–Wales defects, and microvoids [44]. As an example, MWCNTs irradiated with *γ-rays* were subjected to chemical modification with thionyl chloride and decylamine [45]. The results showed that γ-radiation increased the concentration of functional groups bound to MWCNTs, which arose due to the increasing number of defect sites created on the MWCNTs by γ-photons. Compared with untreated MWCNTs, γ-irradiation significantly enhanced the solubility of MWCNTs in acetone and THF.

Also, a method for highly efficient functionalization of SWCNTs by DNA wrapping included exposure of SWCNTs to γ-irradiation (50 kGy), which lowered by one order of magnitude the amount of single-stranded deoxyribonucleic acid (ssDNA) required for SWCNT modification [46]. While γ-irradiation in three different media significantly improved the process of SWCNT dispersion, irradiation in ammonia was the most efficient. The γ-irradiated SWCNTs functionalized with ssDNA were stabilized by electrostatic forces. The authors suggested that γ-irradiation can significantly improve the functionalization of SWCNTs with DNA.

The γ-irradiated CNTs and their composites with elevated dispersibility could have a series of useful applications, for example, as passive dosimeters specifically for the high levels of radiation exposures used in radiation therapy (doses about 2 Gy) [47] and changes in electric properties [48, 49]. In addition, interesting observations for γ-irradiated CNT dispersions are that (1) the effects of γ-irradiation on these dispersions and optical properties of the CNT samples are minimal comparing with the data solid-phase CNT samples (e.g., relative electron irradiation-induced destruction of carbon nanotubes in electron microscopes [50]) and (2) γ-irradiation can be industrially useful for the sterilization of CNT dispersions [51].

A series of radionuclides has been attached to CNTs with distinct purposes. Thus, *carbon-13* was used to enrich CNTs compositions for improved magnetic resonance imaging [52, 53]. An imaging study to determine the tissue biodistribution and pharmacokinetics of prototypical water-soluble DOTA-functionalized CNT labeled with *yttrium-86* (compound **2.1**) and *indium-111* in a mouse model was undertaken [54]. It was noted that the major sites of accumulation of activity resulting from the administration of ^{86}Y–CNT were the kidney, the liver, the spleen, and to a much less extent the bone.

In addition, water-soluble MWCNTs were labeled with *technetium-99m* (this isotope ^{99m}Tc is one of the most used in medicinal chemistry [55]) to study the distribution of MWCNTs modified with glucosamine in mice [56]. It was shown that MWCNTs moved easily among the compartments and tissues of the body, behaving like active molecules although their apparent mean molecular weight is tremendously large. Similar results were obtained for *iodine-125* (^{125}I)-labeled CNTs [57]. In case of *iodine-131*, SWCNTs were coated with a self-polymerized polydopamine (PDA) shell, making them not only highly stable and water soluble but also providing additional possibilities such as chelation with Mn^{2+} to offer enhanced T1 and T2 MR contrast and efficient labeling with ^{131}I (Fig. 2.9) for radioisotope therapy [58].

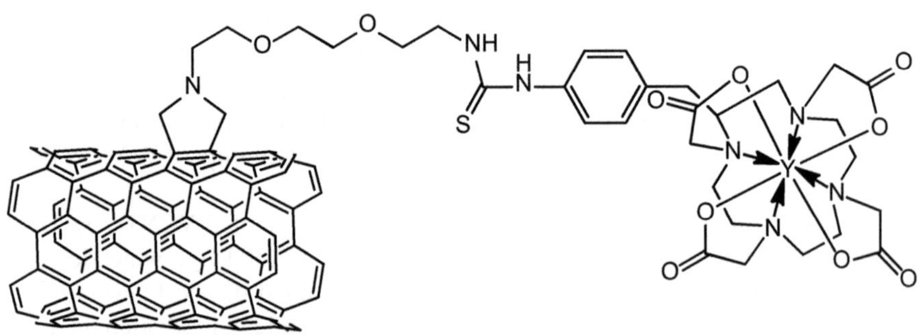

Ittrium-labeled CNTs.

2.4 Mechanical and Related Methods

An interesting approach of attaining the optimum grinding condition of a planetary *ball mill* {Fig. 2.10; the collision medium was monosized (3.00 mm) spherical zirconia (ZrO_2) balls} and simple acid-based method for purifying MWCNTs to investigate the dispersion characteristics of MWCNTs was reported [59]. This experiment was conducted under dry and wet grinding conditions at various rotation speeds (200–500 rpm), with further purification of MWCNTs and their dispersion use of SDS surfactant (see also section about SDS use below). It was shown that the structures of amorphous carbon and carbon particles of MWCNTs were completely eliminated and the tips of nanotubes opened. Moreover, both dry and wet grinding caused the lengths of the MWCNTs to be shortened with increasing rotation speed. Purification of MWCNT removed the impurities and

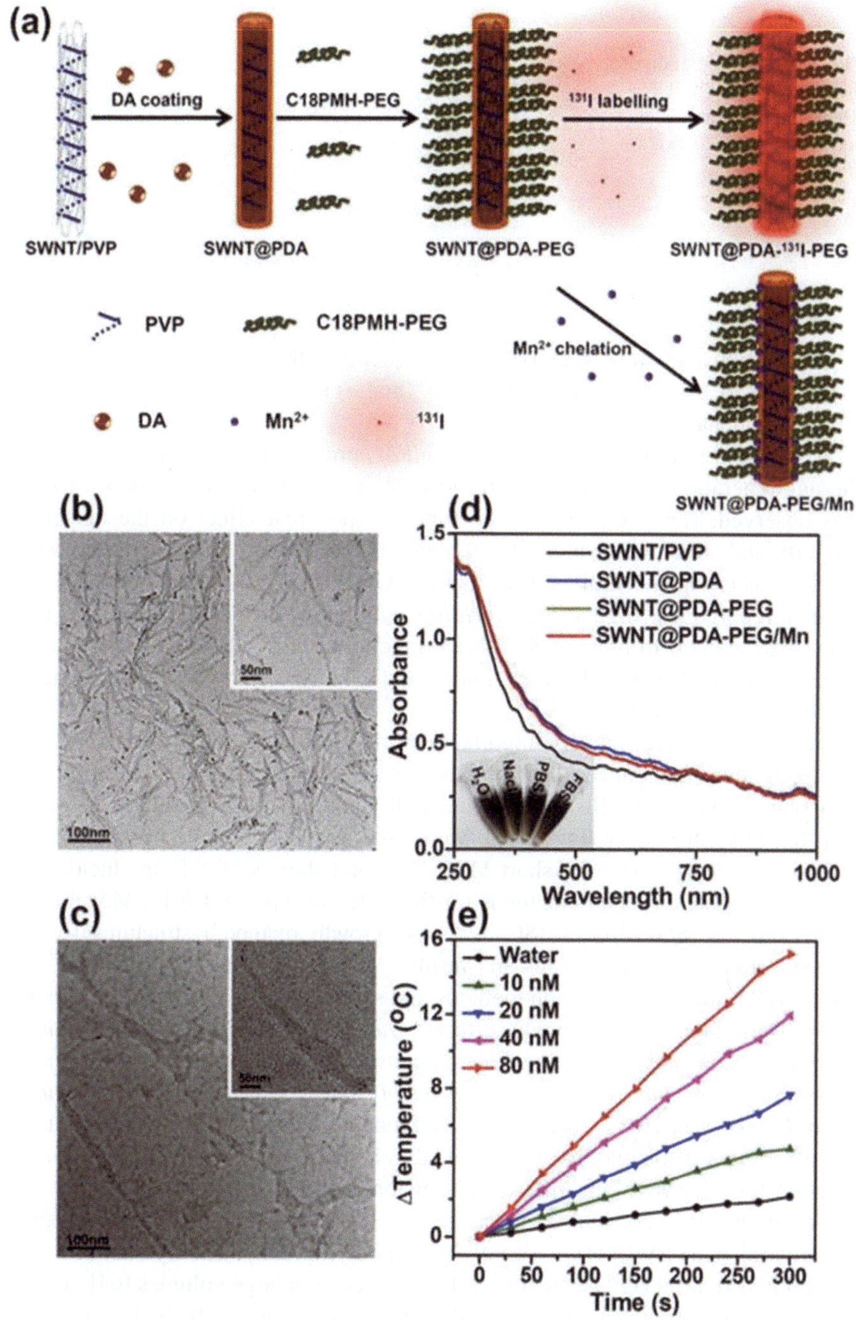

Fig. 2.9 Preparation and characterization of SWCNT@PDA-PEG. (**a**) A schematic illustration for the fabrication of SWCNT@PDA-PEG. (**b**, **c**) TEM images of SWCNT/PVP (**b**) and SWCNT@PDA (**c**). Inset is a TEM image with higher resolution. (**d**) UV-vis-NIR absorbance spectra of SWCNT/PVP, SWCNT@PDA, SWCNT@PDA-PEG, and SWCNT@PDA-PEG/Mn solutions at the same SWCNT concentration. Inset: a photo of SWCNT@PDA-PEG/Mn in various types of physiological buffers. (**e**) Temperature change curves of SWCNT@PDA-PEG solution at different concentrations (i.e., 10, 20, 40, 80 nM) exposed to the 808 nm laser with a power density of 0.7 W cm^{-2} for 5 min (Reproduced with permission of *Ivyspring International Publisher*)

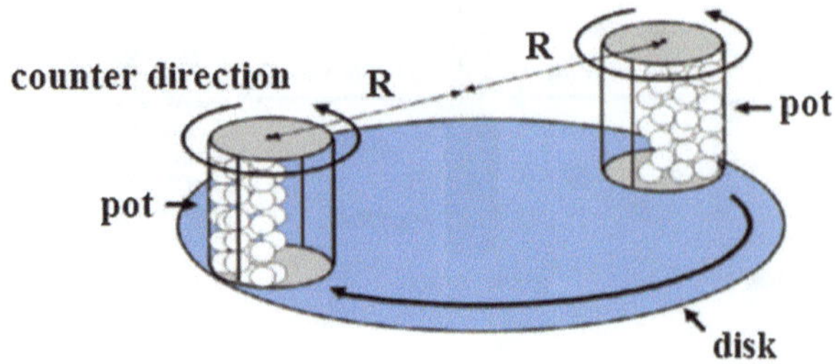

Fig. 2.10 Configuration of the planetary ball mill (Reproduced with permission from *Elsevier Science*)

agglomerations of raw materials; consequently, the pure structure allowed MWCNTs to disperse well in the base fluid. The structure of the MWCNTs remained unchanged after purification.

At a related technique, *cryogenic ball milling* process [60], at high milling speed, the agglomeration of CNTs was considerably reduced, and amorphous structure was observed. However, the milling time has no great effect on the dispersion property and structural change of CNTs compared with milling speed. It is worth noting that cryogenic milling can disperse CNTs in Al2009 matrix too [61]. In another report [62], high-speed *vibration mill* (HSVM) mechanochemical technique was applied to prepare SWCNTs, functionalized with some alkyl and aryl groups. Resulted SWCNTs with long alkyl chains can be dissolved in many common organic solvents.

In a representative and very important report [63] in this area, 11 different dispersion techniques (nanomizer, high-pressure jet mill, probe sonicator, ball mill, bead mill, paint shaker, ball collision mill, cone mill, rotor mill, high-shear batch disperser, and thin-film spin mixer (Fig. 2.11)) were used as a comparative performance for dispersion of long SWCNTs, short MWCNTs, and short SWCNTs to elucidate the most appropriate dispersion methods for the different types of CNTs. Model of the changes in the SG-SWCNT (SG = "super-growth method") structures for two different dispersion mechanisms and resulting structures of CNT dispersions, produced using different dispersion methods, are shown in Figs. 2.12 and 2.13, respectively. The authors found the unique effect ("*when long SWCNTs were dispersed using a turbulent flow method, the resulting composites showed the highest performance compared to 1) composites with other types of CNTs, 2) composites prepared using other dispersion methods*") and proposed a fundamental mechanism to explain it. We can affirm that this is the first fundamental systematic study investigating the comparative performance of the different dispersion methods.

A simple high-speed *shear force mixing* (SFM) was used to disperse nearly monochiral (6,5) SWCNTs with poly[(9,9-dioctylfluorenyl-2,7-diyl)-alt-co-(6,60-{2,20-bipyridine})] (PFO-BPy, Fig. 2.14) in toluene in large volumes [64], leading to average tube length of 1.82 mm. This method was shown to be highly reproducible, scalable, and efficient, allowing to disperse SWCNTs with conjugated polymers and leading to dispersions of exceptionally long (6,5) nanotubes possessing high photoluminescence efficiencies being ideal for optoelectronic applications. Applying this technique, it is possible the cost of sorting SWCNTs without loss of quality.

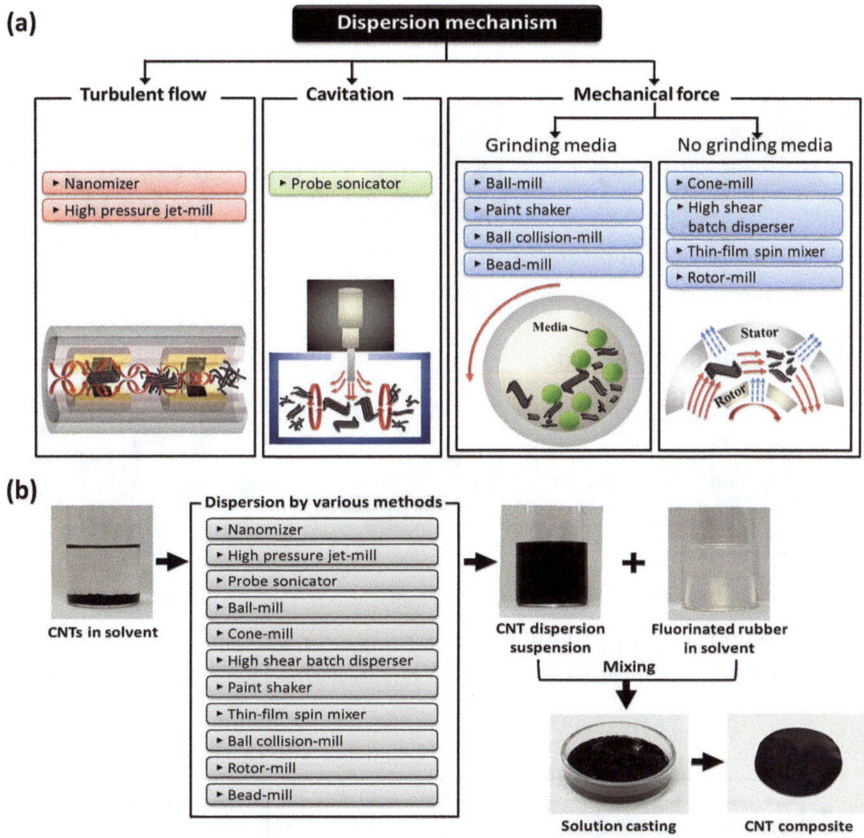

Fig. 2.11 Schematic of methods for dispersing nanotubes in rubbers. (**a**) Various dispersion methods for CNT classified into three general mechanisms: turbulent flow, cavitation, and mechanical force. (**b**) Process for CNT/rubber composite fabrication (Reproduced with permission of *Nature*)

A good dispersion of nanoparticles in aqueous solution is not always achieved by ultrasonication, described in a section above, due to the fact that incoming ultrasonic waves in liquid are usually reflected and damped at the gas/liquid interface. So, various types of ultrasound-combined dispersion methods for CNTs on the basis of *grinding* are offered. Thus, SWCNTs can be universally disperse in water using water-soluble Laponite RD (LRD) vía *grinding–ultrasonication* procedure inspired by Maya Blue [65] due to van der Waals and electron donating–accepting interactions (Fig. 2.15). The formed hybrid is stable over 6 months. In addition, LRD/SWCNT aqueous dispersion can be mixed with other common organic solvents, e.g., DMF, methanol, and DMSO. (It is worth noting that Laponite has been also used for CNT dispersion in liquid crystals: the Laponite plates surrounding (due to strong anchoring to CNT) the CNT, destroying the aggregation, and disintegrating the CNT bundles [66, 67].)

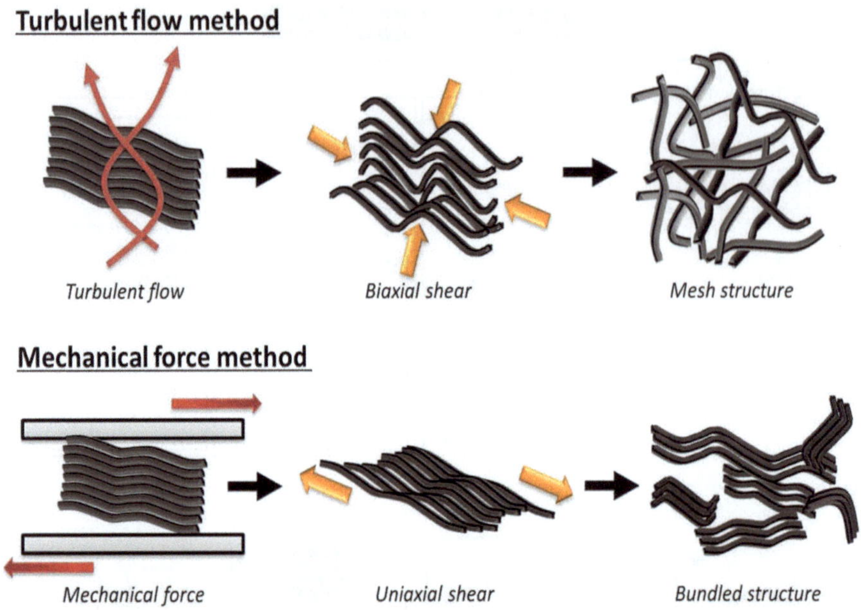

Fig. 2.12 Model of the changes in the SG-SWCNT structures for two different dispersion mechanisms: turbulent flow (*up*) and mechanical force (*down*) (Reproduced with permission of *Nature*)

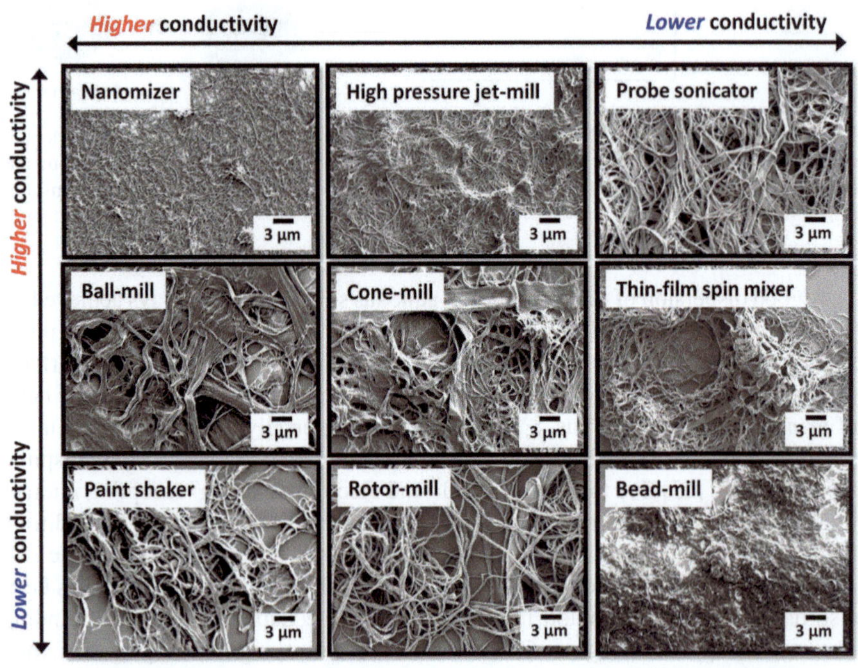

Fig. 2.13 SEM images of the structures of SG-SWCNT dispersions produced using different dispersion methods (Reproduced with permission of *Nature*)

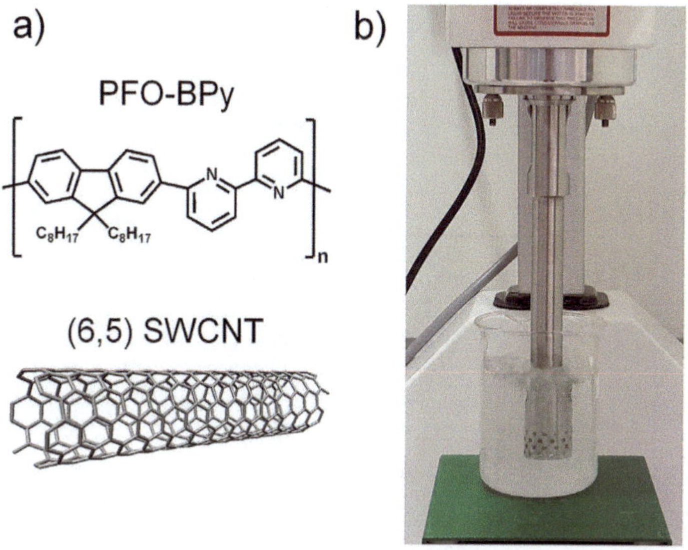

Fig. 2.14 (**a**) Molecular structure of copolymer PFO-BPy and (6,5) SWCNT. (**b**) Lab-scale shear force mixer used for dispersing large volumes of SWCNTs (cooling bath not shown) (Reproduced with permission of *Elsevier Science*)

In a related report [68], a so-called *wet-grinding*-assisted ultrasonication method was offered to improve CNT solubilization, where wet grinding of MWCNTs in chitosan solution is carried out *before* ultrasonication. It was established that the dispersion quality of chitosan/MWCNT suspension prepared by this combined technique is much better than that by wet grinding or ultrasonication alone. It is worth noting that, in certain conditions, the interactions between MWCNTs-OH and MWCNTS-COOH under mechanical grinding, initiated at room temperature in ambient air, can lead to the unzipping of the nanotube [69].

Concluding, the mechanical methods (sometimes the ultrasonic treatment, described above, is also attributed to them), well described as far back as in 2003 [70], are currently not very popular in comparison with sophisticated chemical functionalization techniques below, leading to more "soluble" CNTs. The mechanical methods are usually used in combination with other methods or as the first "pre-synthesis" step for further chemical solubilization of carbon nanotubes.

2.5 Pulsed Streamer Discharge

A few reports are available in the area of discharge methods (it is well known that the arc discharge is one of the production methods for CNTs) for CNT solubilization. Thus, a very interesting, in our point of view, technique "*pulsed streamer discharge*," generated in water, involving chemical reactions between radicals appearing by the pulsed streamer discharge and CNTs, allowed to homogeneously

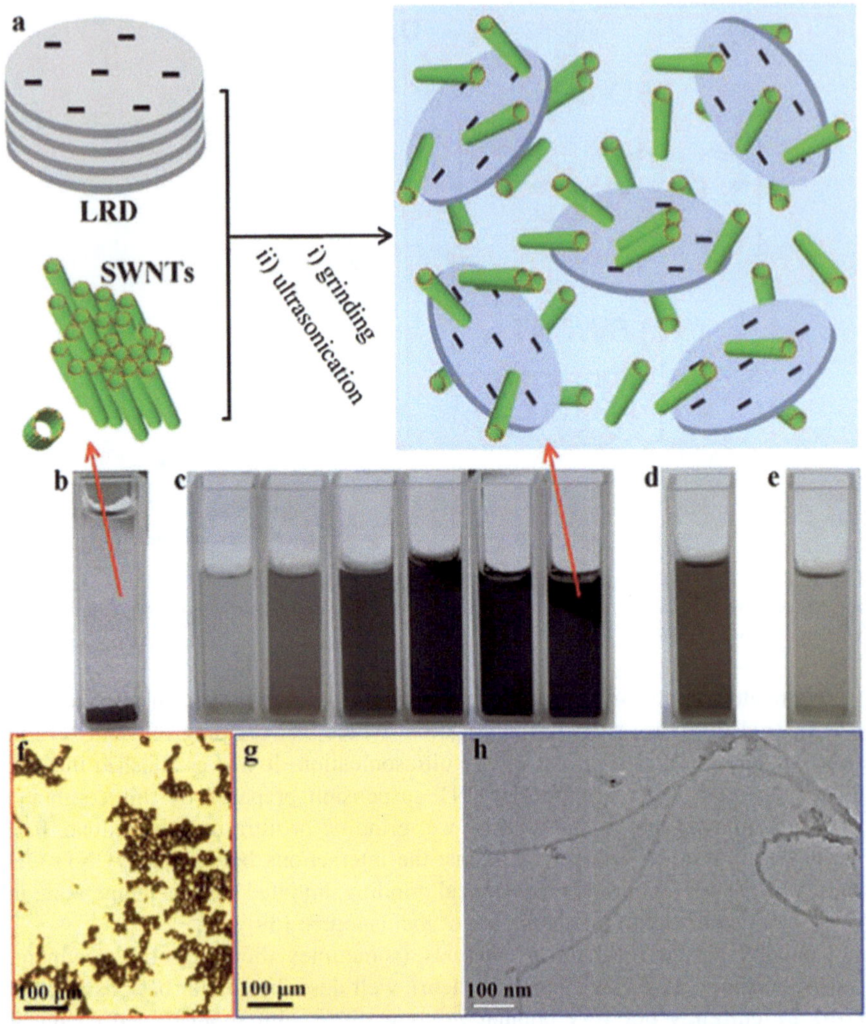

Fig. 2.15 (**a**) Schematic illustration of solid LRD, SWCNT bundles, and homogeneous dispersion of LRD/SWCNTs in aqueous solution. Photographic images of (**b**) pristine SWCNTs in water, (**c**) LRD/SWCNT hybrid of various concentrations in water (from left to right: 0.007, 0.018, 0.037, 0.092, 0.366, and 1.83 mg mL^{-1} SWCNTs), SWCNTs modified with (**d**) SDBS (0.04 mg mL^{-1} SWCNTs in 4 mg mL^{-1} SDBS), and (**e**) SDS (0.02 mg mL^{-1} SWCNTs in 4 mg mL^{-1} SDS) in water. (**f**) Optical micrograph of (**b**). (**g**) Optical micrograph of (**c**) (1.83 mg mL^{-1} SWCNTs). (**h**) TEM image of (**c**) (0.366 mg mL^{-1} SWCNTs) (Reproduced with permission of the *Royal Society of Chemistry*)

disperse them and well solubilize in water for a month or longer [71]. Study of the mechanism revealed that –OH groups, which are known to impart a hydrophilic nature to carbon material, were introduced on the carbon nanotube surface; highly oxidative O* and H* radicals were generated in water and are responsible for the functionalization of the CNT surface by –OH groups. A great advantage of the

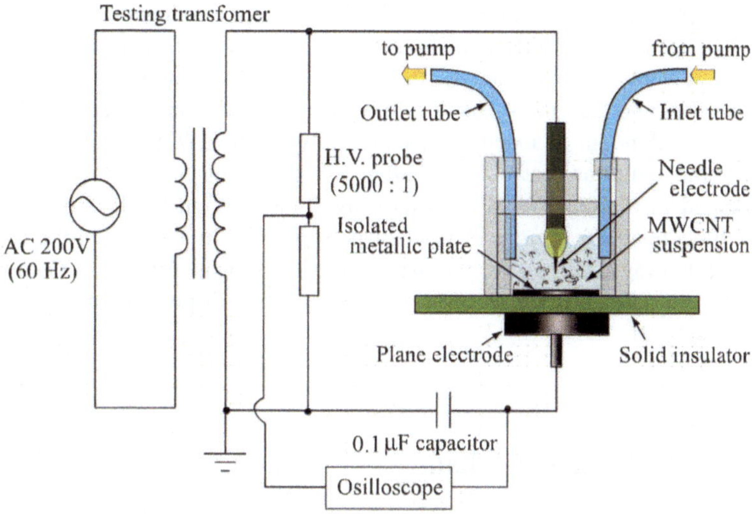

Fig. 2.16 Experimental setup used in the dispersion of CNTs by barrier discharge (Reproduced with permission of the *International Journal of Plasma Environmental Science & Technology*)

proposed method is that there is no need for any chemical agents or additives for solubilization due to their generation from the water itself by the electrochemical reactions induced by the pulsed streamer discharge.

This method was improved by the same researchers [72] by the use of gas bubbling in water. Oxygen, argon, and nitrogen were used as bubbling gas in order to clarify the effects of the gas species on the single-walled CNT (SWCNT) solubilization efficiency, and it was established that gas bubbling has positive effects on microplasma-based SWCNT solubilization as a result of enhanced radical formation and functionalization of the SWCNT surface. In addition, a dispersion method of MWCNTs in organic solvents (octylalcoholes: 2-octanol and 2-ethyl-1-hexanol) using a *barrier discharge* was offered [73], using the needle-to-plane electrode configurations. A barrier discharge was generated between a needle electrode and a grounded plane electrode (Fig. 2.16) which was covered by the solid insulator. The dispersion effect of MWCNTs was observed at more than 60 min. of the discharge.

2.6 Microwave Treatment

Microwave treatment, now a common synthesis technique in chemistry, was successfully applied for chemical functionalization and solubilization of CNTs, for instance, for SWCNT amidation and 1,3-dipolar cycloaddition [74, 75] or preparation in situ of (MWCNT)/polystyrene or poly(methyl methacrylate) composites soluble in common organic solvents such as 1,2-dichlorobenzene, THF, and chloroform [76]. Solubility was a key feature for a successful MW-heated reaction of cycloaddition of 1,3-dipolar azomethine ylides to the sidewalls of MWCNTs, resulting in MWCNTs that contain 2-methylenethiol-4-(4-octadecyloxyphenyl), N-octyl-2-(4-octadecyloxyphenyl), or 2-(4-octadecyloxyphenyl)pyrrolidine units [77]. All these contain the 4-octadecyloxyphenyl substituent that acts as a solubilizing group. Amount of added groups after only 2 h of MW heating at 200 °C was in the same range as that obtained after 100–120 h of conventional heating of soluble and insoluble MWCNTs. Microwave treatment on CNTs "solution" was found to decrease the density of CNT dispersion [78]. In addition to the CNTs, a related technology, *microwave plasma splitting*, of biogas (methane/CO_2 mixture) resulted in solid carbon forms, which can be dispersed in water after purification [79].

2.7 Hydrothermal Method

MWCNTs were hydrothermally (HT) dispersed in organic solvents like dichloromethane, ethanol, isopropyl alcohol, and hexane (Fig. 2.17) [80]. Dichloromethane was found to be the best solvent for the dispersion of MWCNTs.

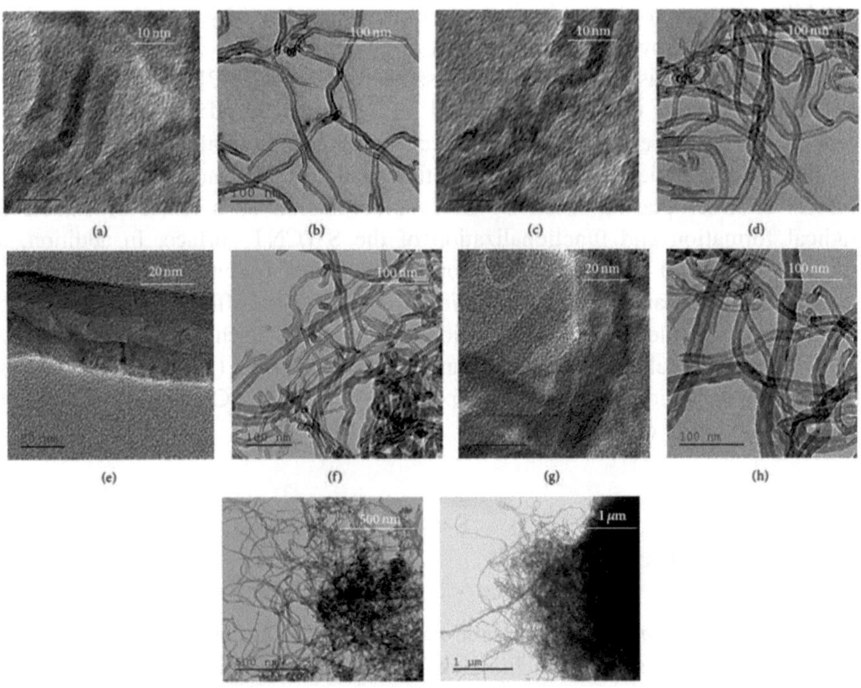

Fig. 2.17 Transmission electron microscope images of dispersed MWCNTs in dichloromethane (**a**, **b**), ethanol (**c**, **d**), isopropyl alcohol, (**e**, **f**) hexane (**g**, **h**), and standard (untreated) MWCNTs (**i**, **j**). (Reproduced with permission of *Hindawi Publishing Corporation*)

A peculiarity of HT method in relation with CNTs is logically explained by authors as follows: different solvents have specific properties under their supercritical conditions, become more fluid, and have greater diffusivity in MWCNTs with mass transformations. MWCNT dispersions showed nanotube diameters of 5–50 nm. Hydrothermal method can be successfully combined with others in order to reach a better CNT solubilization. Thus, nanocrystalline-TiO_2(B)/carbon nanotube hyperdispersion was prepared [81] via an in situ *ultracentrifugation* process coupled followed by *hydrothermal* treatment.

2.8 UV and Visible Light

UV light (and also UV/ozone treatment) influences various carbon materials, in particular CNTs, and is discussed in a comprehensive review [82]. UV irradiation leads to the formation of hydroxyl, carbonyl, and carboxyl groups on MWCNT surface and corresponding change of dispersibility, especially in the conditions of UV/Fenton oxidation treatment (preliminary ultrasonication of MWCNTs and UV irradiation in a photochemical reactor in the presence of H_2O_2 and $FeSO_4$) [83]. In our opinion, UV light is being used quite insufficiently for functionalization of CNTs, despite obvious advantages, availability, and low cost of equipment; a few examples are as follows. Thus, the photochemical reactions involving colloidal dispersions of carboxylated SWCNTs–COOH in sunlight were examined [84]. Production of reactive oxygen species (ROS) during irradiation occurs and is evidence for potential further phototransformation and may be significant in assessing their overall environmental impacts. In aerated samples exposed to sunlight or to lamps that emit light only within the solar spectrum, the probe compounds, furfuryl alcohol (FFA), tetrazolium salts (NBT^{2+} and XTT), and *p*-chlorobenzoic acid (pCBA), were used to indicate production of $^1O_2$, $O_2•^-$, and •OH, respectively. All three ROS were produced in the presence of SWCNTs–COOH and molecular oxygen ($^3O_2$). $^1O_2$ production was confirmed by observing enhanced FFA decay in deuterium oxide, attenuated decay of FFA in the presence of azide ion, and the lack of decay of FFA in deoxygenated solutions.

Also, SWCNTs were functionalized by the covalent attachment of 2-propanol-2-yl radicals, generated by photolysis of 2-hydroxy-2-methyl-1-phenyl-1-propanone under UV light, to their surface in THF solution [85]. A loss of Van Hove singularities and decrease of the intensity ratio of the G band and D band (IG/ID) were observed. The solubility in common organic solvents was improved, and the original electronic structure of the SWCNTs was retained without severe modification that damages the nanotubes. Possible applications of SWCNTs, being modified by photoiniciators, could have applications for creation of new materials.

In addition, the formation of a chemical bond between MWCNTs and side chains of polystyrene was examined using poly(4-chloromethyl)styrene or styrene/4-(chloromethyl)styrene random copolymer, which was mixed with MWCNTs in 1-methyl-2-pyrrolidone and irradiated with UV light [86].

The photo-induced dissociation of the C–Cl bond of the polymer led to appearance of benzyl-type radicals (reaction scheme 2.1), formation of chemical bonds, and deformation of MWCNT carbons. Certain changes in dispersibility of the formed composite in solvents were revealed. It was also found that the dispersant $NH_2OH\cdot\cdot$ HCl could not sufficiently disperse the MWCNTs in the polymer, although it can disperse MWCNTs in organic solvents.

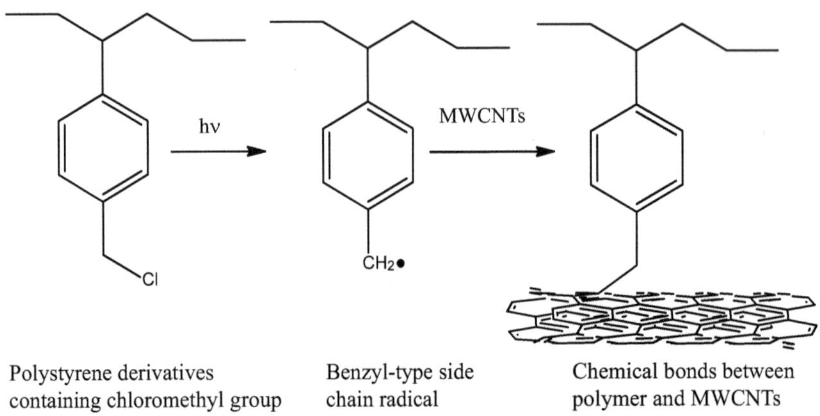

Reaction process of C–Cl dissociation of polymer side chains followed by chemical bond formation between polymer and multi-walled carbon nanotubes

2.9 Other Physical Methods

Among other non-chemical techniques, scarcely used for CNT solubilization, the *cryogenic crushing* CNTs at liquid nitrogen temperature allowed them to be shortened and make them appreciably soluble in a solvent without any dispersant [87]. Typical lengths of less than 500 nm were obtained from 30-min crushing. Laser methods are usually applied for CNTs synthesis [88] or to study the formation of different nanocarbon forms (i.e., nanodiamonds) from solid CNTs [89], but not for their solubilization. On the contrary, a reversible effect was observed: *pulsed laser light* caused flocculation of SWCNTs, previously solubilized individually in dimethylformamide using a polymer with anthracene groups [90]. The SWCNTs absorbed the near-IR light and converted it to heat vía photothermal conversion of SWCNTs. This process did not greatly degrade the polymer wrapping the SWCNTs. In case of SWCNTs/SDS aqueous solution, no such irradiation effect was noted.

Another example is the use of confocal laser scanning microscopy for *visualization* of CNTs: MWCNT-PS (PS = polystyrene) with the highest thickness of 25–60 nm were readily visualized using both reflection mode without necessity of fluorescent labeling and fluorescent mode after addition of free dye to the nanocomposite [91]. In addition, a two-phase liquid–liquid *extraction process* allowed extracting water-soluble SWCNTs into an organic phase [92]. The extraction is based on electrostatic interactions between a common phase transfer agent and the sidewall functional groups on the nanotubes. *Thermal annealing* in air at 350 and 550 °C is a simple and, at the same time, a perspective method to enhance dispersibility of CNTs in different organic solvents, including ethanol, dimethylformamide, chloroform and acetone (Fig. 2.18) [93]. At last, we note the dispersion of CNTs, coated with Fe_2O_3, in polymer matrix via application of *oscillating magnetic field* [94]. This method could be useful for CNT dispersion in solvents too.

As it was shown above, solution-based methods for CNT dispersion are limited by scalability and are not environmentally benign. It was demonstrated [95] that high electric fields can enhance CNT dispersion (Fig. 2.19) and, in addition, allow

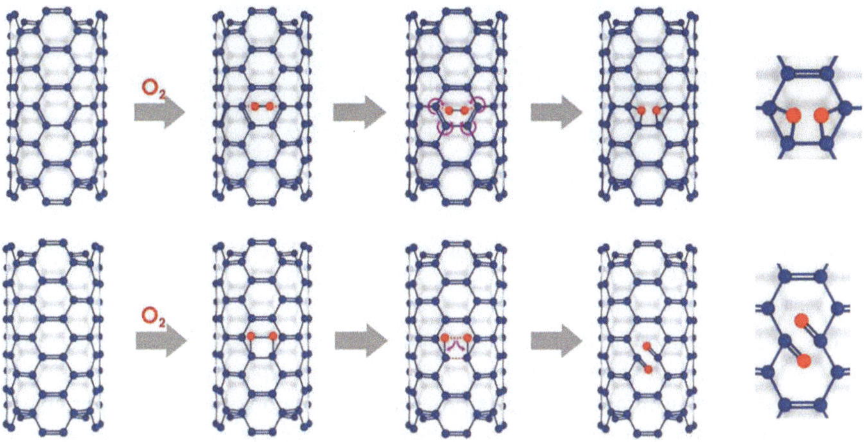

Fig. 2.18 The proposed mechanisms of CNT surface functionalization via thermal annealing. The top part shows the formation of ether functional group, while the bottom part shows the formation of quinone group (Reproduced with permission of *Springer*)

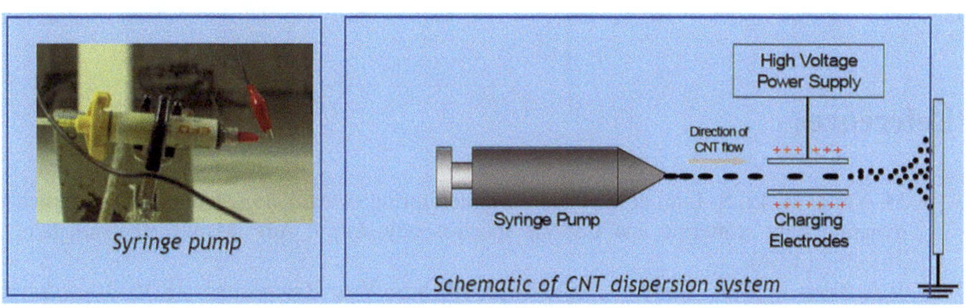

Fig. 2.19 Dispersion and alignment of carbon nanotubes using electric fields (Reproduced with permission of the *Florida State University*)

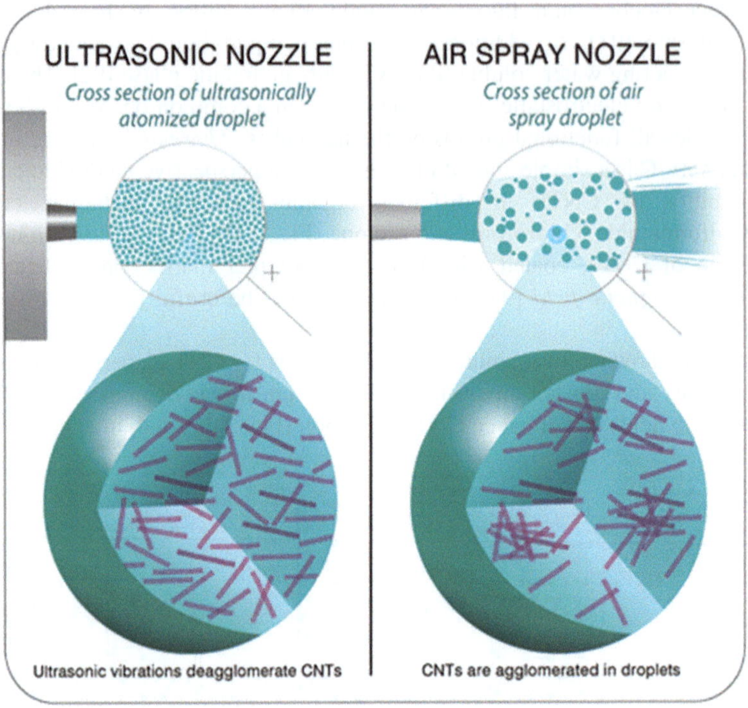

Fig. 2.20 Creation of CNT aerosols in the ultrasonic spray system (Reproduced with permission of *SONO-TEK, NY, USA*)

some level of alignment, despite a little understanding of the response processes of CNTs to electric field. A related *electrospraying method* for CNT dispersion is described in detail in a M.Sc. thesis [96] and other reports [97]. This approach is applied for deposition of CNTs on substrates or for creation of homogeneous composites (in particular, starting from already prepared CNT dispersions [98]) and does not have a direct relation with main idea of this book (solubilization of CNTs in solvents), as well as the formation of CNT aerosols via *aerosolization* with hot water atomization [99] or in the *ultrasonic spray* system (Fig. 2.20) [100].

References

1. M. Yang, M. Li, S. Luo, et al., Real-time monitoring of carbon nanotube dispersion using dynamic light scattering and UV-Vis spectroscopy. Int. J. Adv. Manuf. Technol. **82**(1), 361–367 (2016)
2. V.A. Tarasov, M.A. Komkov, N.A. Stepanishchev, V.A. Romanenkov, R.V. Boyarskaya, Modification of polyester resin binder by carbon nanotubes using ultrasonic dispersion. Polym. Sci. Ser. D **8**(1), 9–16 (2015)

3. K. Yang, Z.L. Yi, Q.F. Jing, R.L. Yue, W. Jiang, D.H. Lin, Sonication-assisted dispersion of carbon nanotubes in aqueous solutions of the anionic surfactant SDBS: The role of sonication energy. Chin. Sci. Bull. **58**(17), 2082–2090 (2013)
4. A. Sesis, M. Hodnett, G. Memoli, A.J. Wain, I. Jurewicz, A.B. Dalton, J. David Carey, G. Hinds, Influence of acoustic cavitation on the controlled ultrasonic dispersion of carbon nanotubes. J. Phys. Chem. B **117**, 15141 (2013)
5. K.L. Lu, R.M. Lago, Y.K. Chen, M.L.H. Green, P.J.F. Harris, S.C. Tsang, Mechanical damage of carbon nanotubes by ultrasound. Carbon **34**, 814–816 (1996)
6. A.D. Willey, J.M. Holt, B.A. Larsen, J.L. Blackburn, S. Liddiard, J. Abbott, M. Coffin, R.R. Vanfleet, R.C. Davis, Thin films of carbon nanotubes via ultrasonic spraying of suspensions in N-methyl-2-pyrrolidone and N-cyclohexyl-2-pyrrolidone. J. Vac. Sci. Technol. B Nanotechnol. Microelectron. Mater. Process. Meas. Phenom **32**, 011218 (2014.) http://avs.scitation.org/doi/abs/10.1116/1.4861370?journalCode=jvb.
7. K. Urbanski, B. Platek, T. Falat, J. Felba, F. Marcq. *Novel Method for CNTs Dispersion in Fluids*. 34th International Microelectronics and Packaging IMAPSCPMT Poland Conference, (Wrocław, 2010), p. 22–25
8. G.T. Caneba, C. Dutta, V. Agrawal, M. Rao, Novel ultrasonic dispersion of carbon nanotubes. J. Miner. Mater. Charact. Eng. **9**(3), 165–181 (2010)
9. T. Yasumitsu, G. Liu, J.-M. Leveque, S. Aonuma, L. Duclaux, T. Kimura, N. Komatsu, A rosette cooling cell: More effective container for solubilization of single-walled carbon nanotubes under probe-type ultrasonic irradiation. Ultrason. Sonochem. **20**(1), 37–39 (2013)
10. Q. Zaib, I.A. Khan, Y. Yoon, J.R.V. Flora, Y.-G. Park, N.B. Saleh, Ultrasonication study for suspending single-walled carbon nanotubes in water. J. Nanosci. Nanotechnol. **12**(5), 3909–3917 (2012)
11. X. Yang, J. Lee, L. Yuan, S.-R. Chae, V.K. Peterson, A.I. Minett, Y. Yin, A.T. Harris, Removal of natural organic matter in water using functionalised carbon nanotube buckypaper. Carbon **59**, 160–166 (2013)
12. B. Koh, W. Cheng, The impact of sonication on the surface quality of single-walled carbon nanotubes. J. Pharm. Sci. **104**, 2594–2599 (2015)
13. H. Yu, S. Hermann, S.E. Schulz, T. Gessner, Z. Dong, W.J. Li, Optimizing sonication parameters for dispersion of single-walled carbon nanotubes. Chem. Phys. **408**, 11–16 (2012)
14. Y. Shi, L. Ren, D. Li, H. Gao, B. Yang, Optimization conditions for single-walled carbon nanotubes dispersion. J. Surf. Eng. Mater. Adv. Technol **3**, 6–12 (2013)
15. M. Mahfuzur Rahman, H. Younes, N. Subramanian, A. Al Ghaferi, Optimizing the dispersion conditions of SWCNTs in aqueous solution of surfactants and organic solvents. J. Nanomater. **2014**, 11 (2014.) Article ID 102621
16. L. Reinert, M. Zeiger, S. Suarez, V. Presser, F. Mucklich, Dispersion analysis of carbon nanotubes, carbon onions, and nanodiamonds for their application as reinforcement phase in nickel metal matrix composites. RSC Adv. **5**, 95149–95159 (2015)
17. Y.Y. Yan Yan Huang, E.M. Terentjev, Dispersion of carbon nanotubes: Mixing, sonication, stabilization, and composite properties. Polymer **4**, 275–295 (2012)
18. L. David, A. Feldman, E. Mansfield, J. Lehman, G. Singh, Evaluating the thermal damage resistance of graphene/carbon nanotube hybrid composite coatings. Sci Rep **4**, Article No. 4311 (2014)
19. B. He, W. Chen, Carboxyl multiwalled carbon nanotubes through ultrasonic dispersing in dimethylfomamide modified electrode as a sensitive amperometric sensor for detection of sulfonamide. Int. J. Electrochem. Sci. **10**, 4335–4345 (2015)
20. V. Oxana, O.V. Kharissova, L.M. Torres Martínez, B.I. Kharisov, Recent trends of reinforcement of cement with carbon nanotubes and fibers, in *Advances in Carbon Nanostructures*, ed. by A. M. Tavares da Silva, S. A. C. Carabineiro (Eds), (INTECH, Amsterdam, The Netherlands, 2016.) http://www.intechopen.com/books/advances-in-carbon-nanostructures.

21. J. Foldyna, V. Foldyna, M. Zelenák, Dispersion of carbon nanotubes for application in cement composites. Procedia Eng **149**, 94–99 (2016)
22. S. Parveen, S. Rana, R.A. Fangueiro, Review on nanomaterial dispersion, microstructure, and mechanical properties of carbon nanotube and nanofiber reinforced cementitious composites. J. Nanomater. **2013**, 19 (2013.) Article ID 710175
23. S. Alrekabi, A. Cundy, A. Lampropoulos, I. Savina, Experimental investigation on the effect of ultrasonication on dispersion and mechanical performance of multi-wall carbon nanotube-cement mortar composites. Int. J. Civ. Environ. Struct. Constr. Archit. Eng **10**(3), 268–274 (2016)
24. L. Cui, Incorporation of multiwalled carbon nanotubes to ordinary Portland cement (OPC): Effects on mechanical properties. Adv. Mater. Res. **641-642**(1), 436–439 (2013)
25. C.S. Davis, J.W. Woodcock, J.W. Gilman, Preparation of multi-walled carbon nanotube dispersions in a polyetheramine epoxy for eco-toxicological assessment. NIST Spec. Publ **1200**(9), 13 (2015.) http://dx.doi.org/10.6028/NIST.SP.1200-9
26. Shahshahan, M., Keinanen, P., Vuorinen, J. *The Effect of Ultrasonic Dispersion on the Surface Chemistry of Carbon Nanotubes in the Jeffamine D-230 Polyetheramine Medium.* 2016 I.E. Nanotechnology Materials and Devices Conference (NMDC), 978–1–5090-4352-1, (Toulouse, France, 2016), p. 2.
27. A. Samuel Afolabi, O. Oluwafolakemi Sadare, M. Olawale Daramola, Effect of dispersion method and CNT loading on the quality and performance of nanocomposite soy protein/ CNTs adhesive for wood application. Adv. Nat. Sci. Nanosci. Nanotechnol. **7**(035005), 8 (2016)
28. http://www.iptonline.com/articles/public/Ultrasonic%20Treatment%20of%20Nanoparticles%20for%20Pharmaceuticals.pdf. Accessed on 6 Jan 2017
29. B. Yonas, G. Tesfaye, Production, purification and functionalization of carbon nanotubes for medical applications. Int. Res. J. Pharm. **7**(7), 19–27 (2016)
30. X. Chang, W.M. Henderson, D.C. Bouchard, Multiwalled carbon nanotube dispersion methods affect their aggregation, deposition, and biomarker response. Environ. Sci. Technol. **49**(11), 6645–6653 (2015)
31. S. Park, H. Kang, Y. Kim, D. Lee, C. Jin, Catalyst-related dispersion of multiwalled carbon nanotubes by simple ultrasonication. Bull. Kor. Chem. Soc. **37**, 174–178 (2016)
32. W. Chen, X. Liu, Y. Liu, H.-I. Kim, Novel synthesis of self-assembled CNT microcapsules by O/W Pickering emulsions. Mater. Lett. **64**(23), 2589–2592 (2010)
33. M. Jellur Rahman, T. Mieno, Water-dispersible multiwalled carbon nanotubes obtained from citric-acid-assisted oxygen plasma functionalization. J. Nanomater. **2014**, 9 (2014.) Article ID 508192
34. N. Hordy, S. Coulombe, J.-L. Meunier, Plasma functionalization of carbon nanotubes for the synthesis of stable aqueous nanofluids and poly(vinyl alcohol) nanocomposites. Plasma Process. Polym. **10**(2), 110–118 (2013)
35. Y.J. Kim, H. Ma, Q. Yu, Plasma nanocoated carbon nanotubes for heat transfer nanofluids. Nanotechnology **21**(29), art. no. 295703 (2010)
36. C. Chen, A. Ogino, X. Wang, M. Nagatsu, Plasma treatment of multiwall carbon nanotubes for dispersion improvement in water. Appl. Phys. Lett. **96**(13), art. no. 131504 (2010)
37. Y. Jiang, Y.-F. Zhu, F.-Y. Wang, W.-J. Chen, Effects of plasma and acid treatment on the dispersion of carbon nanotubes in liquids. Plasma Chem. Plasma Process. **31**(3), 441–448 (2011)
38. Y. Ishibashi, R. Hanaoka, N. Osawa, S. Takata, Y. Kanamaru, H. Anzai, Effect of barrier discharge on homogeneous dispersion of carbon nanotubes in octylalcohols. Int. J. Plasma Environ. Sci. Tech. **5**(1), 62–67 (2011)
39. M. Vesali-Naseh, A.A. Khodadadi, Y. Mortazavi, A.A. Moosavi-Movahedi, K. Ostrikov, H_2O/air plasma-functionalized carbon nanotubes decorated with MnO_2 for glucose sensing. RSC Adv. **6**, 31807–31815 (2016)

40. R. Chetty, K. Kumar Maniam, W. Schuhmann, M. Muhler, Oxygen-plasma-functionalized carbon nanotubes as supports for platinum–ruthenium catalysts applied in electrochemical methanol oxidation. ChemPlusChem **80**(1), 130–135 (2015)
41. A. Merenda, E. des Ligneris, K. Sears, T. Chaffraix, K. Magniez, D. Cornu, J.A. Schütz, L.F. Dumée, Assessing the temporal stability of surface functional groups introduced by plasma treatments on the outer shells of carbon nanotubes. Sci Rep **6**(31565), 12 (2016)
42. Z. Lin, T. Iijima, P. Selvam Karthik, M. Yoshida, M. Hada, T. Nishikawa, Y. Hayashi, Surface modification of carbon nanohorns by helium plasma and ozone treatments. Jpn. J. Appl. Phys. **56**, 6 (2017.) 01AB08
43. B.I. Kharisov, O.V. Kharissova, U. Ortiz Mendez, *Radiation Synthesis of Materials and Compounds* (CRC Press, Boca Raton, 2013), p. 586
44. R. Gorthy. Effects of neutron and gamma radiation on carbon nanotubes and three-dimensional graphene sheets. MS Thesis, University of Cincinnati, 2016. https://etd.ohiolink.edu/pg_10?0::NO:10:P10_ETD_SUBID:112490.
45. J. Guo, Y. Li, S. Wu, W. Li, The effects of γ-irradiation dose on chemical modification of multi-walled carbon nanotubes. Nanotechnology **16**(10), 2385–2388 (2005)
46. S.P. Jovanović, Z.M. Marković, D.N. Kleut, N.Z. Romević, V.S. Trajković, M.D. Dramićanin, B.M. Todorović Marković, A novel method for the functionalization of γ-irradiated single wall carbon nanotubes with DNA. Nanotechnology **20**(44), art. no. 445602 (2009)
47. A. A Alanazi, M. Alkhorayef, K. Alzimami, I. Jurewicz, N. Abuhadi, A. Dalton, D.A. Bradley, Carbon nanotubes buckypaper radiation studies for medical physics applications. Appl. Radiat. Isot. **117**, 106–110 (2016)
48. I. Karbovnyk, I. Olenych, O. Aksimentyeva, H. Klym, O. Dzendzelyuk, Y. Olenych, O. Hrushetska, Effect of radiation on the electrical properties of PEDOT-based nanocomposites. Nanoscale Res. Lett. **11**(84), 5 (2016)
49. T.S. Williams, N.D. Orloff, J.S. Baker, S.G. Miller, B. Natarajan, J. Obrzut, L.S. McCorkle, M. Lebron-Colón, J. Gaier, M.A. Meador, J.A. Liddle, Trade-off between the mechanical strength and microwave electrical properties of functionalized and irradiated carbon nanotube sheets. ACS Appl. Mater. Interfaces **8**(14), 9327–9334 (2016)
50. K. Mølhave, S. Bjarke Gudnason, A. Tegtmeier Pedersen, C. Hyttel Clausen, A. Horsewell, P. Bøggild, Electron irradiation-induced destruction of carbon nanotubes in electron microscopes. Ultramicroscopy **108**(1), 52–57 (2007)
51. J.A. Fagan, N.J. Lin, R. Zeisler, A.R. Hight Walker, Effects of gamma irradiation for sterilization on aqueous dispersions of length sorted carbon nanotubes. Nano Res. **4**(4), 393–404 (2011)
52. R.E. Hurd. Magnetic resonance imaging (MRI) agents: water soluble carbon-13 enriched fullerene and carbon nanotubes for use with dynamic nuclear polarization. US20070025918 (2007)
53. R.E. Hurd. Magnetic resonance imaging (MRI) agents: water soluble carbon-13 enriched fullerene and carbon nanotubes for use with dynamic nuclear polarization. WO08008075, 2008
54. M.R. McDevitt, D. Chattopadhyay, J.S. Jaggi, R.D. Finn, P.B. Zanzonico, C. Villa, D. Rey, M. Juana, B.A. Carl, et al., PET imaging of soluble yttrium-86-labeled carbon nanotubes in mice. PLoS One **2**(9), e907 (2007)
55. M.A. Méndez-Rojas, B.I. Kharisov, A.Y. Tsivadze, Recent advances on technetium complexes: Coordination chemistry and medical applications. J. Coord. Chem. **59**(1), 1–63 (2006)
56. J. Guo, X. Zhang, Q. Li, W. Li, Biodistribution of functionalized multiwall carbon nanotubes in mice. Nucl. Med. Biol. **34**(5), 579–583 (2007)
57. H. Wang, J. Wang, X. Deng, H. Sun, Z. Shi, Z. Gu, Y. Liu, Y. Zhao, Biodistribution of carbon single-wall carbon nanotubes in mice. J. Nanosci. Nanotechnol. **4**(8), 1019–1024 (2004)

58. H. Zhao, Y. Chao, J. Liu, J. Huang, J. Pan, W. Guo, J. Wu, M. Sheng, K. Yang, J. Wang, Z. Liu, Polydopamine coated single-walled carbon nanotubes as a versatile platform with radionuclide labeling for multimodal tumor imaging and therapy. Theranostics **6**(11), 1833–1843 (2016.) http://www.thno.org/v06p1833.htm#coraddress
59. B. Munkhbayar, M.J. Nine, J. Jeoun, M. Bat-Erdene, H. Chung, H. Jeong, Influence of dry and wet ball milling on dispersion characteristics of the multi-walled carbon nanotubes in aqueous solution with and without surfactant. Powder Technol. **234**, 132–140 (2013)
60. J.-H. Lee, K.-Y.A. Rhee, Study on the dispersion characteristics of carbon nanotubes using cryogenic ball milling process. J. Korean Soc. Precis. Eng. **27**(7), 49–54 (2010)
61. T. He, X. He, P. Tang, D. Chu, X. Wang, P. Li, The use of cryogenic milling to prepare high performance Al2009 matrix composites with dispersive carbon nanotubes. Mater. Des. **114**, 373–382 (2017)
62. X. Li, J. Shi, Y. Qin, Q. Wang, H. Luo, P. Zhang, Z.X. Guo, D.K. Park, Alkylation and arylation of single-walled carbon nanotubes by mechanochemical method. Chem. Phys. Lett. **444**(4), 258–262 (2007)
63. H. Yoon, M. Yamashita, S. Ata, D.N. Futaba, T. Yamada, K. Hata, Controlling exfoliation in order to minimize damage during dispersion of long SWCNTs for advanced composites. Sci Rep **4**, Article number: 3907 (2014)
64. A. Graf, Y. Zakharko, S.P. Schießl, C. Backes, M. Pfohl, B.S. Flavel, J. Zaumseil, Large scale, selective dispersion of long single-walled carbon nanotubes with high photoluminescence quantum yield by shear force mixing. Carbon **105**, 593–599 (2016)
65. J. Zhang, B. Li, Universal dispersion of single-walled carbon nanotubes in the liquid phase inspired by Maya blue. J. Mater. Chem. A **1**, 10626–10630 (2013)
66. S. Prakash Yadav, S. Singh, Carbon nanotube dispersion in nematic liquid crystals: An overview. Prog. Mater. Sci. **80**, 38–76 (2016)
67. O. Yaroshchuk, S. Tomylko, O. Kovalchuk, N. Lebovka, Liquid crystal suspensions of carbon nanotubes assisted by organically modified Laponite nanoplatelets. Carbon **68**, 389–398 (2014)
68. C. Tang, T. Zhou, J. Yang, Q. Zhang, F. Chen, Q. Fu, L. Yang, Wet-grinding assisted ultrasonic dispersion of pristine multi-walled carbon nanotubes (MWCNTs) in chitosan solution. Colloids Surf. B Biointerfaces **86**(1), 189–197 (2011)
69. M.A. Kabbani, C. Sekhar Tiwary, P.A.S. Autreto, et al., Ambient solid-state mechano-chemical reactions between functionalized carbon nanotubes. Nat. Commun. **6**(7291), 8 (2015)
70. J. Hilding, A. Eric, E.A. Grulke, Z.G. Zhang, F. Lockwood, Dispersion of carbon nanotubes in liquids. J. Dispers. Sci. Technol. **24**(1), 1–41 (2003)
71. K. Imasaka, J. Suehiro, Y. Kanatake, Y. Kato, M. Hara, Preparation of water-soluble carbon nanotubes using a pulsed streamer discharge in water. Nanotechnology **17**(14), 3421–3427 (2006)
72. K. Imasaka, Y. Kato, J. Suehiro, Enhancement of microplasma-based water-solubilization of single-walled carbon nanotubes using gas bubbling in water. Nanotechnology **18**(33), 335602 (2007)
73. Y. Ishibashi, R. Hanaoka, N. Osawa, S. Takata, Y. Kanamaru, H. Anzai, Effect of barrier discharge on homogeneous dispersion of carbon nanotubes in octylalcohols. Int J. Plasma Environ. Sci. Technol. **5**(1), 62–67 (2011)
74. Y. Wang, Z. Iqbal, S. Mitra, Microwave-induced rapid chemical functionalization of single-walled carbon nanotubes. Carbon **43**(5), 1015–1020 (2005)
75. S. Mitra, Z. Iqbal. Microwave induced functionalization of single wall carbon nanotubes and composites prepared therefrom. WO06099392, 2006
76. H.X. Wu, X.Q. Qiu, W.M. Cao, Y.H. Lin, R.F. Cai, S.X. Qian, Polymer-wrapped multiwalled carbon nanotubes synthesized via microwave-assisted in situ emulsion polymerization and their optical limiting properties. Carbon **45**(15), 2866–2872 (2007)

77. J. Li, H. Grennberg, Microwave-assisted covalent sidewall functionalization of multiwalled carbon nanotubes. Chemistry **12**(14), 3869–3875 (2006)
78. S.-J. Cho, S. Prasad Shrestha, S.-B. Lee, J.-H. Boo, Electrical characteristics of carbon nanotubes by plasma and microwave surface treatments. Bull. Kor. Chem. Soc. **35**(3), 905–907 (2014)
79. F. Hof, K. Kampioti, K. Huang, C. Jaillet, et al., Conductive inks of graphitic nanoparticles from a sustainable carbon feedstock. Carbon **111**, 142–149 (2017)
80. K. Jagadish, S. Srikantaswamy, K. Byrappa, L. Shruthi, M. Ramesh Abhilash, Dispersion of multiwall carbon nanotubes in organic solvents through hydrothermal supercritical condition. J. Nanomater. **2015**, 6 (2015.) Article ID 381275
81. K. Naoi, T. Kurita, M. Abe, et al., Ultrafast nanocrystalline-TiO_2(B)/carbon nanotube hyperdispersion prepared via combined ultracentrifugation and hydrothermal treatments for hybrid supercapacitors. Adv. Mater. **28**(31), 6751–6757 (2016)
82. M. Lung Sham, J. Li, P. Ma, J.K. Kim, Cleaning and functionalization of polymer surfaces and nanoscale carbon fillers by UV/ozone treatment: A review. J. Compos. Mater. **43**(14), 1537–1564 (2009)
83. C. Fan, W. Li, X. Li, et al., Efficient photo-assisted Fenton oxidation treatment of multi-walled carbon nanotubes. Chin. Sci. Bull. **52**(15), 2054–2062 (2007)
84. C.-Y. Chen, C.T. Jafvert, Photoreactivity of carboxylated single-walled carbon nanotubes in sunlight: Reactive oxygen species production in water. Environ. Sci. Technol. **44**(17), 6674–6679 (2010)
85. M.N. Ahmad, J.-Y. Xie, Y.-H. Ma, W.-T. Yang, Surface functionalization of single-walled carbon nanotubes using photolysis for enhanced dispersion in an organic solvent. New Carbon Mater **25**(2), 134–140 (2010.) Xinxing Tan Cailiao
86. T. Takada, T. Baba, S. Abe, Simple process for sidewall modification of multi-walled carbon nanotubes with polymer side chain radicals generated by ultraviolet-induced C–cl bond dissociation of polystyrene derivatives. Carbon Res. **2**, 20 (2016.) 11 pp
87. J. Lee, T. Jeong, J. Heo, S.H. Park, D. Lee, J.B. Park, H. Han, S.M. Yoon, Short carbon nanotubes produced by cryogenic crushing. Carbon **44**(14), 2984–2989 (2006)
88. Y. Yan, J. Miao, Z. Yang, et al., Carbon nanotube catalysts: Recent advances in synthesis, characterization and applications. Chem. Soc. Rev. **44**, 3295–3346 (2015)
89. A. Pérez del Pino, E. György, L. Cabana, B. Ballesteros, G. Tobias, Ultraviolet pulsed laser irradiation of multi-walled carbon nanotubes in nitrogen atmosphere. J. Appl. Phys. **115**(9), 093501 (2014). doi:10.1063/1.4864776
90. K. Narimatsu, Y. Niidome, N. Nakashima, Pulsed-laser induced flocculation of carbon nanotubes solubilized by an anthracene-carrying polymer. Chem. Phys. Lett. **429**, 488–491 (2006)
91. M. Ilčíková, M. Danko, M. Doroshenko, Visualization of carbon nanotubes dispersion in composite by using confocal laser scanning microscopy. Eur. Polym. J. **79**, 187–197 (2016)
92. K.J. Ziegler, D.J. Schmidt, U. Rauwald, K.N. Shah, E.L. Flor, R.H. Hauge, R.E. Smalley, Length-dependent extraction of single-walled carbon nanotubes. Nano Lett. **5**(12), 2355–2359 (2005)
93. X. Chen, X.-Z. Tang, Y.N. Liang, et al., Controlled thermal functionalization for dispersion enhancement of multi-wall carbon nanotube in organic solvents. J. Mater. Sci. **51**(12), 5625–5634 (2016)
94. D. Dima, M. Murarescu, G. Andrei, Dispersion of carbon nanotubes coated with iron (iii) oxide into polymer composite under oscillating magnetic field. Dig. J. Nanomater. Biostruct. **5**(4), 1009–1014 (2010)
95. Dispersion and Alignment of Carbon Nanotubes Using Electric Fields. http://hpmi.research.fsu.edu/. Accessed on 5 May 2017
96. M.-J. Chen. Development and parametric studies of carbon nanotube dispersion using electrospraying. M.Sc. Thesis, 2007, Florida State University. http://diginole.lib.fsu.edu/islandora/object/fsu%3A182086

97. K. Kanakamedala, J. DeSoto, A. Sarkar, T.D. Theda Daniels Race, Study of electrospray assisted electrophoretic deposition of carbon nanotubes on insulator substrates. Electron. Mater. Lett. **11**(6), 949–956 (2015)
98. Y. Meng, G. Xin, J. Nam, S.M. Cho, H.J. Chae, Electrospray deposition of carbon nanotube thin films for flexible transparent electrodes. J. Nanosci. Nanotechnol. **13**(9), 6125–6129 (2013)
99. K.-h. Ahn, S.-M. Kim, I.J. Yu, Multi-walled carbon nanotube (MWCNT) dispersion and Aerosolization with hot water atomization without addition of any surfactant. Saf. Health Work **2**(1), 65–69 (2011)
100. http://www.sono-tek.com/carbon-nanotube-dispersion/. Accessed on 8 May 2017

Chapter 3
Chemical Methods

3.1 Functionalization Leading to Soluble CNTs

3.1.1 Inorganic Compounds

3.1.1.1 Ozone

Ozone can be used alone or with combination with UV light. Thus, chemical functionalization of MWCNTs was carried out [1, 2] by UV/ozone treatment. The presence of oxygen-containing groups (such as carboxylic, quinine, and hydroxyl groups) on the MWCNT surfaces by UV/ozone treatment was confirmed resulting in dispersion stability better than for pristine MWCNTs in polar solvents. Authors predicted the dispersion states of MWCNTs treated with UV/ozone in 32 different solvents in terms of a Hansen solubility parameter sphere, which increased as UV/ozone exposure time increased. The solubility of UV/ozone-treated MWCNTs in polar organic solvents can be increased by as much as 320%, maintaining almost the same thermal stability, compared to raw MWCNTs [3]. The dispersed nanotubes suspension (MWCNTs functionalized by oxygen- and hydrogen-containing groups) can be stable for a few months after treatment [4].

Dry ozone gas flow was also used alone leading to the stable aqueous dispersion of MWCNTs after treatment above a critical level due to appearance of carboxylic groups on MWCNT surface similar to their acidic modification; it could have potential bio-applications [5]. A typical apparatus for ozone generation and SWCNT treatment is shown in Fig. 3.1 [6]. In particular, it was established for SWCNT treatment with ozone that O_3-induced oxidation takes place preferentially on the exposed geometrical outer surface (Fig. 3.2) of CNT conglomerates due to the high reactivity of the O_3 molecule at the sample surface. Also, in case of SWCNT treatment with ozone [7], an increase in the defect density takes place due to the oxidation of the nanotubes, testifying about the removal of π-conjugated electron states near the Fermi level. It leads to the increase in electrical resistance of SWCNTs.

O.V. Kharissova, B.I. Kharisov, *Solubilization and Dispersion of Carbon Nanotubes*, DOI 10.1007/978-3-319-62950-6_3

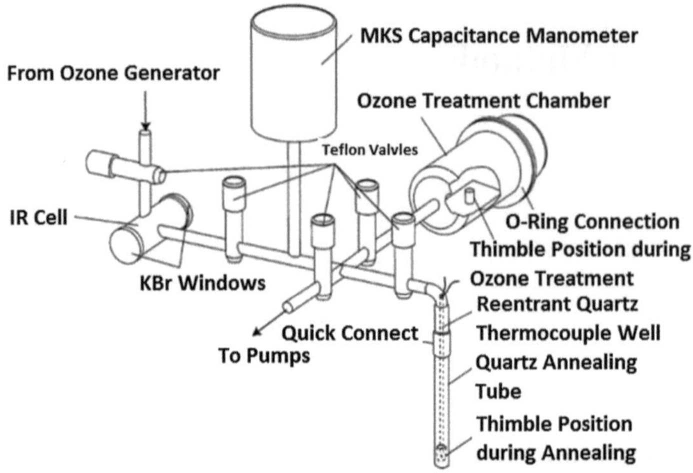

Fig. 3.1 Schematic representation of the glass vacuum system used for ozone purification and control, nanotube oxidative treatment with O_3, and high-temperature annealing (Reproduced with permission of the *American Chemical Society*)

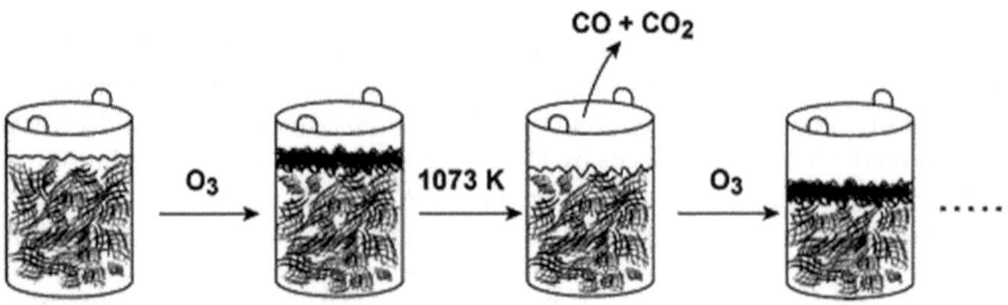

Fig. 3.2 Schematic of O_3 etching of the agglomerated SWCNT sample, showing preferential reaction on the outer geometric surface of the sample (Reproduced with permission of the *American Chemical Society*)

Related applications of these ozone CNT treatments are mainly related with their dispersion in polymers. Thus, an improved dispersion of ozone-treated MWCNTs in polycarbonate matrix was observed, contributing to the enhanced yield strength of MWCNT/PC nanocomposites [8]. Similarly, UV/ozone irradiation allowed an improvement of MWCNT dispersion in polystyrene matrix [9]. Among other applications of ozonated CNTs, we note the catalysis (oxidative dehydrogenation of ethyl benzene to styrene) [10].

3.1.1.2 Liquid Ammonia

In our opinion, the most intriguing example in this section is an unusual application of Na-$NH_{3(liq.)}$ system, which is well known in classic inorganic chemistry courses.

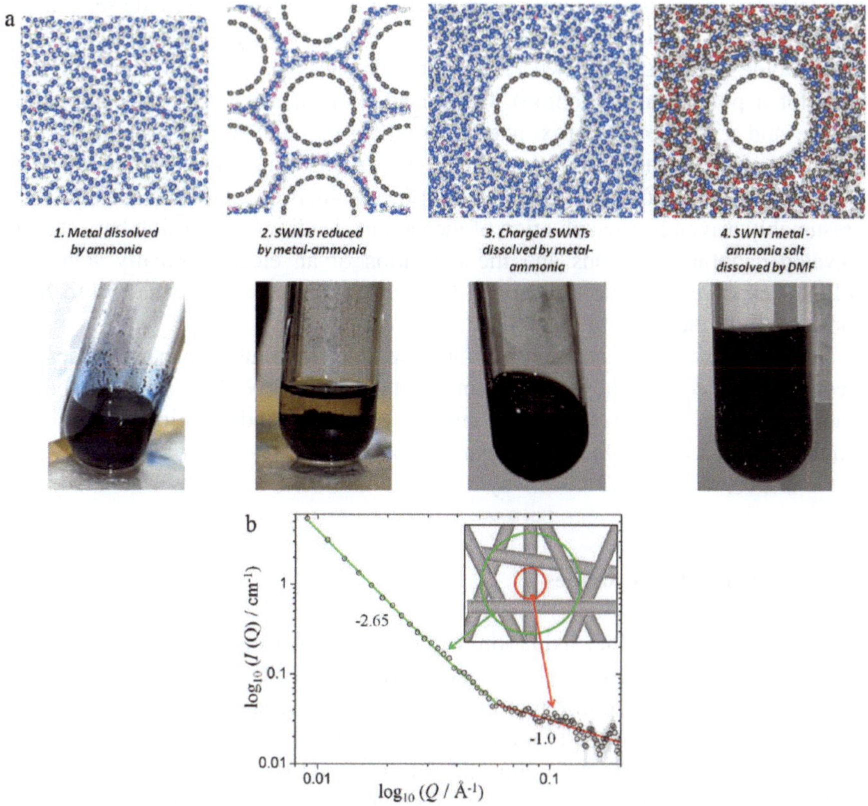

Fig. 3.3 Liquid ammonia reduction of ARC SWCNTs. (**a**) Process scheme for the reduction, solvation, and subsequent dissolution in sodium–ammonia, illustrated by atomistic models (Na ions in pink) and photographs of relevant phases (M:C 1:20). (**b**) SANS pattern from the spontaneously dissolved SWCNTs (M:C 1:24) in sodium–ammonia (Na:ND_3). The red and green lines show power law best fits, giving exponents of −1.0 and −2.65, as marked. The inset shows a schematic of the corresponding nanotube mesh and associated length scales (Reproduced with permission of the *American Chemical Society*)

Thus, an intrinsically scalable method for SWCNT dispersion and separation, using reductive treatment in sodium metal–ammonia solutions, was discussed (Fig. 3.3) [11], optionally followed by selective dissolution in a polar aprotic organic solvent. In situ small-angle neutron scattering demonstrated the presence of dissolved, unbundled SWCNTs in solution, at concentrations reaching at least 2 mg/mL; the ability to isolate individual nanotubes was confirmed by atomic force microscopy. The soluble fraction contained predominately large SWCNTs. SWCNTs can indeed be unbundled to give individual tubes in solution by reductive charging in ammonia. This process requires control of the metal/carbon (M:C) ratio, to much lower values (approximately 1:10) than typical in the literature (>1:1), but does not inherently involve covalent functionalization of the tubes. The initial treatment of

SWCNTs in liquid sodium metal–ammonia solution is a key step in the separation. Following the removal of the liquid ammonia, a dry powder of sodium "nanotubide" was formed (where "nanotubide" was proposed by authors as a term for a pure nanotube anion). On the addition of dry DMF, the nanotubide swells, and a fraction (approximately 40 wt.%) of the SWCNTs spontaneously dissolved. It is worth emphasizing that *no stirring* and, more importantly, *no ultrasound* were used at any stage of the process. This spontaneous dissolution is presumably driven by the solvation of the cations, leading to repulsion between the solvated nanotubide anions and the formation of an electrostatically stabilized colloid (or polyelectrolyte molecule). This separation process was shown to scale consistently to the 100 mg level.

In case of *ammonia aqueous solutions*, we note the solution plasma processing (see also the section above on plasma methods) of MWCNTs in this medium leading to their solubilization [12]. Instead of 7 days and additional chemicals, required for 185 mg in other investigations, only 2 h was necessary to prepare 10 g of the dispersed MWCNTs this way.

3.1.1.3 Inorganic Peroxides

H_2O_2 treatment is a milder oxidation method in comparison with strong acids (see section below) and, as well as acids (HNO_3, H_2SO_4, HNO_3+H_2SO_4, and HNO_3+HCl), is effective in removing non-nanotube carbonaceous species present in the sample, i.e., being used for purification of CNTs [13], or as a pretreatment for further processes in MWCNT composite fabrication [14]. Under H_2O_2 oxidation of MWCNTs, several functional groups such as carboxylic (−COOH), carbonyl (−CO), and hydroxyl (−OH) groups are formed on their surface; however, hydroxyl groups were preferentially formed and reached a maximum atomic concentration of about 46% in 4 days of oxidation [15]. Hydrogen peroxide is frequently used with acids or NH_3 solution [16], resulting different grades of MWCNT purification and concentrations of carboxylic groups (Fig. 3.4), responsible for the "solubility" of CNTs.

An easy strategy was developed for the water-soluble MWCNTs without the length cutting of MWCNTs via the ultrasonic-assisted chemical functionalization with a redox couple ammonium persulfate/diethanolamine at r.t [17]. Diethanolamine groups were found to be covalently attached onto the sidewalls of the MWCNTs; the surface functional group content of the functionalized MWCNTs was 0.66 mmol/g; the MWCNTs were exfoliated into individual nanotubes. The resultant MWCNTs were dispersed easily in water, and the dispersion was stable for more than 4 days. In a relative research [18], MWCNTs with improved dispersion property (Fig. 3.5) were prepared by a mild and fast hydrothermal treatment with the use of potassium persulfate. The modified MWCNTs were highly soluble in polar solvents such as water, ethanol, dimethylformamide, tetrahydrofuran, and dimethylacetamide. These treated MWCNTs showed the highest solubility in water, and the solubility was around 1.16 mg/ml. However,

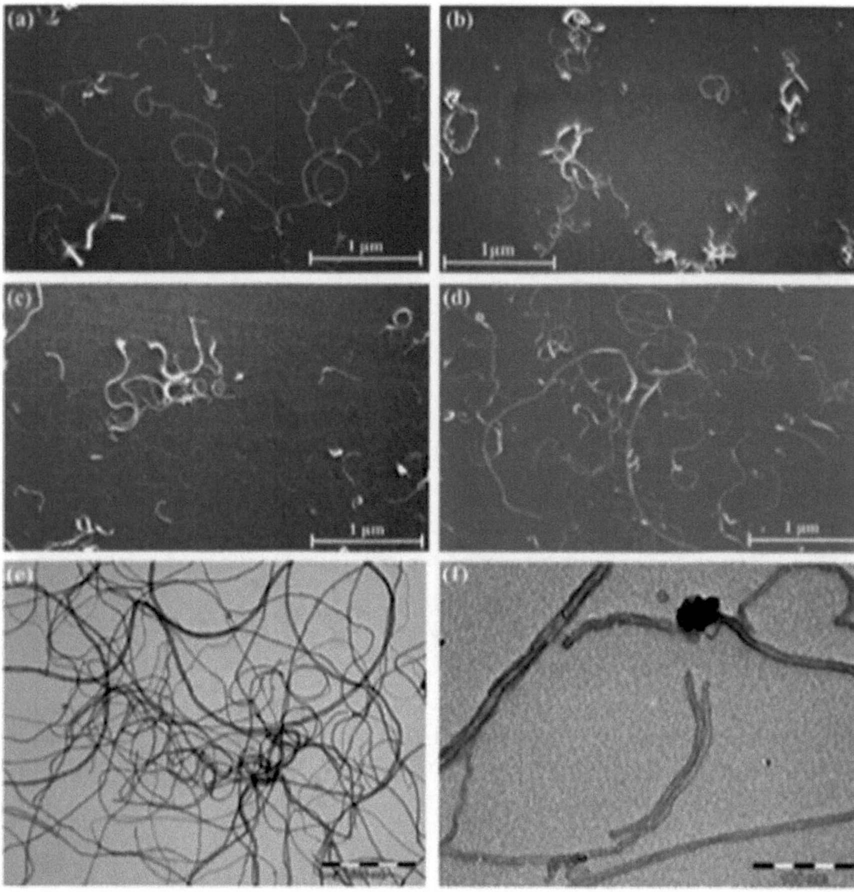

Fig. 3.4 SEM images of the treated MWCNTs at two different magnifications: (**a**) after purification with HCl; (**b**) after nitric acid oxidation; (**c**) after oxidation with piranha; (**d**) after oxidation with ammonium hydroxide and hydrogen peroxide mixture; TEM images after modification with (**e**) ammonium hydroxide and hydrogen peroxide mixture; and (**f**) after nitric acid (Reproduced with permission of the *Elsevier Science*)

they cannot be well dispersed in nonpolar solvents (such as hexane and toluene) and eventually precipitated. This indicated that various hydrophilic groups were created on the treated MWCNTs after hydrothermal modification. We note herein that hydro(solvo)thermal method, in particular, combined with microwave heating (MW-hydro(solvo)thermal technique) is a useful synthesis tool in chemistry and nanotechnology (see also sections on physical methods above) [19].

In addition, dispersive effects of persulfates depend on such important factors as pH and oxidant concentration. Thus, the dispersive effects of sodium persulfate (SPS; its activation modes are shown in reactions (3.1, 3.2, 3.3, 3.4, and 3.5)) are optimal (Fig. 3.6) in a pH range 5–12: in these conditions, the BSPS/CNTs (B – base activated) are well dispersed within 1 month (due to $pH > pK = 4.53$ of

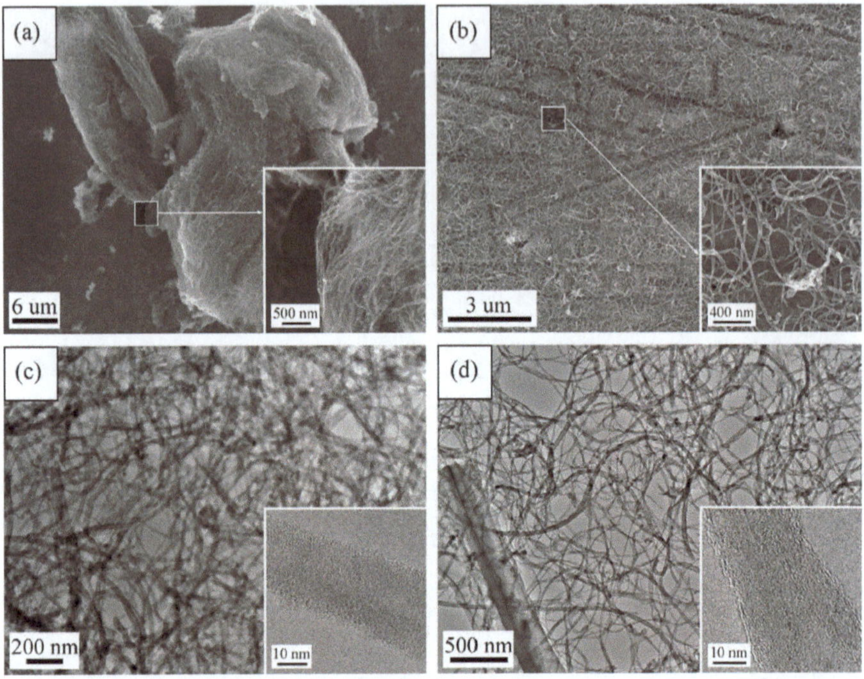

Fig. 3.5 Typical FESEM images of (**a**) pristine MWCNTs and (**b**) treated MWCNTs, with their high magnification images in the inset; typical TEM images of (**c**) pristine MWCNTs and (**d**) treated MWCNTs, with their individual MWCNT images in the inset (Reproduced with permission of the *Elsevier Science*)

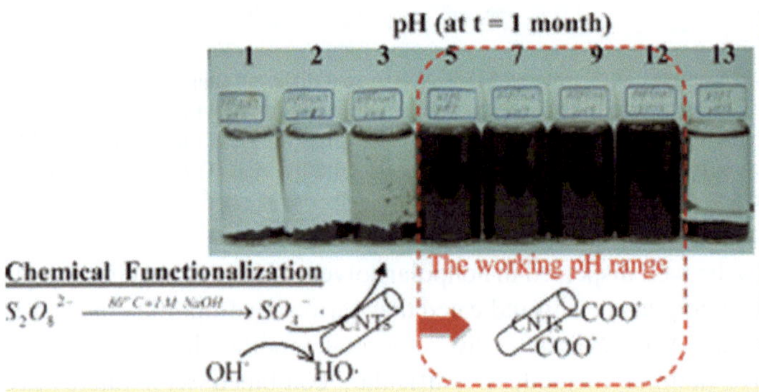

Fig. 3.6 Dispersion of MWCNTs at different pH magnitudes under SPS activation (Reproduced with permission of the *American Chemical Society*)

–COONa) [20]. However, at extreme alkaline conditions (e.g., pH > 13), the stronger negatively charged $-COO^-$ groups on CNTs would be affected by the increased sodium cations, resulting in compression of the electrical double layer of the charged surface and inducing the destabilization of dispersed particles. Dispersion stability of the MWCNTs, oxidized using $K_2S_2O_8$, increases due to the generation of sulfate groups on surface of the MWCNTs, but higher concentration of SO_4^{2-} groups has a detrimental effect on dispersion stability (the optimal concentration of potassium persulfate for the best dispersion stability of the oxidized MWCNTs was 0.01 M) [21].

Thermal SPS activation:

$$S_2O_8^{2-} + \text{heat} \rightarrow 2SO_4^{2-} \tag{3.1}$$

Base SPS activation:

$$S_2O_8^{2-} + OH^- \rightarrow HSO_4^- + SO_5^{2-} \tag{3.2}$$

$$SO_5^{2-} + H_2O \rightarrow HO_2\text{-} + SO_4^{2-} + H^+ \tag{3.3}$$

$$HO_2\text{-} + S_2O_8^{2-} \rightarrow SO_4\text{-} + SO_4^{2-} + H^- - O_2\text{-} \tag{3.4}$$

Radical interconversion:

$$SO_4\text{-} + OH^- \rightarrow HO + SO_4^{2-} \tag{3.5}$$

3.1.1.4 Elemental Metals

There are a lot of reports on the decoration of carbon nanotubes with elemental metal particles (generalized in a recent review [22]), but a few such metal dopations of CNTs have been studied in a point of view of improvement of CNT dispersion in solvents. The effect of metal dopation on CNT surface is itself weak in respect of CNT "solubilization," so doped CNTs are normally dispersed using an additional force, i.e., ultrasound, linkers, or surfactants. Thus, highly conducting crystalline metal nanoparticles (Pd, Au, Ag) decorated the MWCNT surface (Fig. 3.7), and further dispersion of these metal–MWCNTs in polar base fluids like deionized water (DI water) and ethylene glycol (EG) without the addition of any surfactant was carried out [23]. Metal-decorated MWCNTs were ultrasonically dispersed homogenously in the base fluid, forming nanofluids.

As an example of stabilizers, three elemental metals (Au, Fe, Co) have been used in the form of nanoparticles or solutions in liquid ammonia (alkali metals) through organic linkers. In those experiments, water-soluble hybrids of MWCNT and *gold* nanoparticles (Au@MWCNTs) were fabricated via the in situ solution method using an optoelectronic-active compound of *N,N′*-bi(2-mercaptoethyl)-perylene-3,4,9,10-tetracarboxylic diimide as interlinker and stabilizer [24]. It was found that the formed hybrid exhibited strong visible luminescence under UV lamp irradiation, which might extend its potential applications to biological labeling.

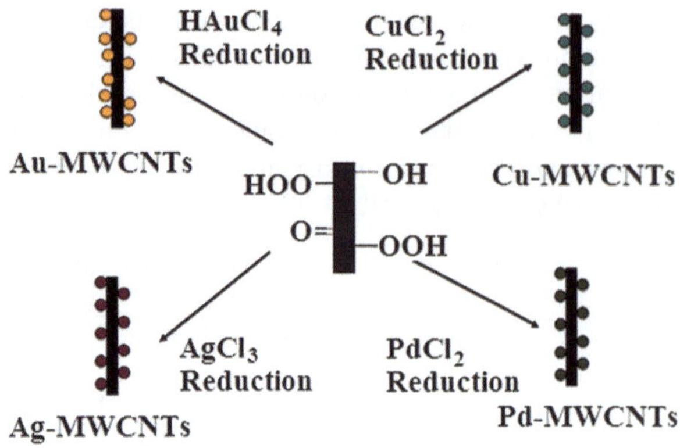

Fig. 3.7 Schematic representation of metal decoration on MWCNT surface (Reproduced with permission of the *AIP Publishing*)

In another report, dedicated to gold–CNT hybrids, the efficient aqueous dispersion of pristine HiPco SWCNTs with ionic liquid surfactants 1-dodecyl-3-methylimidazolium bromide (D) and 1-(12-mercaptododecyl)-3-methylimidazolium bromide (M), the thiolation of nanotube sidewalls, and the controlled self-assembly of positively charged SWCNT–D,M composites on gold were studied [25]. *Iron*-filled Fe@MWCNTs were surface modified with various functionalities via a rapid, single-step process involving ultrasonication-assisted and microwave-induced radical polymerization reactions [26]. The offered process is universal for both hydrophobic (e.g., polystyrenes and polymethyl methacrylate) and hydrophilic (e.g., polyacrylamide, polyacrylic acids, and polyallyl alcohols) polymer chains (see also polymer section below), which can be chemically grafted onto the surface of MWCNTs in the same conditions within ~10 min. The solubilities of the formed functionalized MWCNTs are in the range of 1200–2800 mg/l in solutions. An intriguing method for CNT derivatization with magnetic nanoparticles is described in [27]. Capped iron oxide or *cobalt and cobalt/platinum* magnetic nanoparticles were attached to carbon nanotubes by means of an interlinker molecule, a carboxylic derivative of pyrene (reaction scheme 3.6) (see also sections below on the functionalization with aromatic molecules). The available carboxylic groups of pyrene derivative can be further linked to metal or metal oxide nanoparticles. The formed composites were highly soluble in organic solvents, such as chloroform, toluene, and hexane.

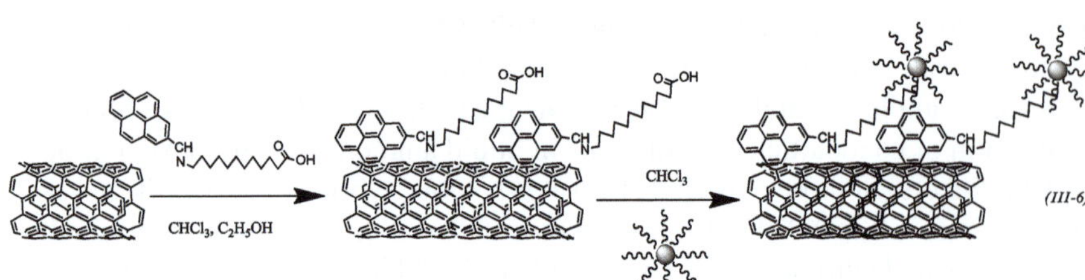

(III-6)

Modification of CNTs by capped magnetic nanoparticles.

Sidewall-functionalized nanotubes, soluble in organic solvents, were prepared by alkylation of nanotube salts obtained using *lithium*, *sodium*, or *potassium* in liquid ammonia (*Billups* reaction) [28]. Such reactions can produce different types of derivatized CNTs. It was shown that the alkali metal intercalates into the SWCNT ropes (in case of further reaction of nanotube salts Na–CNTs with PhI, the arylated CNTs can be produced); alkali metals used to behave differently requiring distinct temperature ranges (the least range corresponds to Li). The *Billups* reaction protocol involving dissolving metal reduction of MWCNTs and their subsequent alkylation or arylation was shown to produce functionalized MWCNTs that were soluble in either organic or aqueous solvents. This method allows for the attachment of alkyl or aryl pendent groups, using either lithium or sodium, and has been used to produce gram quantities of alkylated MWCNTs [29] (reaction scheme 3.7).

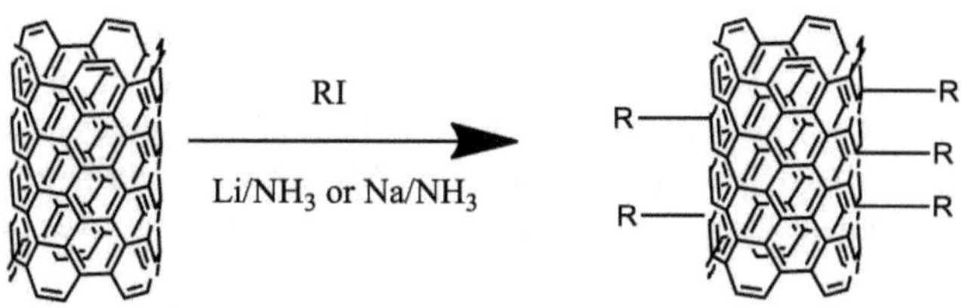

(III-7)

R = p-$C_6H_4NH_2$, Ph, p-C_6H_4COOH, p-$C_6H_4CH_2OH$, p-$C_6H_4C(CH_3)_3$, p-$C_6H_4OCH_3$, $C_{12}H_{25}$

Functionalization of MWCNTs by *Billups* reaction.

The use of Li as the mostly used alkali metal in these processes was in situ Raman studied in detail [30]. It was shown that addition of 1-iodododecane to the lithiated SWCNTs resulted in the covalent attachment of dodecyl groups. The intercalation of lithium throughout the SWCNT ropes led to complete dodecylation of all individual SWCNTs. Lithium was also applied to yield a water-soluble polyethylene glycol PEG–CNT conjugate in a two-step process, including the reaction of lithium, SWCNTs, and 11-bromoundecanoic acid in liquid ammonia forming carboxylic acid derivatized SWCNTs (reaction scheme 3.8) and further interaction of the obtained product with H_2N-PEG-OMe in mixture of solvents [31, 32].

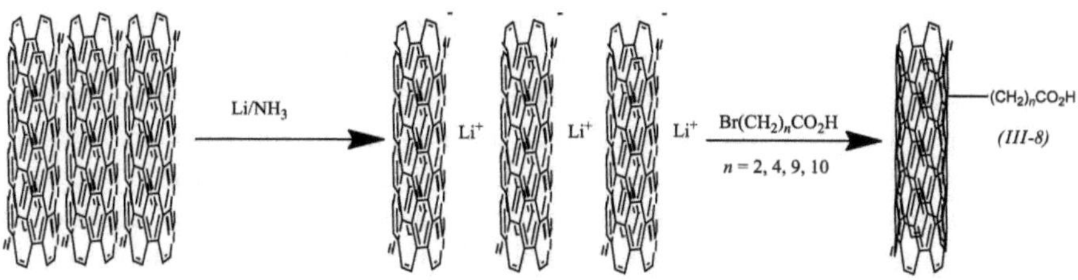

(III-8)

General procedure for the synthesis of carboxylic acid derivatized SWCNTs.

3.1.1.5 Elemental Nonmetals

Nonmetals in elemental form, applied for precisely enhancing CNTs "solubility," are also scarce, although they are widely used for CNT decoration. Thus, doping of low-dimensional graphitic materials, including graphene, graphene quantum dots, and SWCNTs with *boron*, *sulfur*, or *nitrogen*, carried out by simple fluorination followed by annealing in a dopant source, can considerably change their properties [33]. *Iodine* doping into SWCNTs can be effectively done using an electrochemical method [34] and can be easily and finely controlled by changing the polarity. Iodine molecules were found to be mainly inserted into the hollow core of SWCNTs, and these iodine-doped SWCNTs can be homogeneously dispersed in water at low temperature (ca. <15 °C). The enhanced dispersion stability of MWCNTs in a fluorinated refrigerant (FC-72) was reached by a combined process of oxidation and fluorination using (tridecafluoro-1,1,2,2-tetrahydrooctyl)trichlorosilane as a *fluorine* source and reactant with –OH groups on MWCNT surface, obtained by chemical oxidation [35]. An excellent colloidal stability of thus obtained MWCNT–F in FC-72 is explained by the chemical affinity between FC-72 and functional groups (-CF_n) on the sidewalls of MWCNTs. Fluorine-treated CNT solutions are known long ago to be metastable, persisting over a week in dispersion [36].

Other *carbon allotropes* and their derivatives, the *nanodiamonds* [37] (NDs), *graphene* [38] and *graphene oxide*, were applied for improvement of CNT dispersion. Thus, nanodiamond particles were used [39] to disperse CNTs, allowing the formation of their stable colloidal suspensions. Both MWCNTs and SWCNTs could be suspended in deionized water using either detonation NDs or high-pressure, high-temperature NDs. Negatively charged NDs were found to suspend CNTs in deionized water more favorably as compared to positively charged particles, possibly caused by electrostatic interactions. Graphene oxide (GO) sheets, considered as "soft" 2D macromolecules containing multiple aromatic regions and hydrophilic oxygen groups, can adsorb the pristine MWCNTs through the π stacking interaction, thus causing pristine MWCNTs to stably disperse and fractionate in aqueous media (Figs. 3.8 and 3.9) [40], forming optically transparent solution-like aqueous suspensions [41]. The water-soluble and water-insoluble complexes can be obtained by changing the initial proportion of MWCNTs to GO sheets, which are important for noncovalent approaches toward solubilizing CNTs and open a new way for GO applications in colloidal chemistry. Solubility results indicated that the GO sheets lean to stabilizing MWCNTs with larger diameters, mainly depending on whether the CNTs are inclined to form bundles, twisted structures, or CNT–GO complexes. Dispersed GO/CNT composites can have applications for electrochemical detection of a variety of chemicals, in particular sodium nitrite, hydroquinone, and catechol [42], and for water purification [43]. In addition, stable aqueous dispersion of polymer-bound graphene sheets, produced via in situ reduction of exfoliated graphite oxide in the presence of cationic PEI, was further used for self-assembly with MWCNTs, resulting hybrid carbon films (Fig. 3.10) [44], which are promising as supercapacitor electrodes.

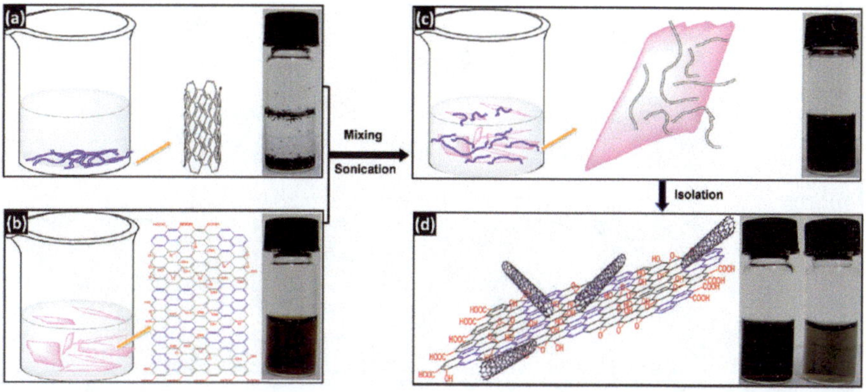

Fig. 3.8 Schematic mechanism for formation of the MWCNT–GO complexes (Reproduced with permission of the *American Chemical Society*)

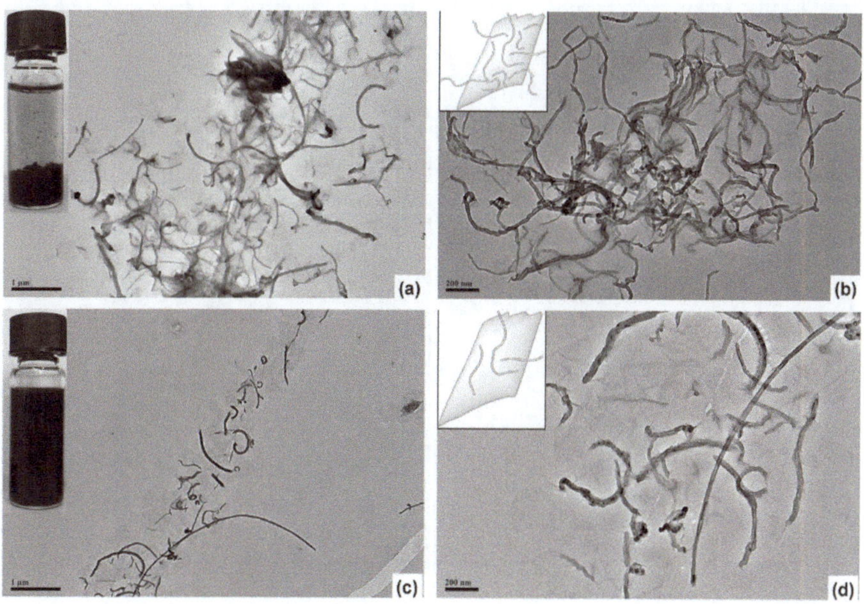

Fig. 3.9 TEM images of interactions formed between MWCNTs and GO sheets by changing the initial proportion of MWCNTs to GO sheets with 2:1 at different magnifications (**a**, **b**) and the initial proportion of MWCNTs to GO sheets with 1:2 at different magnifications (**c**, **d**). The inset in (**b**, **d**) is the schematic description of the MWCNT–GO complexes (Reproduced with permission of the *American Chemical Society*)

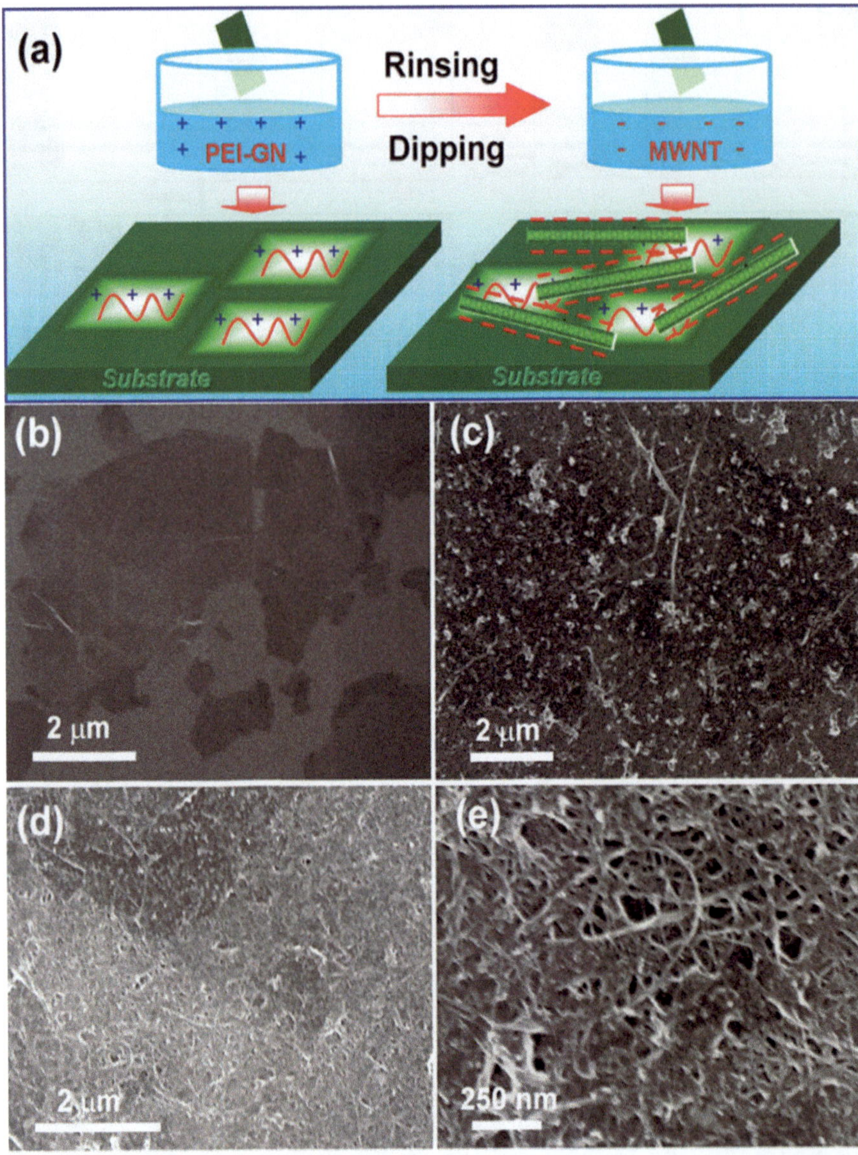

Fig. 3.10 (**a**) Illustration of positively charged PEI(cationic poly(ethyleneimine))-GN (graphene nanosheets) and negatively charged MWCNT film deposition process on an appropriate substrate (e.g., silicon wafer, ITO glass). SEM images of (**b**) the first layer PEI-GN and (**c**) the first bilayer [PEI-GN/MWCNT–COOH] film deposited on a silicon substrate. (**d**, **e**) SEM images of the [PEI-GN/MWCNT–COOH] film after the ninth deposition cycle under different magnifications (Reproduced with permission of the *American Chemical Society*)

3.1.1.6 CO_2 and SiO_2

Simple inorganic compounds are capable to solubilize CNTs even without formation of direct chemical bonds "CNTs substrate." Thus, the dispersion of MWCNTs at the very low milligram level was achieved by effervescence due to the in situ generation of carbon dioxide like in some traditional pharmaceuticals [45]. A special tablet containing the effervescence precursors (sodium carbonate as carbon dioxide source and sodium dihydrogen phosphate as proton donor) and the MWCNTs was applied for MWCNT solubilization. Another reported type of carbon dioxide action was completely different. Thus, an efficient method using supercritical (SC) CO_2 to help a *p*-type conducting polymer, conjugated polymer MDMO-PPV wrapping on CNTs [46]. 1D CNTs were periodically decorated with the congeries of MDMO-PPV molecular chains forming functional nanohybrid structures. The solubility and light emission of the hybrid MDMO-PPV/CNTs can be controlled by varying SC CO_2 pressure. Chemical interaction between MDMO-PPV and CNTs was confirmed, and the interaction not only bestowed CNTs desired solubility but also added functionality of light emission to CNTs. For DMSO as the solvent, the higher SC CO_2 pressure ensures the excellent dispersion and solubility of MDMO-PPV/CNTs. A relative research was carried out for a typical piezoelectric polymer, poly(vinylidene fluoride) (PVDF), which was wrapped on SWCNTs using a simple supercritical carbon dioxide (SC CO_2) antisolvent-induced polymer epitaxy method [47] and for preparation of CNT/poly (phenyl sulfone) composites (Fig. 3.11) [48].

In case of SiO_2 [49], nanoscale silica functional groups were covalently bonded to CNTs to improve their dispersion in water while providing interfacial bond sites for the proposed matrix material [50]. At the nanoscale, interparticle spacing was greatly increased, and, as a result, dramatically improved dispersion stability in water (for 6 months) was observed. Mechanism of dispersion of SWCNTs with

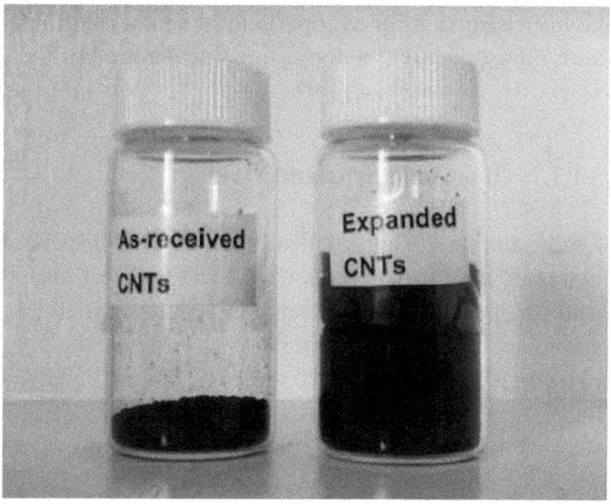

Fig. 3.11 Photopicture of carbon nanotubes with the same weight (0.10 g) before (*left*) and after (*right*) expansion using supercritical CO_2 (Reproduced with permission of the *Wiley*)

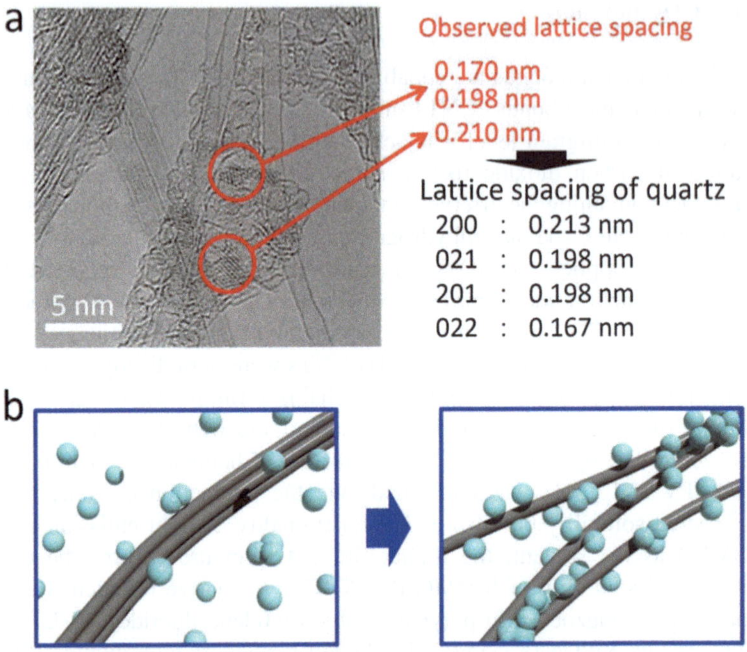

Fig. 3.12 Mechanism of dispersion of SWCNTs with nanosilicas. (**a**) TEM image of SWCNTs with nanosilica particles along with the lattice spacings of the lattice fringes and quartz segments. The lattice spacings of nanosilica adsorbed on the SWCNTs were determined to be 0.17, 0.198, and 0.21 nm, which are in good agreement with the 200, 021, 201, and 022 planes of quartz crystal. (**b**) Nanosilica adheres to the surface of SWCNTs, leading to debundling SWCNTs (Reproduced with permission of the *American Chemical Society*)

nanosilicas and view of suspensions are shown in Figs. 3.12 and 3.13 [51]. In addition, MWCNTs were dispersed in aqueous solutions through *alumina-coated silica* (ACS) nanoparticle halos [52]. MWCNTs were directly dispersed into a highly charged ACS nanoparticle aqueous solution, which was stable for weeks after ultrasonication, without functionalization of their surfaces.

3.1.1.7 Inorganic Salts and Oxides

Few examples of salt applications for CNT dispersion are known. The effects of inorganic monovalent salts (NaCl and NaI) on dispersion stability of CNTs were studied, performing all-atom molecular dynamics simulations (MDS) using nonpolarizable interaction models to compute the potential of mean force between two (10,10) SWCNTs in the presence of NaCl/NaI and compare to the potential of mean force between SWCNTs in pure water [53]. Addition of salts enhances stability of the contact state between two SWCNTs on the order of 4 kcal/mol. The ion-specific spatial distribution of different halide anions gives rise to starkly

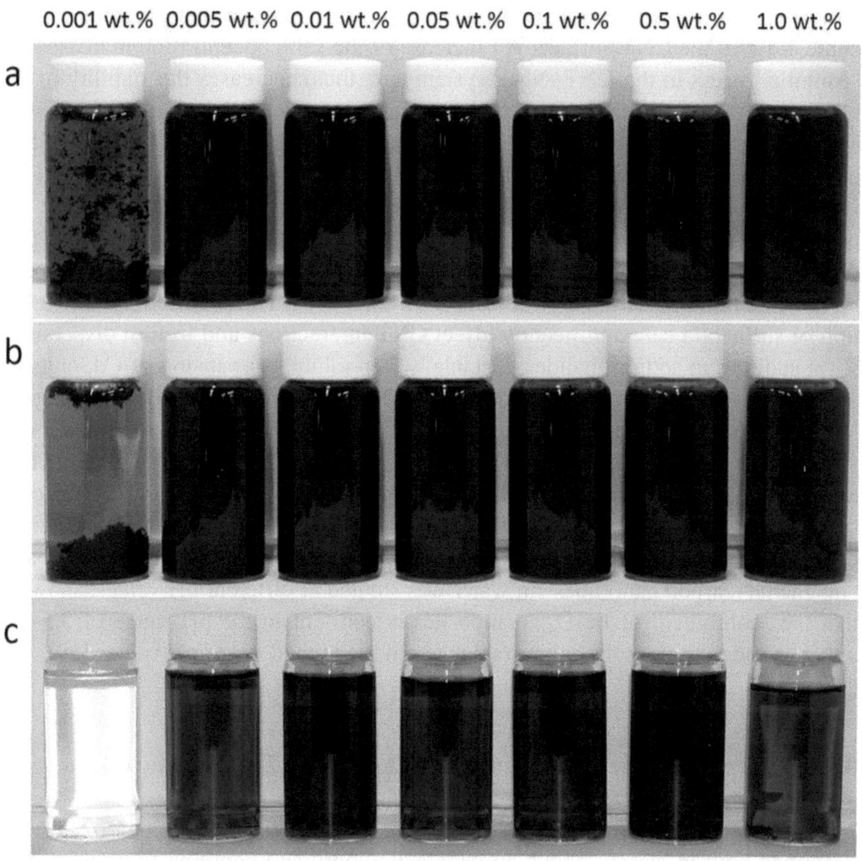

Fig. 3.13 Dispersion states of SWCNTs in aqueous solutions with differing nanosilica concentrations after (**a**) 30 min, (**b**) 1 day, and (**c**) centrifugation for 60 min at 4000 rpm (Reproduced with permission of the *American Chemical Society*)

different contributions to the free energy stability of nanotubes in the contact state. Iodide anion directly stabilizes the contact state to a much greater extent than chloride anion. The enhanced stability arises from the locally repulsive forces imposed on nanotubes by the surface-segregated iodide anion. Within the time scale of these simulations, both NaI and NaCl solutions stabilize the contact state by equivalent amounts. So, fine control over the SWCNT aggregation state is possible with salt additions, which result in selective aggregation of a very narrow diameter range [54].

In a related report [55], the effects of salt (NaI) on pristine CNT dispersions in N-methyl-2-pyrrolidone (NMP) were also MDS studied, revealing another group of conclusions. It was revealed that both ions (Na^+ and I^-) are depleted from the CNT surface in the CNT–NMP dispersions mainly due to the two reasons: (1) there is a high energy penalty for the ion partial desolvation at the CNT surface; (2) NMP molecules form a dense solvation layer at the CNT surface that prevents ions to come close to the CNT surface. An increase of the salt concentration increases the "osmotic" stress in the CNT–NMP system and, thus, decreases the stability of the CNT dispersions in NMP. Direct experiments confirmed the simulation results: addition of NaI salt into the NMP dispersions of pristine CNTs led to precipitation of CNTs (bundle formation) even at very small salt concentration ($\sim 10^{-3}$ mol·L^{-1}). It is necessary to mention that the MDS method was also used to compute for the potential of mean force as a function of the distance and orientation between a pair of SWCNTs in water [56]. An adaptive biasing force method was applied to speed up the calculations. Simulation results showed that CNT orientation and the addition of surfactant can significantly affect CNT interactions and in-turn dispersion.

In addition to sodium halides, a stable, water-soluble suspension of CdS/aligned MWCNT heterostructures with electron transfer from CdS nanoparticles to aligned MWCNTs, implying potential applications in photovoltaic cells, photocatalysis, and solar energy conversion, was reported [57]. CdSe quantum dots were noncovalently attached to SWCNTs through an intermediary 1-pyrenebutyric acid *N*-hydroxysuccinimide ester molecule [58]. A classic example of metal oxides for CNT applications is the use of iron oxides. Thus, the dispersing ability of magnetic composites "magnetite–carbon nanotubes" in water was studied at different pH values, showing that the "solubility" of Fe_3O_{4o} composites is constant in the pH range of 3.5–10, although these composites are unstable at both higher and lower pH values [59].

3.1.1.8 Strong Acidic Media (Inorganic Acids and Their Mixtures)

Mineral acids were extensively used in first attempts to functionalize CNTs, described now in classic books on nanotechnology, and continue to be used now too, generally as a first simple step of attachment of polar groups to CNT surface with their further interchange to others. Generally, suitable acid media include any acidic medium capable of dispersing CNTs in a substantially individualized state. By first dispersing CNTs in an acidic medium, bundled CNTs can be separated as individual CNTs, affording exposure of the CNT sidewalls and thereby facilitating the functionalization of such CNTs, wherein functional groups are attached to the subsequently exposed sidewalls of these individualized CNTs. Strong acids, oxidizing CNT surface, lead to formation of -COOH groups, so the use of sulfuric and/or nitric acid can be a first step of CNT functionalization, followed by further reaction of –COOH groups with, for example, $SOCl_2$ or other substances which are present in solution.

Among acids, the nitric acid is the most common, leading to –OH and –COOH attachment to CNT surface. Thus, the CNTs, CVD synthesized from heavy petroleum fractions like furnace oil and light diesel oil (~70 nm in the first case and ~90 nm in the second case), were purified, oxidized with dilute nitric acid, and dispersed in different solvents [60]. It was shown that these –COOH- and –OH-functionalized CNTs were highly dispersed in acetone and distilled water as compared to ethanol and methanol and stable even up to 60 °C. Also, combination of HNO_3 application with further physical treatments has also been developed. For instance, individual electric arc-produced SWCNTs were prepared in aqueous solution on a large scale through three steps of processing: refluxing in concentrated HNO_3, low speed centrifugation, and high-speed centrifugation [61]. The bulk production (10 g of starting SWCNTs) resulted in a concentration of 0.2 mg/mL individual SWCNTs stably dispersed in DI-H_2O without any external protection. It was shown that the aqueous dispersion contained approximately 80% individual SWCNTs with lengths ranging from 500 nm to 1 mcm and an absolute zeta potential value of ~72 mV with a concentration of 0.05 mg/mL at pH 5. The authors believed that it is this high zeta potential resulting from an electrical double layer which produces the repulsion to overcome the van der Waals attraction, thereby keeping the SWCNTs individually dispersed. Nowadays, despite the existence of many other softer functionalization methods, the concentrated nitric acid (as well as the combination "nitric acid – ultrasound" [62]) continues [63] to be useful as a first and/or principal step for CNT functionalization, although "milder" alkaline mixture (NH_4OH+H_2O_2) is expected to be less destructive and causes minimal structural damage on CNT surface [64]. Additionally, a nondestructive method, for the preparation of stable aqueous colloidal solutions of graphene nanoribbons and carbon nanotubes included sonication of carbon nanomaterials in hypophosphorous acid (H_3PO_2), filtration accompanied by washing the solids with water, and dispersion of the solids in a fresh portion of water to form colloidal solutions, was described [65].

A few patents and articles are dedicated to use oleum in order to functionalize CNTs. Thus, the functionalization of SWCNTs in oleum via *diazonium species* generated in situ (reaction scheme 3.9) was reported in [66, 67]. In a typical example, purified SWCNTs were dispersed in oleum (20% free SO_3) at 80 °C; then sulfanilic acid was added to the dispersion followed by sodium nitrite and 2,2′-azo-*bis*-isobutyrylnitrile (AIBN), and the mixture was stirred for 1 h being further filtered and washed. In other reports [68, 69], the phenylated SWCNTs, obtained from SWCNTs and distinct quantities of benzoyl peroxide to reach most, medium, or least functionalized level, were further sulfonated by reaction of phenylated SWCNTs with sulfuric acid (simply dispersing the phenylated SWCNTs in oleum), wherein -SO_3H substitutes for -H on the phenyl groups at a position para to their attachment to the SWCNT (reaction scheme 3.10). Phenyl sulfonated SWCNTs are found to be true water-soluble carbon nanotubes and can serve as a platform technology for the development of SWCNTs for several industries including pharmaceutical, energy, and electronics. Degree of functionalization was determined both qualitatively using Raman spectroscopy and quantitatively using thermogravimetric analysis and X-ray photoelectron spectroscopy. Similar derivatization of SWCNTs using the diazonium-species technique was reported in [70, 71].

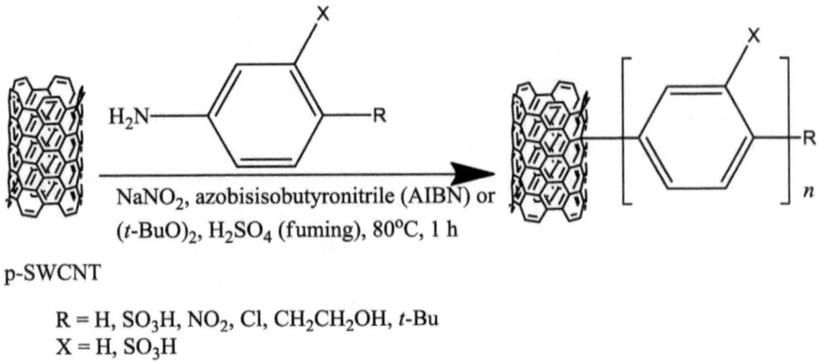

(III-9)

Functionalization of SWCNTs in oleum *via* diazonium species generated *in situ*.

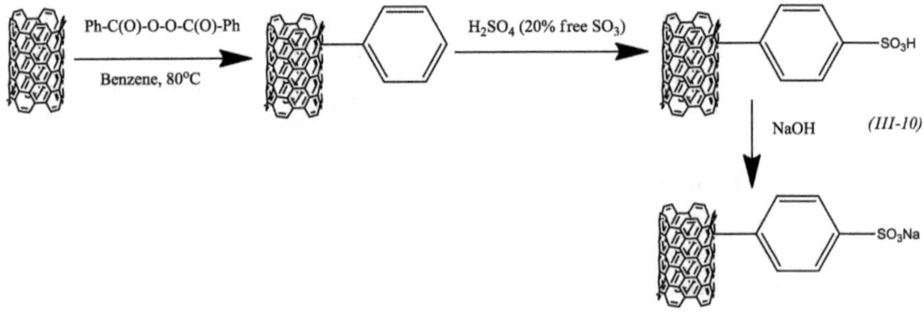

(III-10)

MWCNTs can also be purified and functionalized using oxidizing mixture of acids [72] (generally sulfuric and nitric acids, although a mixture $H_2SO_4/HNO_3/HCl$ was also used [73]). It was revealed that the length of MWCNTs decreased when the purification and functionalization time was increased; also, by functionalization the solubility and colloidal stability of MWCNTs were increased in polar media like water and methanol. A comparison of CNT treatments with individual acids and their mixtures also led to interesting conclusions. Thus, the effect of five different chemical treatments (HNO_3, H_2SO_4, H_2O_2, HNO_3+H_2SO_4, and HNO_3+HCl) on the homogeneity, surface chemistry, structure, and dispersibility in water of iron- and iron oxide-filled MWCNT samples was evaluated by a series of methods [74]. It was indicated that the chemical treatments are generally effective in removing non-nanotube carbonaceous species present in the sample. Obviously, with the exception of the H_2O_2 treatment, the chemical treatments offered good removal of free iron species. Besides the increase in the sample homogeneity, the chemical treatments promoted an increase in the carboxyl and hydroxyl groups at the carbon nanotube surface, which directly affects the dispersibility of these carbon nanotubes in water. In particular, dispersions of

2.24×10^{-1} g·L^{-1} were obtained for the treated samples with a mixture of nitric and sulfuric acid. Also, MWCNT particles were dispersed in *n*-hexadecane (used to decrease supercooling) [75] through surface modification of the MWCNT particles with strong acids H_2SO_4 and HNO_3, plus the addition of 1-decanol as a surfactant to the organic liquid. The supercooling of *n*-hexadecane was significantly decreased with the concentration of 0.1 and 0.5% but only slightly with the concentrations over 1.0%. The authors suggested that well-dispersed nanoparticles provided stable foreign nuclei of proper size to promote the heterogeneous nucleation process and accelerate crystallization process; thus, the supercooling was significantly reduced. Generally, the use of HNO_3+H_2SO_4 [76] and other acid mixtures is the first step of functionalization; further steps lead to attachment of various functional organic groups.

3.1.2 Organic Compounds

3.1.2.1 Aliphatic, Aromatic, and Polyaromatic Compounds

Alkyl and aryl groups General recent chapters and reviews [77–80] describe covalent and noncovalent functionalization of CNTs leading, in several cases, to an increasing "solubility" of nanotubes. Simple alkyl, aryl, and poly-aryl groups, attached to the CNTs, due to their hydrophobic nature, enhance their dispersibility in organic medium. Thus, high-speed vibration mill (HSVM) mechanochemical technique, described above (section *Mechanical methods*), was applied to prepare single-walled CNTs, functionalized with some *alkyl* and *aryl* groups (reaction scheme 3.11). As a result, SWCNTs with long alkyl chains can be dissolved in many common organic solvents [81]. In addition, halogenated aromatics (fluorinated liquid crystalline surfactants) represent a good approach for dispersion and alignment of CNTs [82].

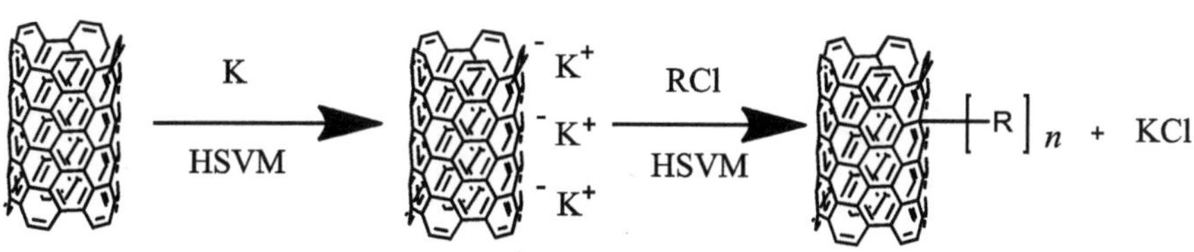

SWCNTs

$R = CH_2(CH_2)_{12}CH_3, C_6H_5$

SWCNTs, functionalized with alkyl and aryl groups.

Polyaromatic compounds Polyaromatic hydrocarbons are also very common for CNT noncovalent functionalization and enhanced dispersion purposes. Thus, the synthesis of perylene-based (**3.1**) SWCNT surfactants and the dispersion and exfoliation of SWCNTs in water by a variety of designed surfactants are investigated [83]. Significantly, the dispersion efficiency can be increased at higher pH, as water solubility of the surfactants is ensured by peripheral derivatization with carboxyl-functionalized first- and second-order Newkome dendrimers. Even at very low perylene concentrations of 0.1 $g{\cdot}L^{-1}$ and a nanotube-to-surfactant ratio of 1:1, the nanotube supernatant after centrifugation contained up to 73% of the pristine material with exfoliation degrees (the number of fractions of individualized nanotubes N1/NT) of up to 76%. In a related report [84], bolaamphiphilic perylene bisimides (PBI) were used for CNT functionalization and dispersion water. PBI functionalized with Boc-lysine and phenylalanine led to very good suspension of CNTs in water for 24 h. Useful information on related perylene derivatives is also given in [85].

A water-soluble stilbene (**3.2**) derivative with cationic charges enables tuning of the dispersibility of the SWCNTs in aqueous solution by photoirradiation [86]. The nearly coplanar core of the stilbene dispersant interacts with the SWCNTs via π–π interactions, resulting in a stable dispersion of the SWCNTs. Photoinduced cyclization for the dispersant-triggered reprecipitation of the SWCNTs is due to detachment of the dispersant from SWCNT surfaces. Stilbene moieties, as well as those of 1,2,3-triazole, azomethine, ethynyl, biphenyl, and azobenzene, can also serve as lateral units in polymers, used for SWCNT dispersion [87]. In addition, it is known that several stilbene derivatives are ones of most widely used photochromic molecules, whose hybrids with CNTs can serve as optically responsive materials [88].

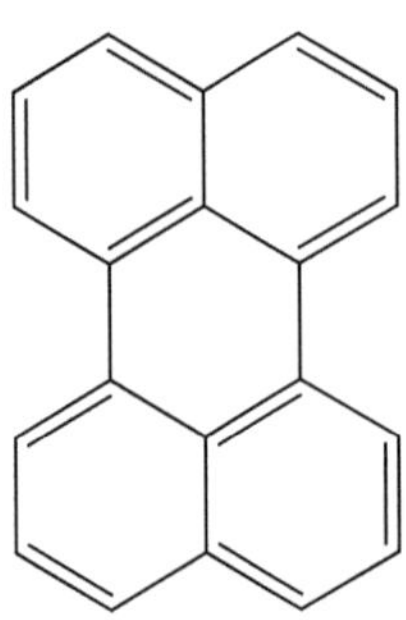

3.1

Perylene

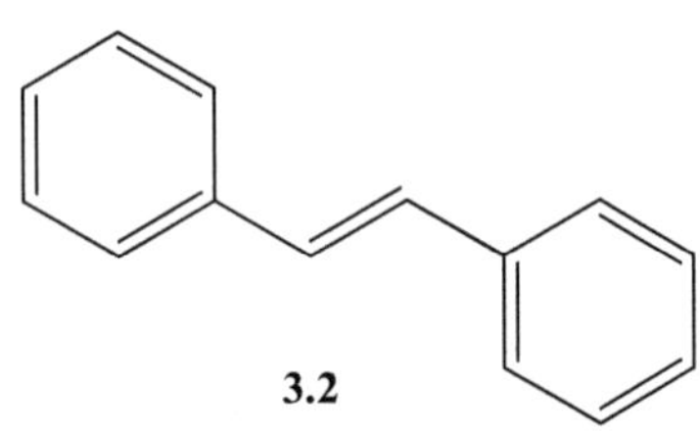

3.2

(E)-Stilbene

A preparation method for individually dispersed SWCNTs in both liquid and dried solid states was developed using triphenylene **3.3** derivatives [89]. On its basis, highly concentrated solution of SWCNTs (0.7–0.8%) was prepared, and also the SWCNTs were dispersed well after the addition of methanol (up to 70%) for a long time. These properties (redispersion ability and dispersion to organic solvent) were not observed in typical dispersing surfactants, such as SFBS and sodium cholate. Also, triphenylene-based discotic ionic liquid crystals with six imidazolium ion pendants can disperse pristine SWCNTs [90].

Triptycene **3.4** (Trp) (as well as its sugar-functionalized derivatives [91]) was designed with the ability to solubilize SWCNTs and C_{60} in water through noncovalent interactions [92]. The prepared compounds were designed with either two ionic or nonionic tails to ensure a large number of supramolecular interactions with the solvent, thereby promoting strong solubilization. The surfactants produced stable suspensions in which the SWCNTs were dispersed, and the surfactant/SWCNT complexes formed are stable for more than 1 year. In a related research [93], supramolecular surface modification of SWCNTs using an amphiphilic molecules containing a bent triptycene moiety and a hydrophilic oligo(ethylene glycol) chain **3.5** and **3.6** was described. The surface modification was realized through the binding of the triptycene moiety onto the sidewall of SWCNTs through a π–π stacking interaction, and the oligo(ethylene glycol) chains extend into the water and act as dispersing agents, thus yielding an aqueous SWCNT dispersion. This dispersion was found to be stable for more than 6 months and contained a high concentration of SWCNTs. Based on shape fitting of SWCNTs and the triptycene moiety, the stacking of triptycene moieties on the SWCNT sidewall showed a nice selectivity for SWCNTs with a diameter of 1.0 nm. The stacking of Trp-TEG molecules onto the SWCNT sidewall is illustrated in Fig. 3.14.

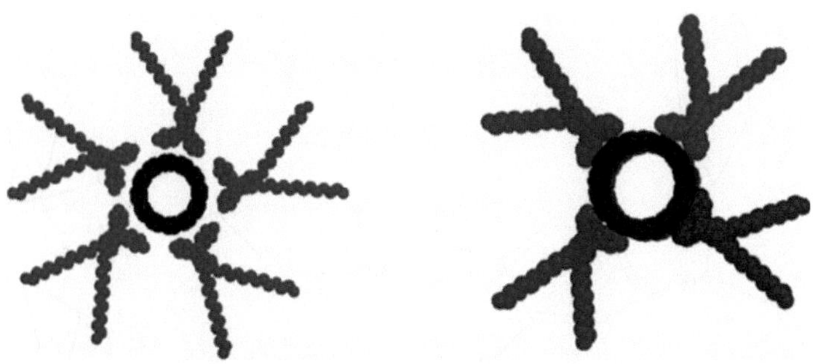

Fig. 3.14 Illustration of noncovalent surface modification of SWCNTs by amphiphilic Trp-TEG molecules (Reproduced with permission of *Elsevier Science*)

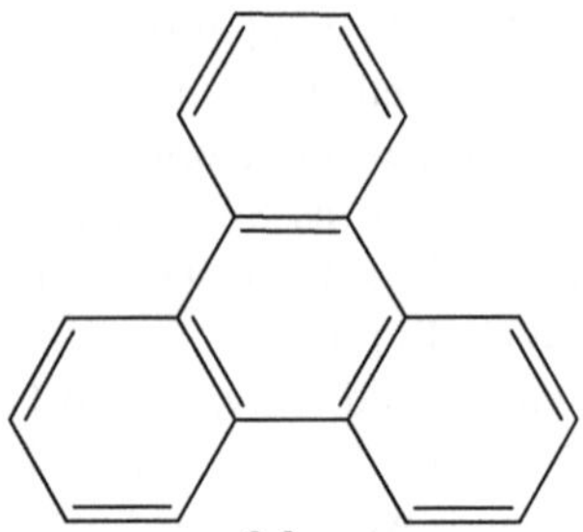

3.3

Triphenylene

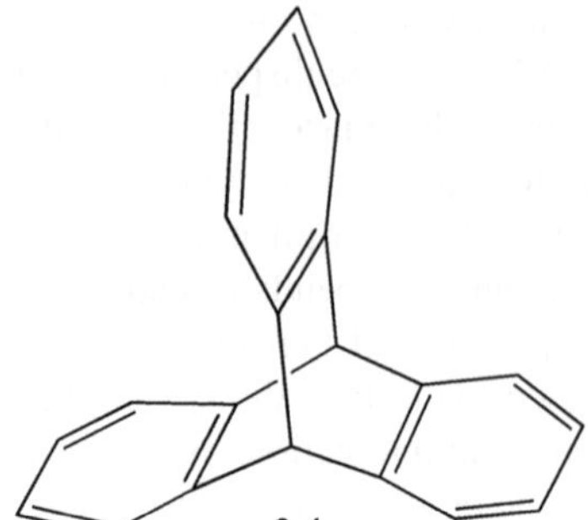

3.4

Triptycene

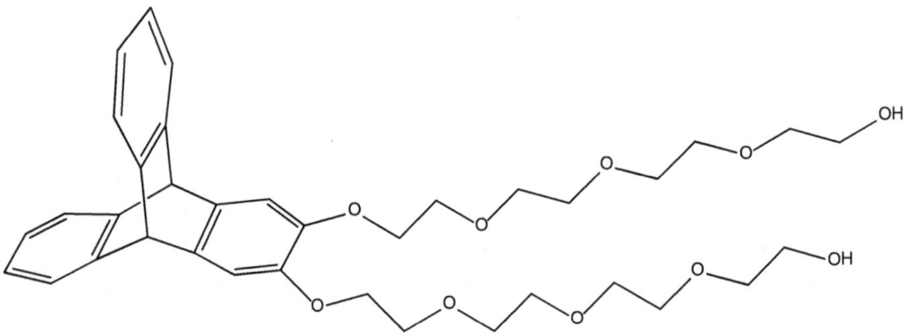

3.5

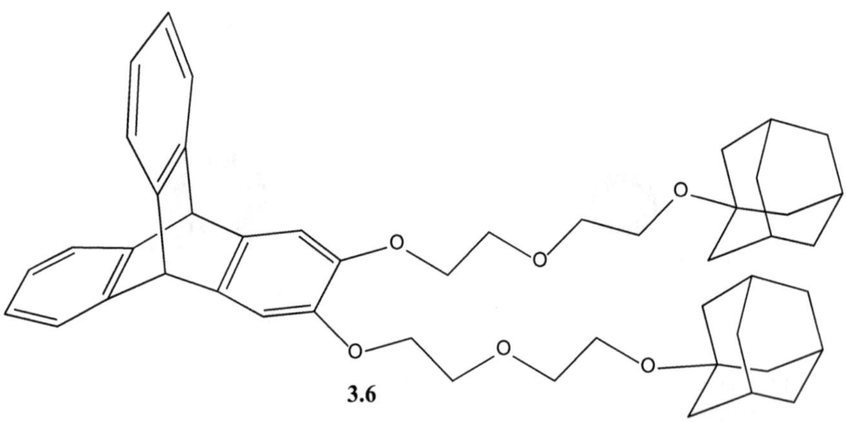

3.6

Chemical structures of the synthetic bifunctional molecules.

Nanohybrid dispersant molecules based on pyrene **3.7** (classic agent for noncovalent functionalization of CNTs) and polyhedral oligomeric silsesquioxanes were created for noncovalent functionalization of MWCNTs [94]. The prepared dispersant improved the dispersion of MWCNTs in organic solvents with very different polarities such as tetrahydrofuran, toluene, and *n*-hexane. The functionalized MWCNTs were used to introduce conductivity into polydimethylsiloxane membranes which can be used for electrostatic discharge applications. We note that silsesquioxanes were also used without pyrene moieties: covalently attaching polyhedral oligomeric silsesquixoane **3.8** (POSS) to the surface of MWCNTs can increase [95] their dispersion both in solution and in polymer matrices. Three types of POSS with different organic substituent groups were used as dispersing agents to study the effect of R group and spacer length between POSS and the MWCNT surface on MWCNT dispersion. It was found that POSS molecules with longer spacer length and less bulky R groups facilitated CNT modification process better, while CNTs modified by the POSS with shorter spacer length and less bulky R group showed better dispersion in THF.

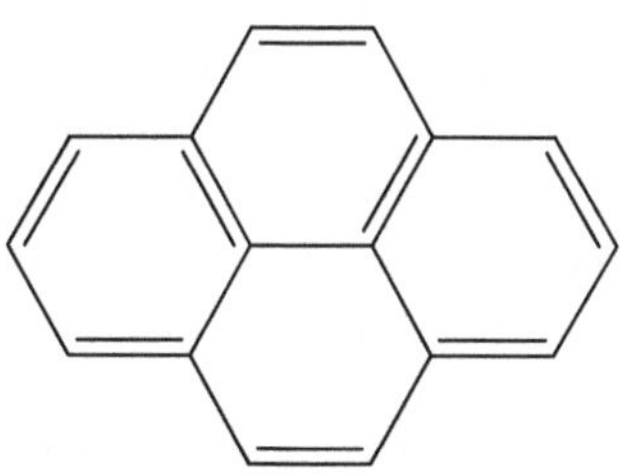

3.7

Pyrene

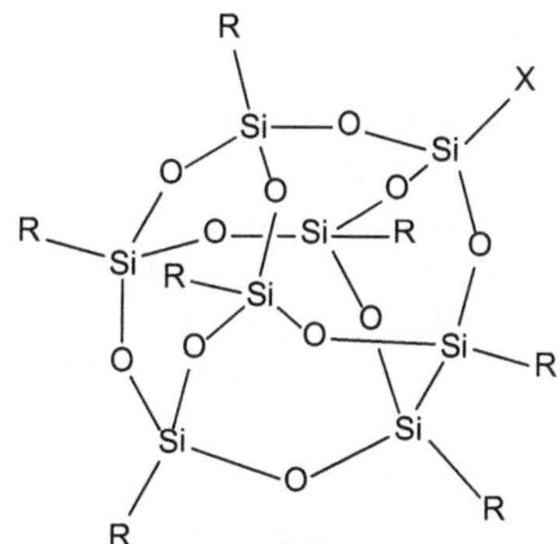

3.8

Silsesquioxane cage structure

In addition, pyrene can be used together with coordination compounds (see also the section below), although the last in the example below was used for another purpose, not for solubility increase. Thus, the synthesis of a multifunctional block copolymer incorporated with pyrene and ruthenium terpyridyl thiocyanato complex moieties by reversible addition–fragmentation chain transfer polymerization [96]. The pyrene block in the copolymer facilitates the dispersion of MWCNTs in DMF solution because of the strong π–π interaction between the pyrene moieties and nanotube surface. On the other hand, the ruthenium complexes greatly enhance the photosensitivity of the functionalized nanotubes in the visible region. Selected important results on application of pyrene derivatives for CNT dispersion are described in [97–99]. Among them, we especially emphasize CO_2-responsive "smart" SWCNTs on the basis of the following pyrene derivatives [100], showing reversible (switchable) dispersion in water, N,N-dimethyl-N′-(pyren-1-ylmethyl) acetimidamidinium [101], β-cyclodextrin modified with a pyrene group followed by further functionalization with ferrocene-terminated water-soluble poly(ethylene glycol) [102], as well as exhibiting a solvent-dependent fluorescence property using pyrene-derivatized hydrolyzed poly(styrene-co-maleic anhydride) [103].

3.1.2.2 Organic Acids, Their Salts, and Other Compounds Containing O-Containing Moieties

A few organic acids have been applied for CNT functionalization with the objective to increase their solubility. Thus, preparation of stable dispersion of MWCNTs using carminic acid (**3.9**, CA, 7-d-glucopyranosyl-3,5,6,8-tetrahydroxy-1-methyl-9,10-dioxo-9,10-dihydroanthracene-2-carboxylic acid) [104] as a dispersing agent revealed that MWCNTs are well dispersed in CA aqueous solution, and CA was well adsorbed at MWCNT walls. The prepared CA–CNTs aqueous dispersion was found to be highly stable for 4 months. The π–π stacking interactions between CA and MWCNTs provided good stability to the CA–CNT dispersion. On the basis of this composite, selective amperometric biosensor for H_2O_2 and iodate was developed, which exhibited excellent amperometric response toward H_2O_2 and iodate in the linear concentration range between 10 M–3.2 mM and 0.01–2.16 mM (Fig. 3.15). CA–CNTs could be used as an effective platform for immobilizing vital proteins and biomolecules to explore their direct electrochemistry. Functionalization of MWCNTs with benzenetricarboxylic acid (BTC) was conducted [105, 106] by using direct Friedel–Crafts acylation in an optimized PPA/P_2O_5 medium. It was shown that MWCNTs were functionalized effectively and without damaging the CNTs compared to mineral acid oxidation method (using HNO_3 and H_2SO_4), described in the previous section, and the dispersion and compatibility of thus obtained MWCNTs in organic medium was improved. In addition, benzyldimethylhexadecylammonium chloride (BAC, **3.10**) surfactant and caffeic acid (CA, **3.11**) were reported [107] to be selectively adsorbed on MWCNTs and MnO_2, respectively (Fig. 3.16), resulting stable aqueous suspensions of positively charged MWCNT and negatively charged MnO_2.

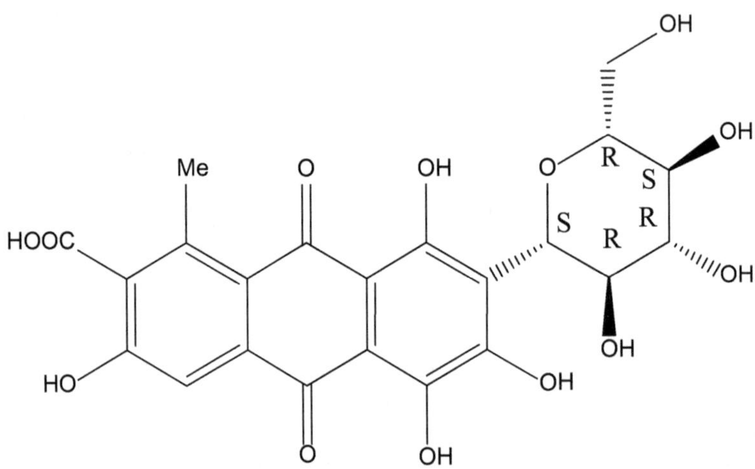

3.9

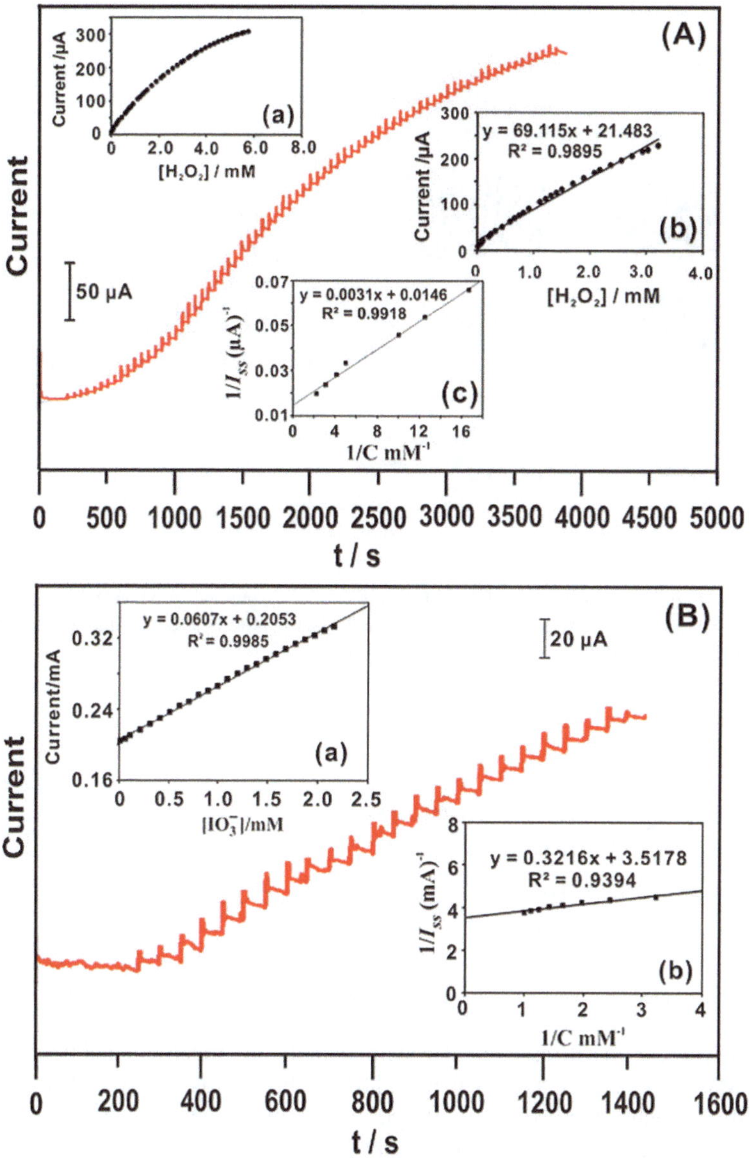

Fig. 3.15 (**a**) Amperometric response at CAT/CA–CNT modified RDE upon the addition of 1 μM to 5.8 mM H_2O_2; applied potential, −0.4 V. The inset (*a*) response current vs. [H_2O_2] plot, (*b*) linear calibration plot, and (*c*) Lineweaver–Burk plot used for K^{app}_M determination. (**b**) Amperometric response at CAT/CA–CNT modified RDE upon the addition of 0.01–2.16 mM IO_3^-; applied potential, −0.75 V. The insets are (*a*) the linear calibration plot and (*b*) Lineweaver–Burk plot used for K^{app}_M determination. Supporting electrolyte: N_2-saturated PBS; rotation rate, 1200 rpm (Reproduced with permission of *Elsevier Science*)

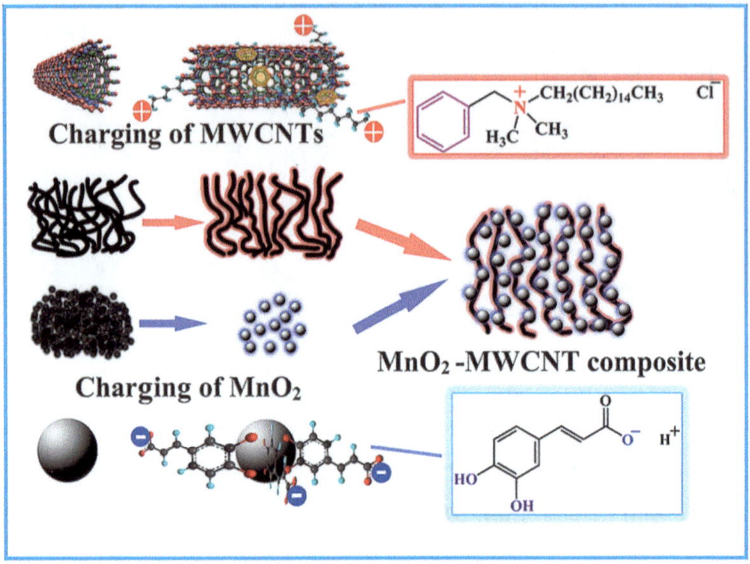

Fig. 3.16 Fabrication of MnO_2–MWCNT composite by dispersion of MWCNT using cationic BAC, dispersion of MnO_2 using anionic CA, and electrostatic heterocoagulation (Reproduced with permission of the *Royal Society of Chemistry*)

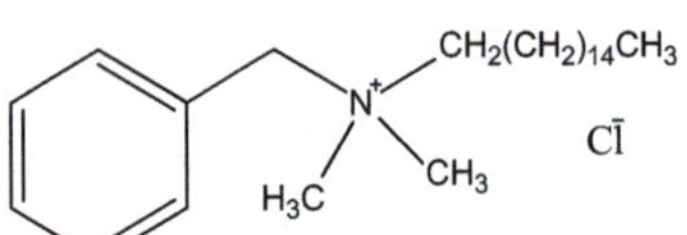

3.10

Benzyldimethylhexadecylammonium chloride

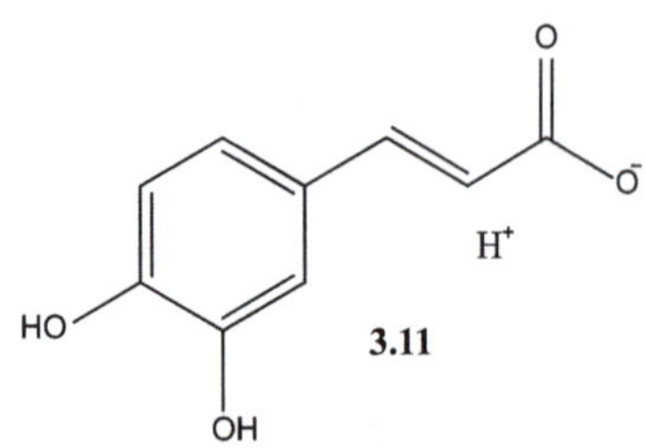

3.11

Caffeic acid

In a synthesized self-assembled system, namely, hydrogen-bonded liquid crystal **3.12**, MWCNTs were readily dispersed [108]. This self-assembly system of liquid crystal was realized by formation of hydrogen bond between levo tartaric acid and undecyloxy benzoic acid. A small amount (0.05 wt.%) of MWCNTs was dispersed in the liquid crystal, and the resultant mixture in its isotropic state is ultrasonicated for 2 h so as to obtain monodispersion of MWCNTs without any aggregation. Another example is acid yellow 9 (4-amino-1-1′-azobenzene-3,4′-disulfonic acid, AY), which was found to be a good stabilizing agent for MWCNTs [109]. MWCNTs dispersed in AY solution were remained stable about 3 months and even remained stable after centrifugation at 10,000 rpm for 30 min.

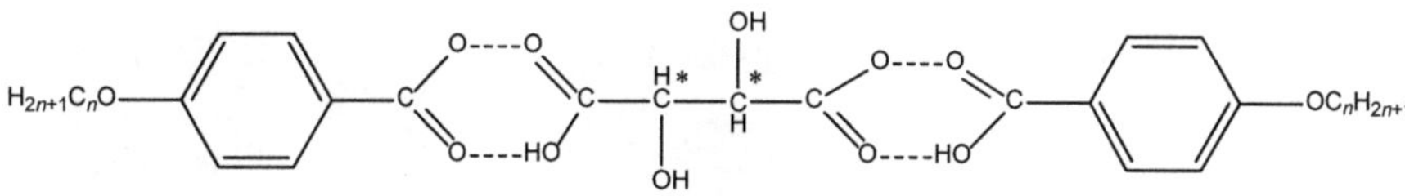

3.12

Molecular structure of the hydrogen bonded liquid crystal.

Among other compounds with O-containing organic groups, CNTs were solubilized by functionalization with *ester* moieties, for instance, esters with *n*-pentyl [110], other alkyl chains (*n*-butyl, *n*-hexyl, *n*-octyl, *n*-dodecyl, and *n*-hexadecyl) [111], or dendron-type moieties with long alkyl chains [112] (**3.13, 3.14**), *crown ethers* (4-′-aminobenzo-15-crown-5-ether [113] **3.15** or 4′-aminobenzo-18-crown-6-ether [114] **3.16** (the CNTs composites are "soluble" in propanol (Fig. 3.17)), phosphate esters [115] **3.17, 3.18**, hyperbranched *polyglycerol* [116], or *aldehydes* (anisaldehyde together with 3-methylhippuric acid) [117]. Functionalization of SWCNTs with *phenol* groups by 1,3-dipolar cycloaddition reaction and further derivatization of the products with 2-bromoisobutyryl bromide resulted in the attachment of atom transfer radical polymerization initiators (active in the polymerization of *t*-butyl acrylate from the surface of the nanotubes) to the sidewalls of the nanotubes (**3.19**) [118, 119]. The obtained SWCNTs, functionalized with poly(*t*-butyl acrylate), were soluble in a variety of organic solvents (see section on polymers below).

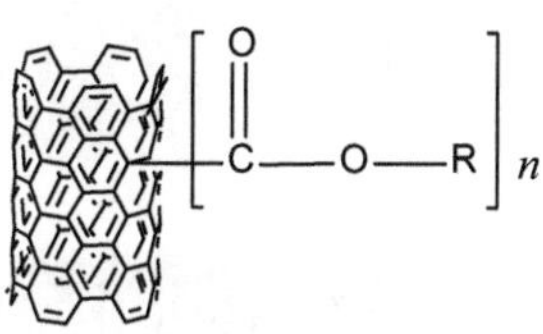

$R = -(CH_2)_nCH_3$
$n = 3, 5, 7, 11, 15$

3.13

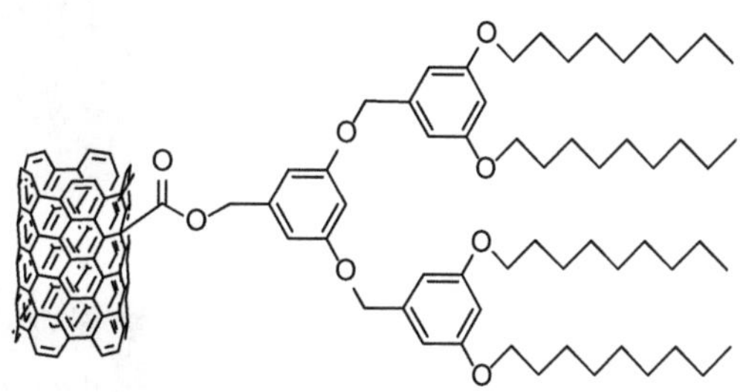

3.14

Ester-functionalized CNTs.

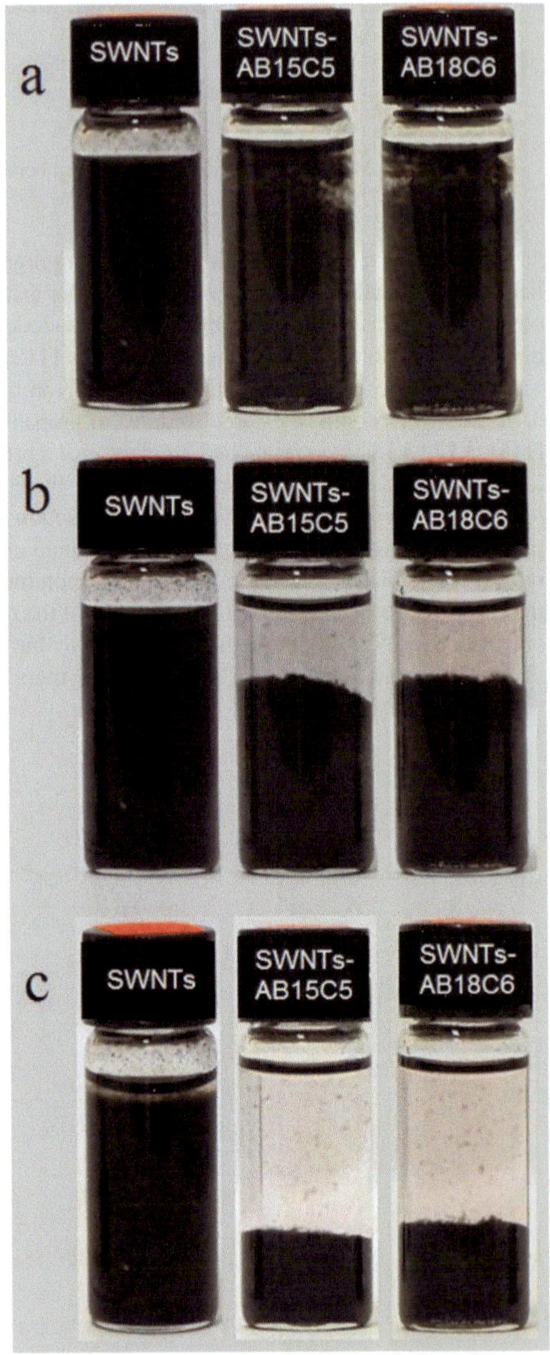

Fig. 3.17 Solubility test in propanol for oxidized SWCNTs before and after functionalization with crown ethers at elapsed time of (**a**) 0, (**b**) 1, and (**c**) 15 h after ultrasonic bath treatment for 5 min (Reproduced with permission of *Taylor and Francis*)

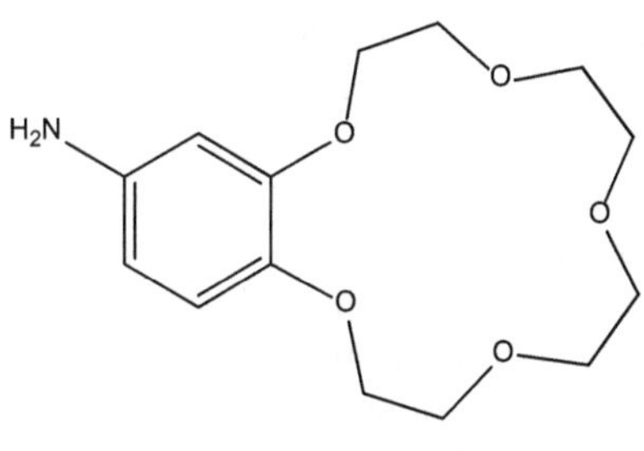

3.15

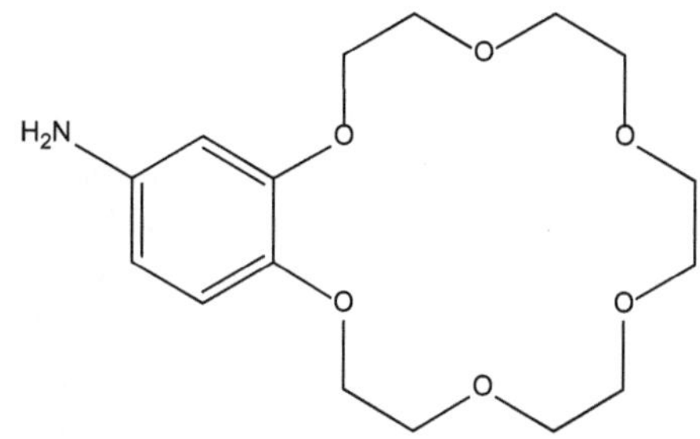

3.16

40-aminobenzo-15-crown-5 **3.15** and 40-aminobenzo-18-crown-6 **3.16**.

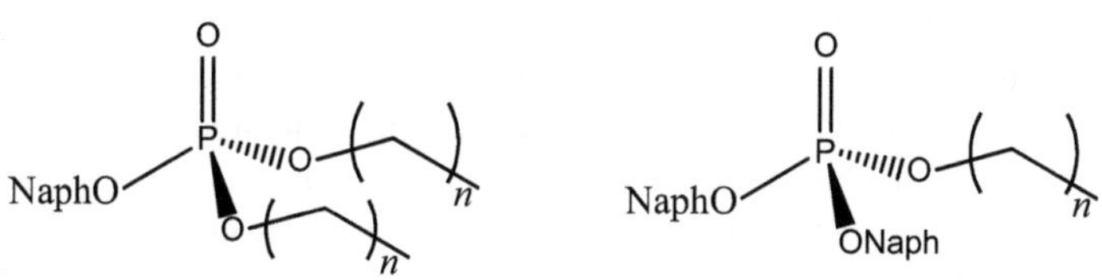

$n = 7, 9, 11, 13, 15$

3.17 **3.18**

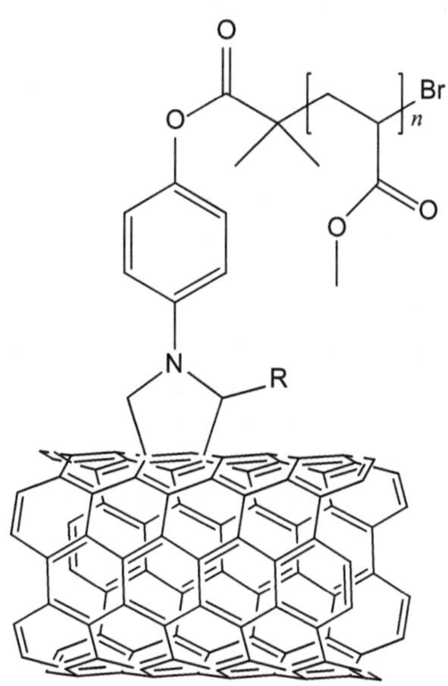

3.19

CNTs, obtained by 1,3-dipolar cycloaddition reaction.

Among a series of ionic salts of organic acids, sodium dodecylbenzene sulfonate **3.20** (SDBS) and sodium dodecyl sulfate **3.21** (SDS) are very popular and mostly frequently used [120–125] due to their outstanding properties for CNT dispersion, first of all dispersion stability in distinct temperature conditions and solvents. Thus, as a representative example, stability of dispersions of SWCNTs stabilized by SDBS in binary polar solvents "water + antifreeze" (glycerol, polyethylene glycol) with eutectic compositions was studied [126, 127]. The absorption spectra of the suspensions demonstrated no changes during 1 year storage with temperature spanning from −40 to +40 °C. Aqueous dispersions of nanotubes exhibited considerable enhancement of optical limiting parameters alongside an increase of the bundled material populace. Using SDBS, the layered double hydroxide LDH/CNT composites with interlayer anion of dodecylbenzene sulfo group (DBS-LDH/CNTs) were obtained [128] and dispersed in several organic solvents. The suspension of DBS-LDH/CNTs in 1-butanol, ethanol, and xylene kept stable in a long duration while sedimentated quickly in tetrahydrofuran. In addition, the delamination extent for LDH in these solvents ranked as follows: 1-butanol > ethanol > xylene > tetrahydrofuran. Among other important studies, the aggregation kinetics for SWCNT and MWCNT carbon nanotubes dispersed using SDBS were investigated [129] using time-resolved dynamic light scattering (DLS), in the presence of several electrolytes, and humic acid (HA). The CNTs could be effectively suspended in aqueous solution using the SDBS and that increased electrolyte concentrations induced aggregation. Increases in the solution pH from 3 to 10 led to a significant decrease in CNT aggregation, indicating the presence of functional groups on the CNT surface. In a NaCl solution, the stability of the CNTs was increased. A pseudostationary phase of MWCNTs dispersed with SDBS was used [130] for the EKC separation of nucleotides. SDBS has a long hydrophobic chain and a benzylsulfonate group. It suspends more MWCNTs (about 100-fold) than SDS, and the π–π interaction between the benzene ring of SDBS and MWCNTs prolongs the slurry suspension time. Using SDBS as a surfactant can reduce the required amount of MWCNTs and decrease the baseline noise. In addition, for the case of SDBS, it was found [131] that during its adsorption on SWCNT Coulombic forces do not play a central role but are overcome by the hydrophobic interactions between the surfactant tail and the nanotube walls. The hydrophobic forces between the surfactant tail and the nanotube determine the structure of the surfactant-stabilized nanotubes (Fig. 3.18). In such a structure, each nanotube is covered by a monolayer of surfactant molecules in which the heads form a compact outer surface, while the tails remain in contact with the nanotube walls. It is important to note that although the final configuration can be described as a cylindrical micelle with a nanotube in the center.

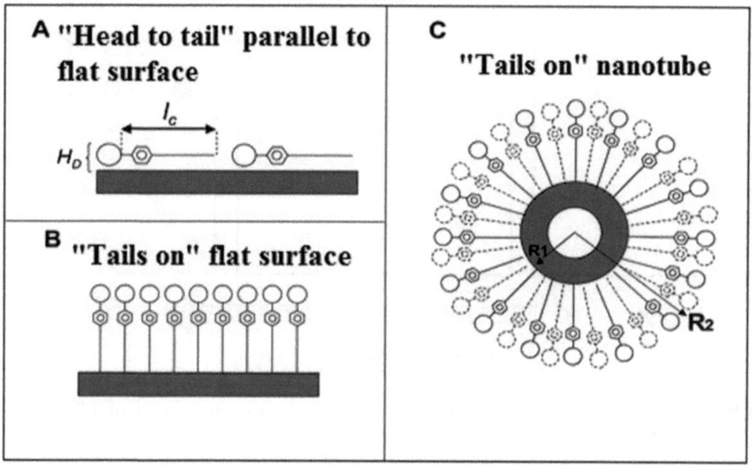

Fig. 3.18 Model for the surfactant–nanotube interaction. Head cubic packing density = 1.45 molecules/nm^2. Head diameter (H_D) = 0.69 nm. Chain length (l_c) = 1.5 + 1.265n (Å). n = number of carbons; l_c = 24.27 (Å) (Reproduced with permission of the *American Chemical Society*)

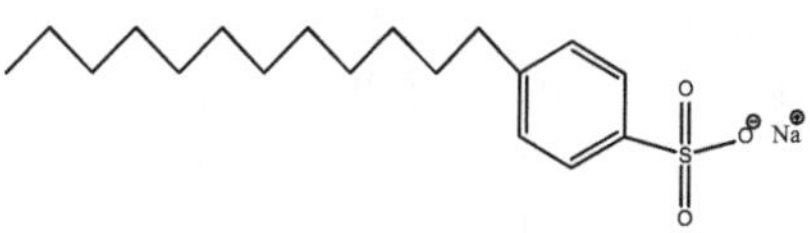

3.20, SDBS[1]

3.21, SDS

As an example of SDS applications, CNTs were dispersed in a SDS/decanol/water nematic fluid [132, 133]. The long-term stability of the dispersions was ensured by the small density gradients existing between nanotubes and the nematic fluid and by its viscosity, as well. Presumably, surfactant or nematic micelles adsorb onto nanotubes and concur to stabilize them. It was quantified how changing the counterion (Cs^+ instead of Na^+) affected the morphology of dodecyl sulfate surfactants adsorbed on CNTs [134]. Using atomistic molecular dynamics, aqueous cesium dodecyl sulfate (CsDS) adsorbed on (6,6), (12,12), and (20,20) SWCNTs at r.t. were simulated. It was suggested that CsDS should be more effective than SDS at stabilizing aqueous carbon nanotubes dispersions. More importantly, these results were obtained only for the (6,6) nanotubes simulated. The analysis of surface tension of SDS-stabilized dispersions of CNTs was carried out [135]. It was shown that surface tension is very sensitive to small changes of nanotube and surfactant concentrations. It was suggested that, at moderate concentrations, surfactant displaces carbon nanotubes from the air–water interface, and the nanotubes are mostly moved into the bulk of the liquid. Also, the effect of CNT diameter

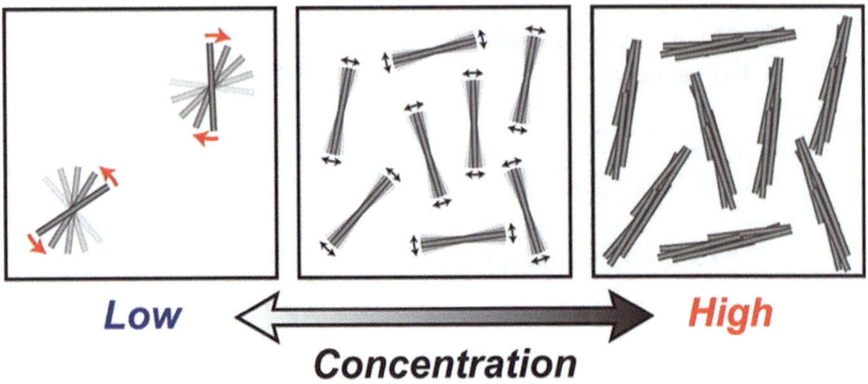

Fig. 3.19 Proposed concentration dependence for the aggregation state of CNT dispersions (Reproduced with permission of the *American Chemical Society*)

(10–500 nm) and CNT length (1–40 μm) in the presence of 1% sodium dodecyl sulfate (SDS) on suspension of CNTs in water was studied [136]. It was also emphasized that CNT dispersions showed a dispersion-state transition as a form of rotation restriction at a certain concentration (Fig. 3.19) [137].

In addition, a series of comparative studies for SDBS **3.20** and SDS **3.21** between each other, as well as with frequently used non-sulfonate surfactants, have been carried out. Thus, the SDBS contributed to a better dispersibility and electrical conductivity of SWCNTs than SDS **3.21**, sodium cholate **3.22** (SC), and cetyltrimethylammonium bromide **3.23** (CTAB) [138]. Similar conclusion was also made [139] on the basis of molecular dynamic simulations comparing SDBS and SDS, concluding that SDBS surfactant contributed weaker interactions between CNTs as compared with that of SDS surfactant by a factor of about 10. A series of surfactants (SDBS, CTAB, etc.) were used for CNT dispersion studying the effects of surfactant properties (chain length, aromatic rings, headgroup charge, etc.) and nanotube morphology [140]. SDBS was clearly observed as the best dispersant for CNTs, and the following order of performance was established: SDBS **3.20** > CTAB **3.23** ≈ CPyCl (cetylpyridinium chloride **3.24**) > STS **3.25** > TTAB **3.26** ≈ SDS **3.21** > DTAB **3.27**. Among other results, it was established that the presence of micelles in the exfoliation process is not a key factor either for starting CNT dispersibility or attaining its saturation value. Additional effects of CNT dispersion process in surfactants under ultrasonication, such as sonication duration, SDBS/CNT ratio, and dispersant agents, were also studied [141].

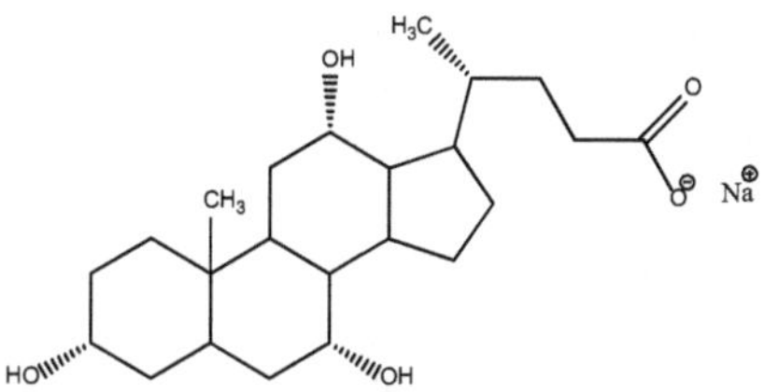

3.22, SC

3.23, CTAB

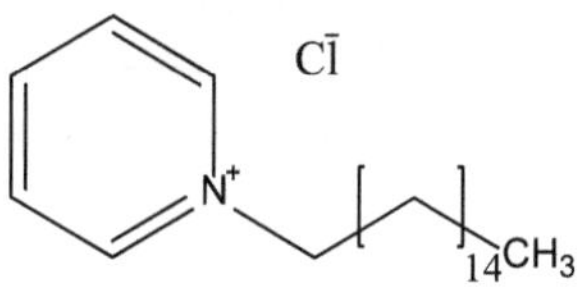

3.24

CPyCl

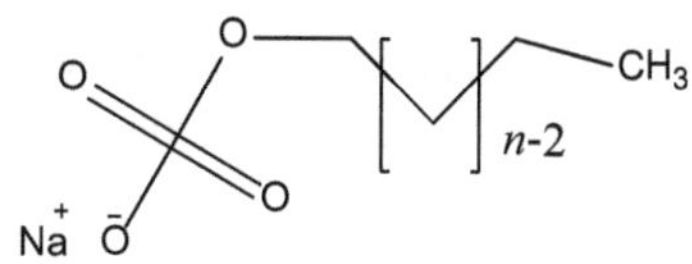

$n = 14$

3.25

STS

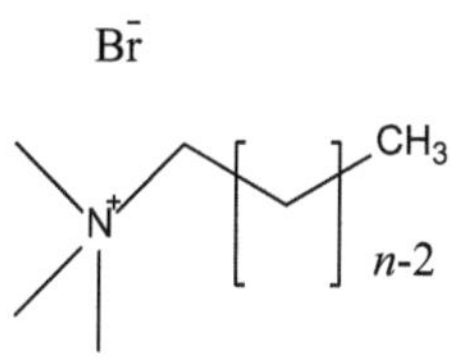

3.26, TTAB n=14

3.27, DTAB n=12

Adsorption isotherms of four different surfactants, SDBS **3.20**, SDS **3.21**, benzethonium chloride **3.28**, and Triton X-100 **3.29**, were measured on MWCNT in water [142]. The nature of the polar head had only little influence on adsorption which was mainly driven by hydrophobic interactions. However, the outcome of the dispersion experiment was dependent on the purity of the carbon nanotubes. Using these results, it was possible to prepare concentrated colloidally stable dispersions of MWCNTs in water (c = 32 g/L). Other comparative studies on interaction and dispersion of MWCNTs using two different surfactants (SDS **3.21** and polyoxyethylene sorbitan monooleate **3.30**, Tween 80) showed [143] that, for effective dispersion, the minimum weight ratio of MWCNTs to surfactant was 1:41 and 1:3 for SDS and Tween 80, respectively. Other Tween-type surfactants are also frequently applied. Thus, in case of Tween 65 (compound **3.31**), it could suspend high-loading nanotubes in ethanol [144]. MWCNT suspensions in ethanol with Triton X-100 and Tween 65 initial supernatant concentrations of 1.0 $g{\cdot}L^{-1}$ were prepared, and their concentrations were more than 0.50 and 0.35 $g{\cdot}L^{-1}$ after 240 h, respectively. These concentrations for MWCNT suspensions, which are stable over the long term in ethanol without polymer or covalent functionalization, are higher than others reported in literature. The slow diffusion of Tween 80 (see also the previous section) surfactant molecules in SWCNT aqueous dispersion was directly observed using the pulsed field gradient nuclear magnetic resonance method [145]. The slow diffusion of Tween 80 molecules was attributed to the strongly adsorbed molecules on the SWCNTs in the aqueous dispersion. The amount of bound Tween 80 molecules was estimated to be approximately 12% of the total amount of Tween 80 molecules, contributing to the stability of the SWCNT aqueous dispersion. This SWCNT/Tween 80 aqueous dispersion was found to be very stable for at least 3 weeks. The observed zeta potentials of this SWCNT dispersion are between −10 and 0 mV, indicating that the stability of the SWCNTs in the Tween 80 solution was maintained by steric interactions between the small amount of adsorbed Tween 80 molecules on the SWCNTs, while the effect of electrostatic interactions between adsorbed Tween 80 was minimal.

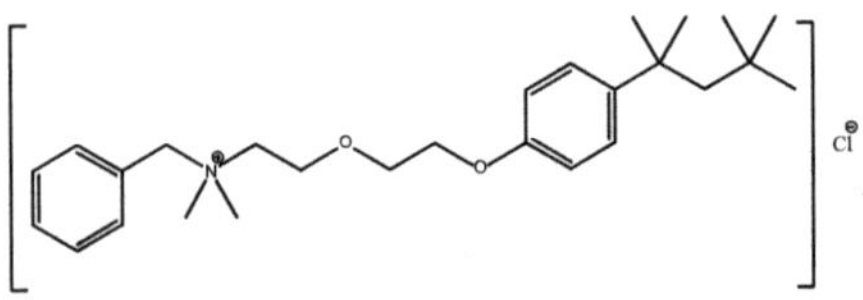

3.28, Benzethonium chloride

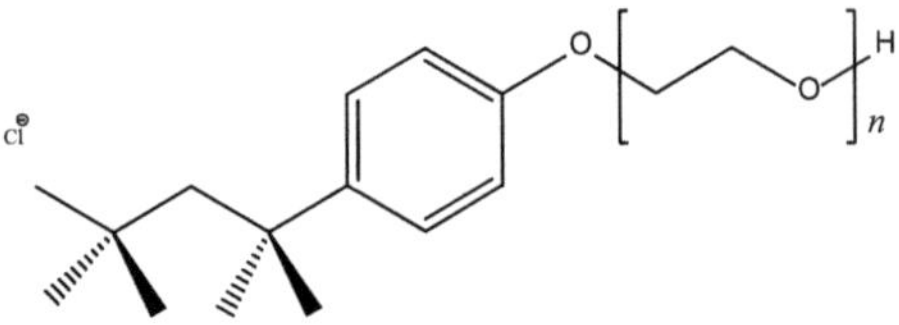

3.29, Triton X-100

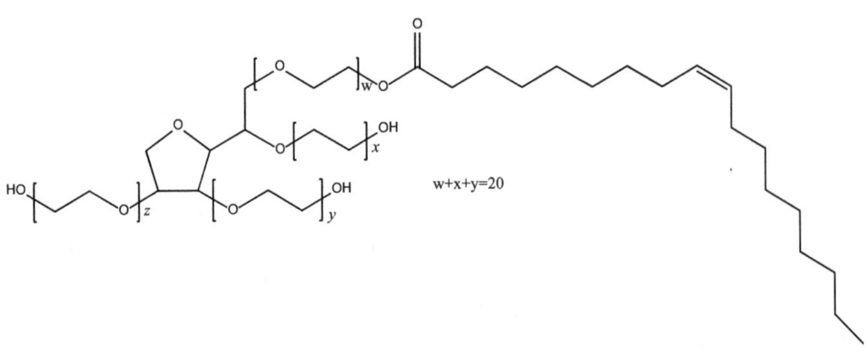

3.30, Tween 80

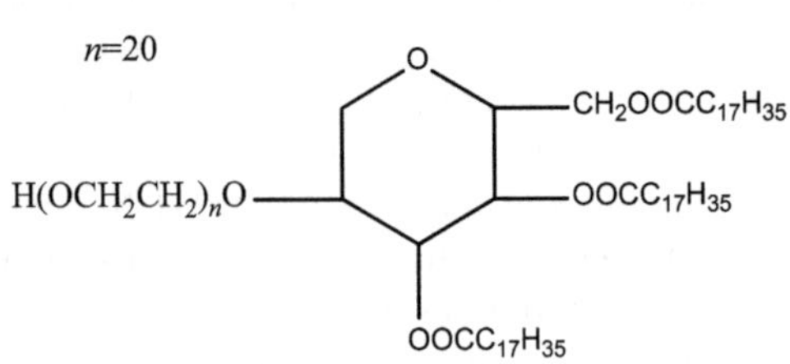

3.31

Tween 65

Other sulfonate-containing surfactants are rare; [146] among them, we note water-soluble polyaniline blend poly(sodium 4-styrenesulfonate), (PANI·PSS), which was used [147] to disperse MWCNTs by noncovalent surface modification. The anionic of PANI·PSS solution was prepared by interfacial polymerization of aniline monomer in the presence of PSS as the blending reagent to provide water solubility. The MWCNTs dispersed with PANI·PSS were found to be highly dispersible in water, which open up the new possibilities for the fabrication of composite films. Related poly(3,4-ethylenedioxythiophene) poly(styrenesulfonate) (PEDOT:PSS, **3.32**) also showed its effective properties as dispersant for SWCNTs, enabling uniform aqueous suspensions to be obtained at weight loadings of up to 0.23 mg/ml (>1% by weight relative to PEDOT:PSS) without necessity of additional surfactants [148]. We note that PEDOT/PSS/CNT composites have important applications [149, 150]. Water-soluble sodium *p*-aminobenzene sulfonate-grafted MWCNTs (MWCNTs-CO-$NHC_6H_5SO_3Na$) were synthesized by an amidation reaction of sodium *p*-aminobenzene sulfonate with MWCNT–COCl (prepared by MWCNT oxidation [151] by concentrate HNO_3 and then reacted with $SOCl_2$). The MWCNTs-CO-$NHC_6H_5SO_3Na$ showed a good solubility and dispersion in water and had an improved adsorptive performance for the removal of low concentrations of Cu^{2+} in water. MWCNTs were also functionalized [152] with sodium lignosulfonate (**3.33**, SLS) at various SLS/MWCNT ratios. It was shown that the solubility of the MWCNTs did not increase significantly with further increases in SLS after the appropriate amount of SLS was employed. Excessive use of SLS rather decreased the conductivity of functionalized MWCNTs. Also, the solubility of MWCNTs was influenced by dispersing method even when the same amount of SLS was used.

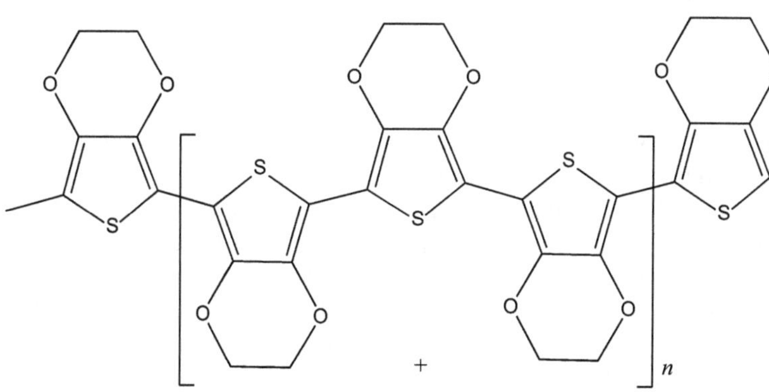

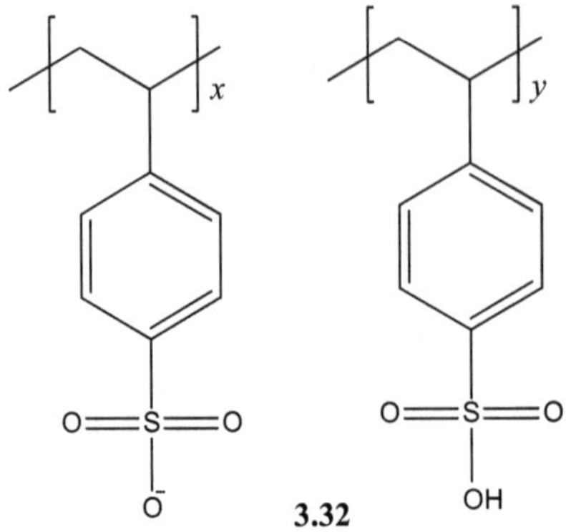

3.32

PEDOT:PSS

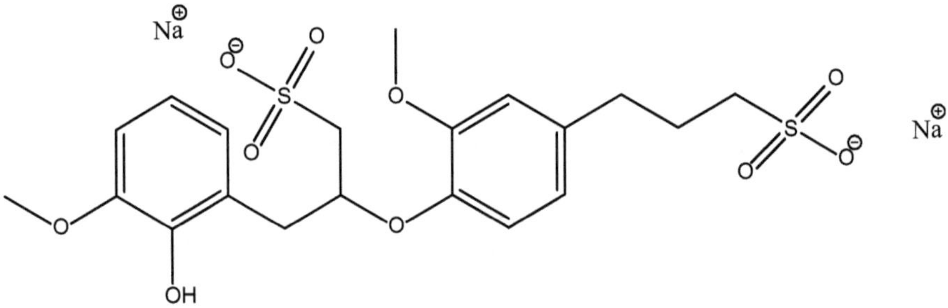

3.33, SLS

Several other organic salts, applied for CNT dispersion, are also known, for instance, dodecyl quaternary ammonium bromides [153]. Thus, dodecyl trimethylammonium bromide (**3.27**, DTAB) and sodium octanoate (**3.34**, SOCT) were found [154] to form exceptionally stable MWCNT dispersions. Stable dispersions were obtained at low total surfactant concentration as compared to their concentration when used alone. A waterproof biomaterial, stearic acid (STA), which is one of components of the wax present on the lotus leaf surface, was used as the material with low surface energy to fabricate superhydrophobic MWCNT hybrids through a solution method. This method involved preparation [155] of a sodium stearate (SST)-stabilized MWCNT dispersion (Fig. 3.20), followed by a precipitating process. STA was assembled on the MWCNT–SST hybrid surface by a reaction of SST with acetic acid. With increasing the SST/MWCNT weight ratio, the water CA of MWCNT hybrid increased and then decreased after a maximum value of 163 °C at the ratio of 1:1. It was interesting that the wetting property of MWCNT hybrids (SST/MWCNT = 0.5/1 and 1/1) was tunable between superhydrophobicity and superhydrophilicity by changing temperature.

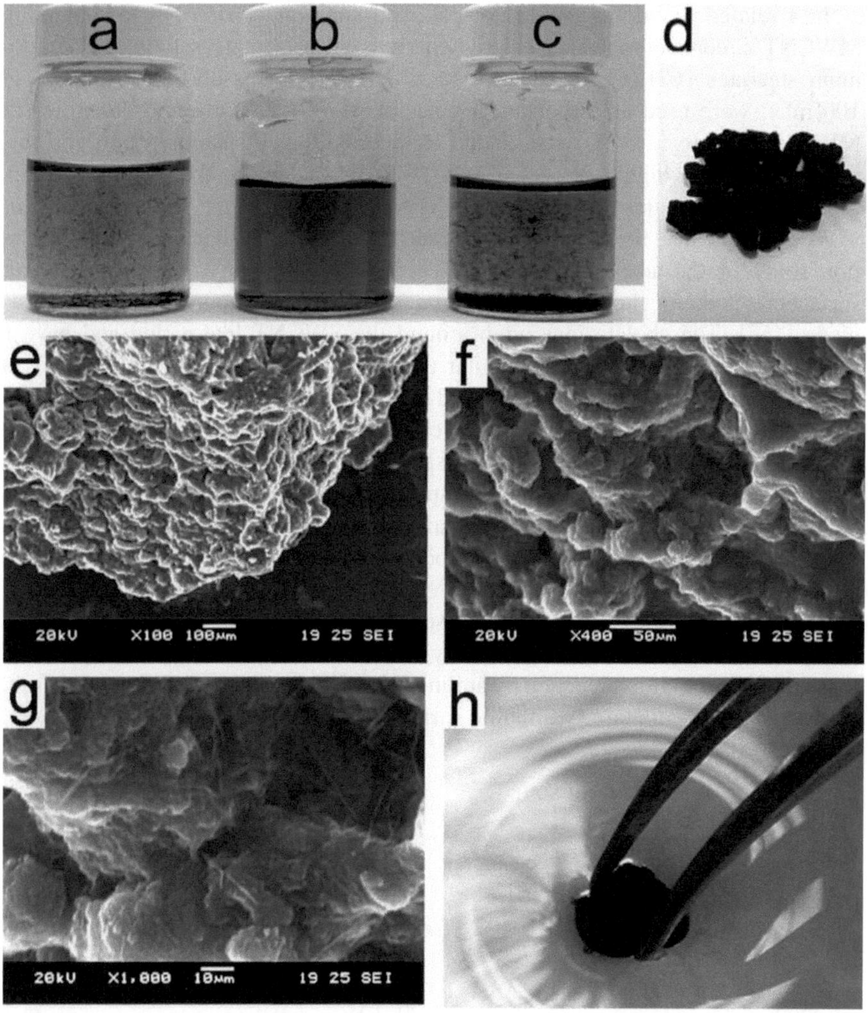

Fig. 3.20 Images of MWCNT aqueous suspensions: (**a**) MWCNT (2×10^{-4} g/ml), (**b**) SST/MWCNT (1/1), (**c**) SST/MWCNT (1/1) after addition of acetic acid; (**d**) image of MWCNT hybrid (SST/MWCNT = 1/1); (**e–g**) SEM images of MWCNT hybrid (SST/MWCNT = 1/1); (**h**) image of a water concave formed when a MWCNT hybrid (SST/MWCNT = 1/1) was touched to water (Reproduced with permission of *Elsevier Science*)

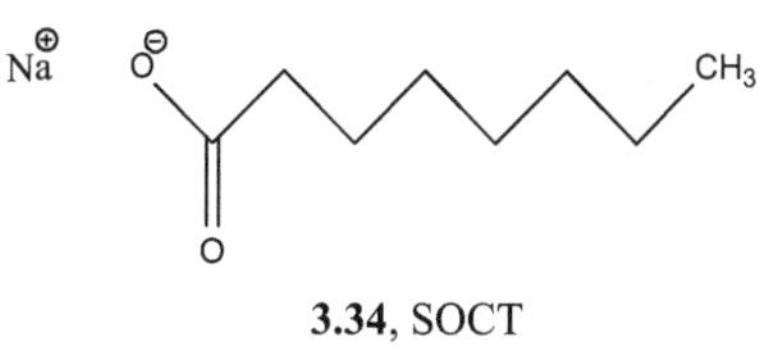

3.34, SOCT

In a related research [156], a sol-gel synthesis was carried out for obtaining MWCNT/zirconia composites, where two kinds of surfactants, sodium and ammonium stearates {$CH_3(CH_2)_{16}COOM$, M = Na, NH_4^+} dissolved in water (1 g/ 100 mL), were used as dispersant agents for MWCNTs. Different quantities of MWCNTs (up to 5 wt.%) were added in the solution of stearate/water, and this solution with the highly dispersed MWCNTs was added to the zirconia sol-gel, producing composites of MWCNTs/zirconia with different concentrations of MWCNTs. Finally, the detailed large-scale all-atom molecular dynamics simulation study of the adsorption and surface self-assembly of a common bile salt surfactant, sodium cholate **3.22** (SC) on a SWCNT in aqueous solution, was made [157]. The cholate ions wrap around the SWCNT like a ring and have a small tendency to orient perpendicular to the cylindrical axis of the SWCNT, a unique feature that has not been observed for conventional linear surfactants such as SDS. At the saturated surface coverages, SC is a better stabilizer than SDS, a finding that is consistent with the widespread use of SC to disperse SWCNTs in aqueous media. Indeed, the superior dispersion-induced stability of SC over SDS results from a higher repulsive energy barrier and a shallower attractive energy well induced by SC in the PMF profile. The sodium cholate-assisted CNT dispersion can be stable and clear for 2 months [158]. We note that sodium cholate-assisted density gradient ultracentrifugation can be used to obtain small diameter SWCNTs [159]. Also, a good water dispersibility was reached by MWCNT modification through radical addition reaction by aniline 2,5-double sulfonic acid diazonium salt, leading to the creation of hydrophilic surface (reaction Scheme 3.12) [160].

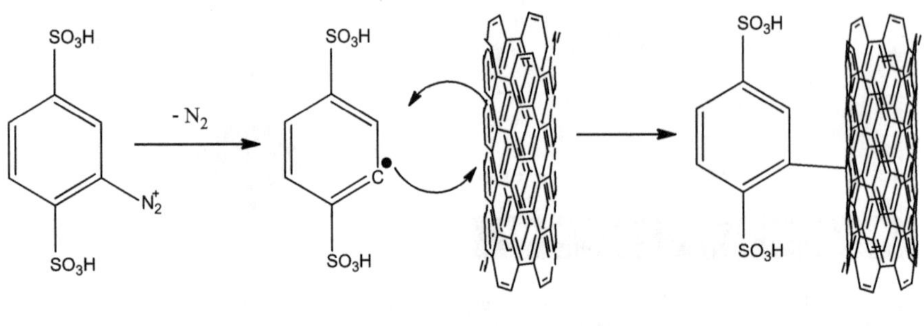

The hydrophilic modification process of MWCNTs.

In case of using *mixed surfactants* [161], CTAB and SDS, and their molecular dynamics simulation studies (Fig. 3.21), their mixtures were found to be more effective with the lower concentrations and more individual CNTs, reflecting a synergistic effect, which is mainly due to the electrostatic attractions between surfactant heads (see also the section below on special studies of CNT dispersion). The mixtures of 90:10 cationic rich and anionic rich of CTAB and SDS are able to disperse the MWCNTs well at lower concentration as compared to pure CTAB or SDS.

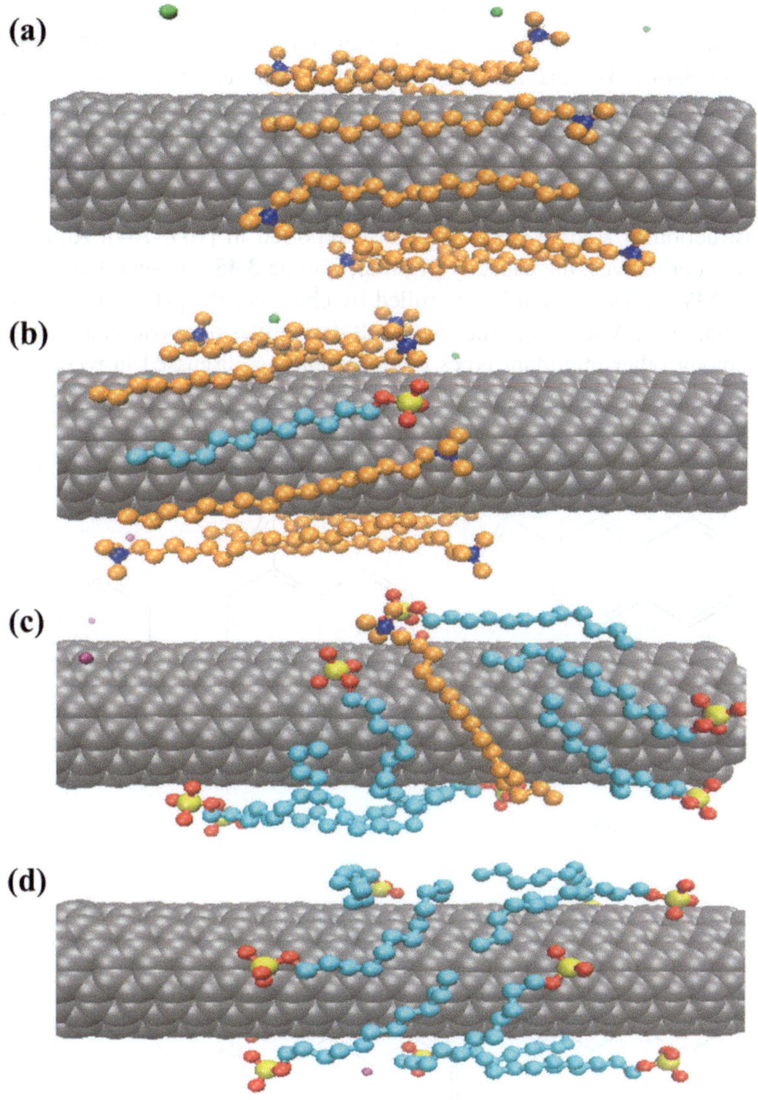

Fig. 3.21 Representative simulation snapshots of a (7,7) SWCNT in aqueous solutions of (**a**) pure cetyltrimethylammonium chloride, (**b**) a 90:10 mixture of cetyltrimethylammonium chloride/sodium dodecyl sulfate, (**c**) a 10:90 mixture of cetyltrimethylammonium chloride/sodium dodecyl sulfate, and (**d**) pure sodium dodecyl sulfate. Water molecules are not shown for clarity. Color code: blue, nitrogen; green, chloride counterion; orange, carbon atoms in cetyltrimethylammonium ion; red, oxygen; yellow, sulfur; purple, sodium counterion; cyan, carbon atoms of dodecyl sulfate ion; silver, carbon atoms in the SWCNT. All snapshots are at the 23 ns in the MD simulations (Reproduced with permission of the *American Chemical Society*)

Pyrene and naphthalene hydrocarbons were discussed above (see the section *Polyaromatic hydrocarbons*). In case of their *ionic derivatives* (**3.35**, **3.36**, **3.37**, **3.38**, **3.39**, **3.40**, **3.41**, **3.42**, **3.43**, and **3.44**), their ionic salts were used for obtaining water-soluble SWCNT polyelectrolytes (SWCNT-PEs), which are analogous to polyanions and polycations [162]. The nanotube–adsorbate interactions consist of π–π stacking interactions between the aromatic core of the adsorbate and the nanotube surface and charge transfer between them. Pyrene-containing moieties in construction of soluble CNTs were also reported in [163, 164]. Among other arene salt derivatives, the capacity of pillar[6]arene **3.45** host–guest complexes to disperse MWCNTs, reversibly controlled by changing the pH, should be emphasized [165, 166]. A series of other ionic salt derivatives of polyaromatic hydrocarbons, among other important types of surfactants, is discussed in recent reviews [78, 167].

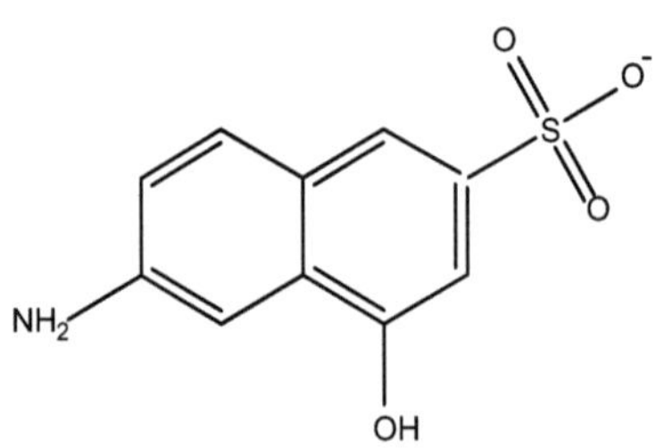

3.35

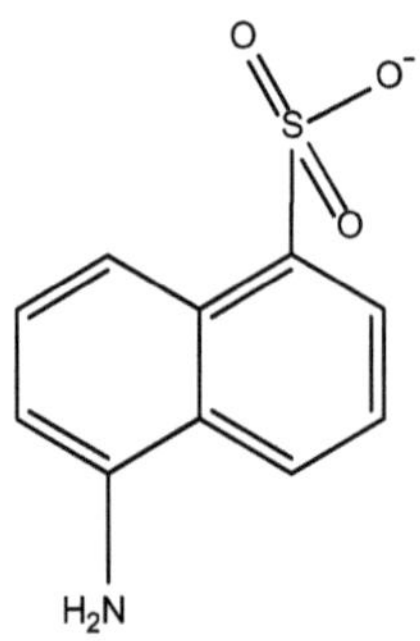

3.36

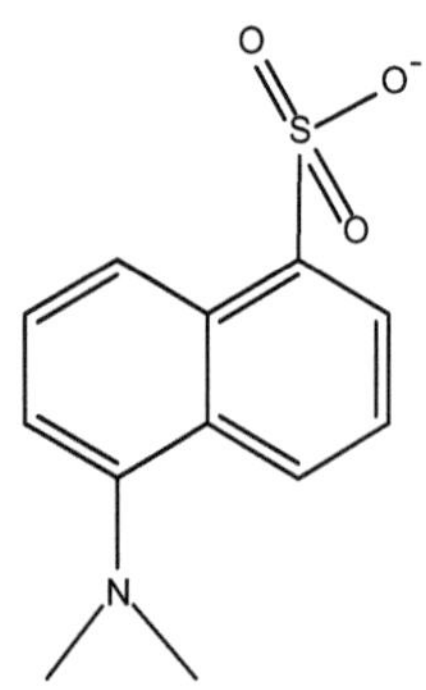

3.37

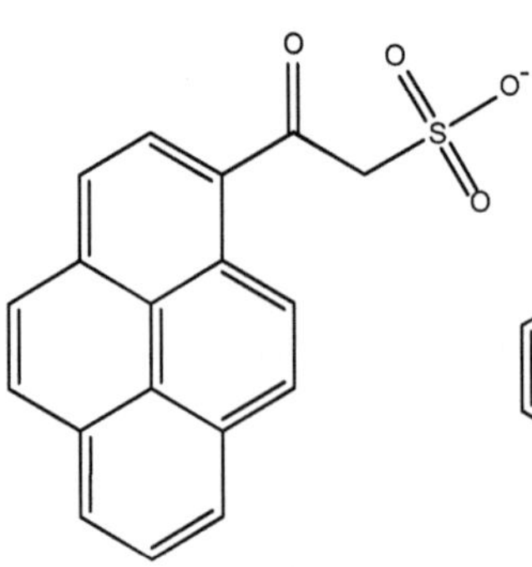

3.38

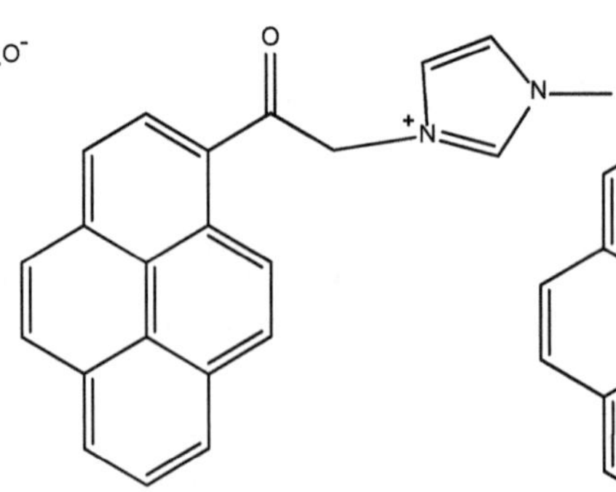

3.39

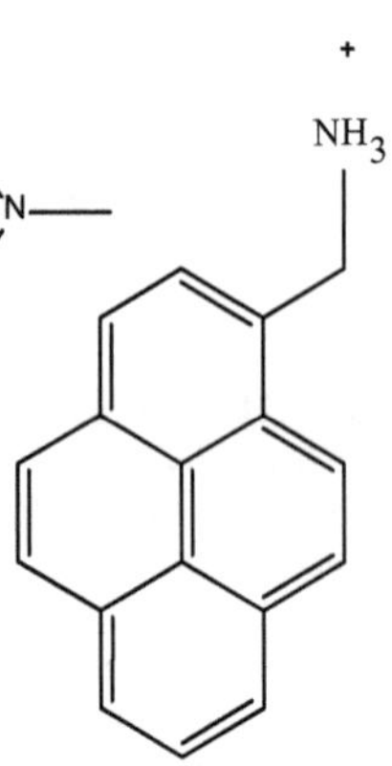

3.40

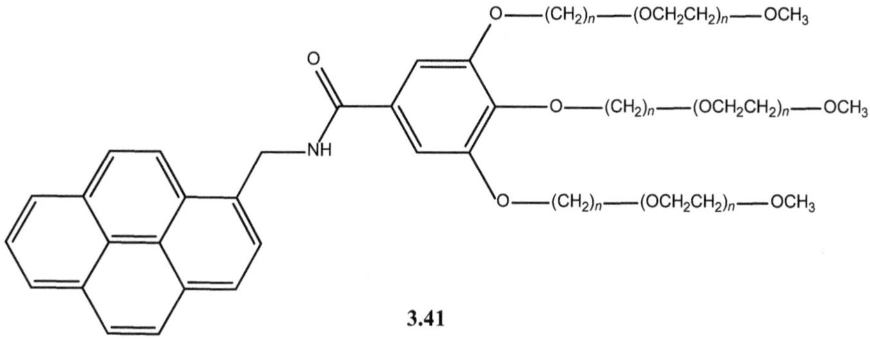

3.41

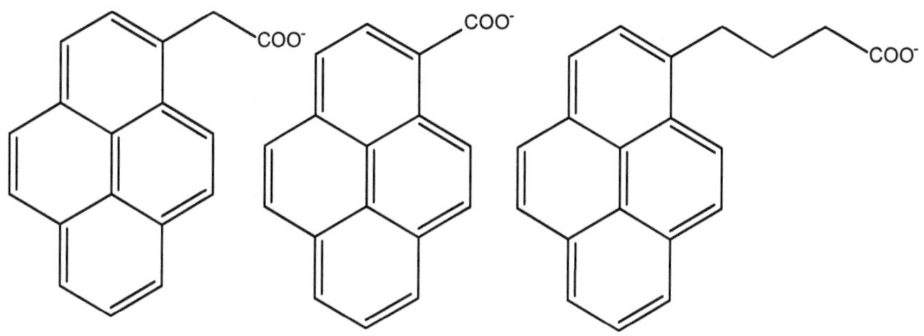

3.42 **3.43** **3.44**

Polynuclear aromatics, used for obtaining water-soluble SWCNT polyelectrolytes.

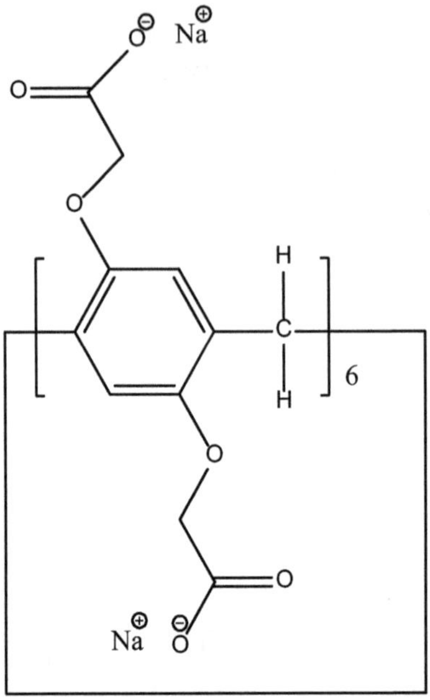

3.45

At last, we note microwave-assisted chemical functionalization of SWCNTs with undecyl groups decomposed from lauroyl peroxide **3.46** [168]. This rapid efficient procedure reduced the reaction time to 10 min and obtained the products with higher functionalized degree than that by the conventional refluxing method. Longer treatment time leads to partial defunctionalization, and higher microwave power (higher than 900 W) can reduce the functionalized degree by removing some initially attached functional groups. The resulting SWCNT had enhanced dispersivity in organic solvents compared to the pristine nanotubes.

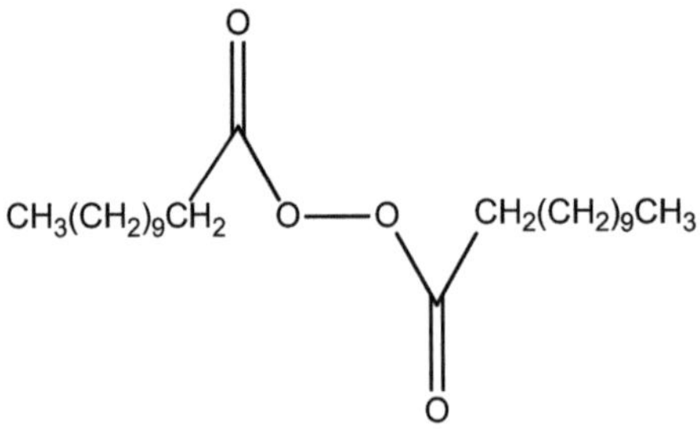

3.46, lauroyl peroxide

3.1.2.3 Compounds with N-Containing Moieties

Functionalization with aliphatic and aromatic amine(amido)-containing moieties N-containing compounds represent a majority of compounds having dispersion capacity for CNTs. Despite that many existing complex amines or amides below with various functional groups are used for CNT dispersion, a simple compound, one of classic solvents in organic and coordination chemistry, dimethylformamide (DMF), under ultrasonication [169], after strong acid treatment of SWCNTs [170] or in the conditions of photooxidation using singlet oxygen [171], can be used for successful debundling and dispersion of CNTs, being adsorbed on the CNT surface through a hydrophobic interaction. It was shown that this solvent works better than mixture of acetone and ethanol, N-methyl-2-pyrrolidone (NMP), and distilled water [172, 173]. One-step dissolution of SWCNTs was demonstrated through the use of dimethylacetamide (DMA) as a stable solvent for both sodium naphthalide and reduced SWCNTs, enabling the synthesis of concentrated solutions of nanotubide [174]. Dimethylacetamide–cholic acid mixture was also successfully applied for SWCNT dispersion [175].

However, more sophisticated compounds are generally applied instead of those simple solvents above, for instance, or a series of amines {octadecylamine (ODA), 2-aminoanthracene, 1-H,1-H-pentadecafluorooctylamine, 4-perfluorooctylaniline and 2,4-bis(perfluorooctyl)aniline (see also aniline derivatives below)}, reported to interact with SWCNTs, previously thermally been treated in air and optionally purified with nitric acid [176]. The formed products exhibit different solubility in dependence of amines and solvents used. For one of the most frequently applied amines, octadecylamine [177–180], it was revealed that the ODA chains grafted on MWCNT are partially crystallized [181]. Similar functionalization of SWCNTs with octadecylamido moieties allowed to create soluble nanotubes (s-SWCNT-$CONH(CH_2)_{17}CH_3$), where the mentioned groups are attached in the end groups and at defect sites; the weight percentage of the octadecylamido functionality in the s-SWCNTs is about 50% [182]. A representative reaction between ODA y SWCNTs is shown in the reaction scheme 3.13 [183]. Octadecylamine-functionalized MWCNT–ODA was also used [184] as the model compound (studying the possibility of antisolvent precipitation of hydrophobic, organic soluble-functionalized carbon nanotubes (f-CNTs), where water acts as an antisolvent is presented), and it was found to form highly stable dispersions in different water/solvent systems, and the particle sizes ranged from 170 to 400 nm. Colloidal behavior was studied using dynamic light scattering, and particle aggregation was found to increase with the addition of electrolytes, with THF and ethanol showing the maximum effect. Based on this reported study, it is evident that hydrophobic f-CNTs are potential water pollutants.

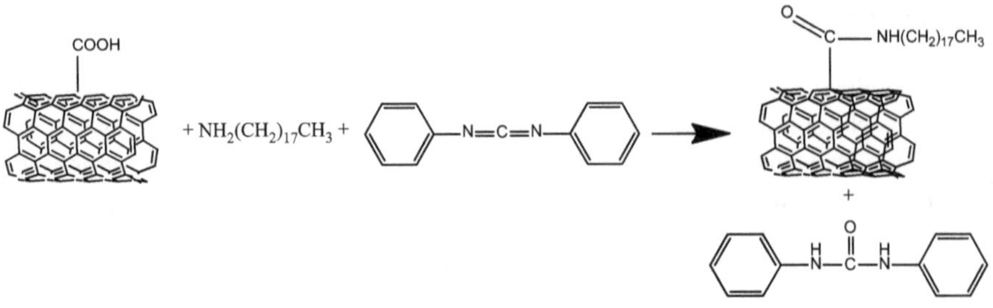

Amidation of SWCNTs with octadecylamine.

An organogelator, N,N′-*bis*(octadecyl)-l-(1-pyrenebutyric acid)-glutamic diamide (LPG), whose interaction with pristine SWCNTs in the gel state was investigated [185]. It was found that LPG can form organogels with various kinds of organic solvents, and SWCNTs can be well dispersed into LPG gel. The gelation process and the properties of the resulting nanocomposites were found to be closely related to the pyrene group in the gelator. The textures of the nanocomposites were altered from layer structure to intertwined fibers upon the incorporation of SWCNTs, which also supported the effective mixing of SWCNTs into LPG organogels. In addition to this LPG gel, hybrid organogels, where SWCNTs were incorporated [186] into organogel fibers, were prepared. The SWCNTs were covalently functionalized with organic branches {1,8-*bis*[3,4,5-*tris*(decyloxy) benzoylamino]octane and N-(8-aminooctyl)-3,4,5-*tris*(decyloxy)benzamide} (reaction scheme 3.14) that had a similar structure to the organogelator. The functionalized SWCNTs in the hybrid organogel formed in decane were mainly located inside or on the surface of the organogel fibers, while the f-SWCNTs in the hybrid organogel formed in DMF were distributed evenly over the sample. When an organogelator had a different chemical structure to that of an organic functional group on the SWCNT surface, SWCNTs existed as large aggregates, or long bundles, which were not incorporated inside of the organogel fibers. The dispersion properties of the f-SWCNTs in the organogels are greatly dependent on their relative interactions with the solvent and the organogelator and can be exploited to optimize the properties of hybrid gels bearing CNTs. Different chemical strategies to aminate oxidized CNTs are extensively discussed [187]. Generally, the amino(amido) functionalization of MWCNTs can improve their dispersion in H_2O; however, other reactions may also have occurred, which influenced their dispersity in organic solvents [188]. In case of the use of N,O-containing dendrimer-like dispersants as surfactants, the PAMAM *dendrimer* (**3.47**) was found to be an “ideal” candidate for SWCNTs separation [189].

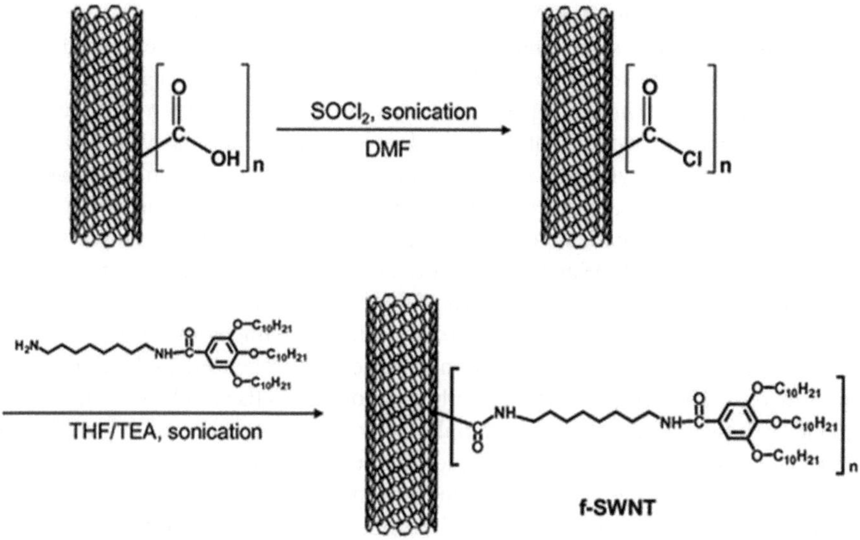

Covalent functionalization of a SWCNT with N-(8-aminooctyl)-3,4,5-*tris* (decyloxy)benzamide

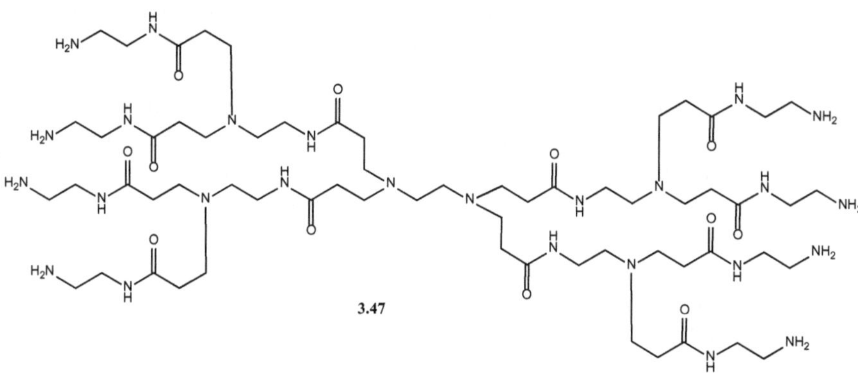

3.47

PAMAM dendrimer

Aniline and its derivatives Such classic chemical as aniline $PhNH_2$ was found to act as a very efficient dispersing agent to debundle CNTs from 200 to 50 nm at a very low concentration of 0.5% in an IPA/water solution [190]. The aniline-stabilized CNTs had a larger pore volume and larger amount of mesopores than pristine CNTs, and the debundling of CNTs by aniline appears to be a physical rather than a chemical process. In addition, the aniline is an efficient dispersant and stabilizer for the preparation of Pt nanoparticles deposited on CNTs. Aniline was also used as a solvent for MWCNTs, grafted by the carboxylic acid group [191]. These nanotubes were rapidly (30 min.) dissolved in aniline under microwave treatment. The solubility of SWCNT in aniline is up to 8 mg/mL [192]. A series of *aniline derivatives* (4-pentylaniline, 4-dodecylaniline, 4-tetradocylaniline, 4-pentacosylaniline, 4-tetracontylaniline, 4-pentacontylaniline), as well as amines (octadecylamine, nonylamine, dodecylamine, pentacosylamine, tetracontylamine, pentacontylamine, see also the section above on amines), etc., and mixtures thereof were used for functionalization of CNTs [193, 194]. The formed composites are soluble in carbon disulfide and common organic solvents as chlorobenzene, dichlorobenzene, trichlorobenzene, tetrahydrofuran, chloroform, methylene chloride, diethylene glycol dimethyl ether, benzene, toluene, tetrachlorocarbon, pyridine, dichloroethane, diethyl ether, xylene, naphthalene, nitrobenzene, ether, and mixtures thereof. The solubilities of the carbon nanotubes in these solvents range from about 0.01 to 5.0 mg/ml. For a completely soluble optically active polyaniline (PANI)-multiwalled carbon nanotube composite, it was found [195] that the polymer's optical activity was retained in the presence of carbon nanotubes. Solutions were found to be easily processable into thin films, which exhibited dendritic structures only in the presence of nanotubes. We note that PANI–CNT interactions are being intensively studied [196–199].

Pyridine derivatives Highly soluble *pyridyl*-functionalized SWCNTs were obtained by a 1,3-dipolar cycloaddition of a nitrile oxide on the SWCNT walls (reaction scheme 3.15), similar to 1,3-dipolar cycloadditions that are common for fullerene functionalization (reaction scheme 3.16), and characterized by NMR, FT-Raman, and electron microscopy [200]. The CNTs here were *doubly functionalized*, at the tips with pentyl esters (to provide sufficient solubility in organic solvents) and on the walls by pyridyl isoxazoline groups (they are capable to coordinate the metalloporphyrin to the pyridyl group) (see also the section on macrocycles below). This composite was further used as a precursor in the synthesis of CNTs-Zn–porphyrin analogue of fullerene-C_{60}-Zn–porphyrin. The occurrence of this complex is clearly revealed by optical spectroscopy and by the shifts in CV potentials of Zn–porphyrin in the presence of Py-SWCNT, similar but larger than for the corresponding fullerene analogues. Pyridine-based polymers (see also the section below on polymer-assisted "dissolution" of CNTs) are also applied, for example, poly(4-vinylpyridine) (P4VP) and P4VP/pentacyanoferrate (II) metallopolymer, which were used together to suspend up to 1 mg mL^{-1} of MWCNTs in EtOH/water mixtures, resulting noncovalent decoration of MWCNT with electroactive iron complex [201]. The peculiarity of this system is that the major interaction between polymer side chains and CNTs is via π–π stacking, rather than charge transfer interaction reported for a lot of other N-containing interacting molecules.

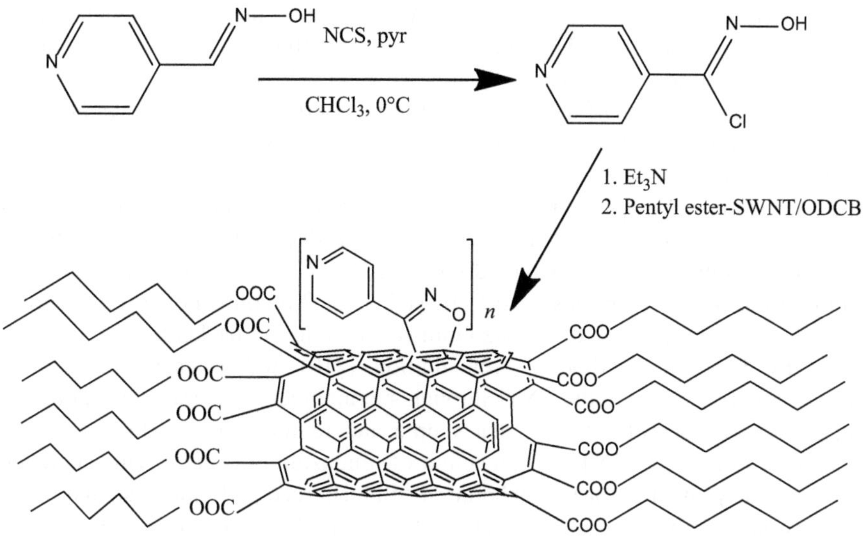

Synthesis of pyridyl-functionalized SWCNTs.

(III-16)

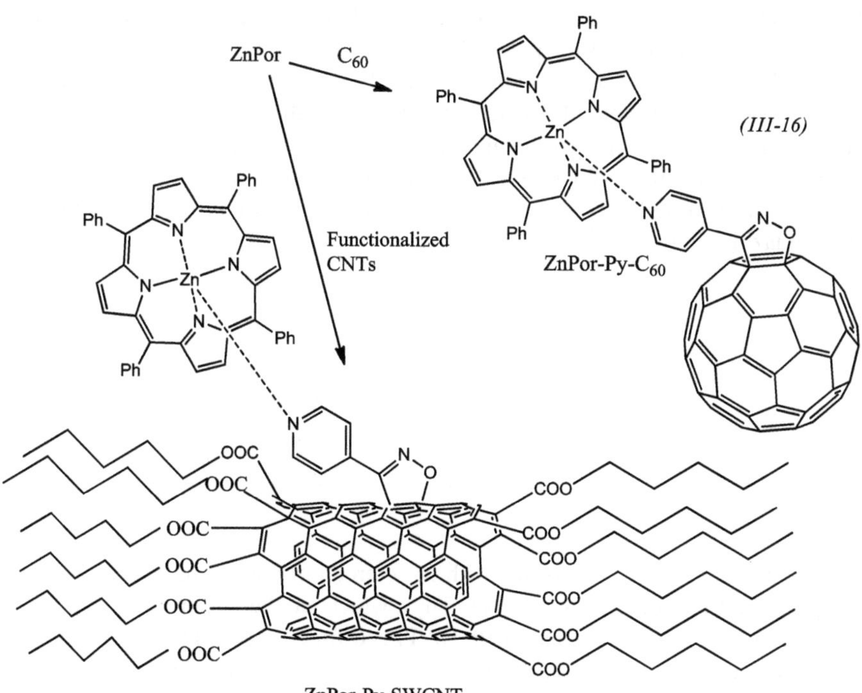

A comparison of Zn-porphyrin complex of pyridyl-functionalized SWCNTs and its fullerene analogue.

Other N-Heterocycles The dispersion of SWCNTs in the presence of water-soluble *polypyridyl* complexes of the general formula $[Ru_x(bpy)_yL]^{2+}$ (L = dppz, dppn, tpphz) was reported [202]. These ligands have extended planar π systems, which aid in the solubilization of SWCNTs via π–π interactions. A series of *imidazolium* derivatives have been reported as good surfactants for CNTs, leading to composites with useful properties and possible applications [203–208]. Thus, an approach for the stabilization of dispersion based on the use of cationic surfactants (derivatives of imidazolium iodide and bromide **3.48**) synthesized [209] from raw materials was offered. Resulting homogeneous dispersions of CNTS with sonication with imidazolium iodide were observed to be stable for 2 months. Polyesters incorporating a variable quantity of imidazolium rings along the backbone were synthesized by Huisgen 1,3-dipolar cycloaddition [210] and revealed an efficient dispersing ability toward the CNTs through supramolecular interactions in organic media. These compounds offered a simple and reliable strategy to control the quantity of dispersed CNTs as a function of imidazolium content. This approach is particularly suitable for the elaboration of biosourced and biodegradable materials based on poly(butylene succinate) with high-performance properties. A core–shell carbon nanotube/conducting polymer hybrid, using an imidazolium salt-based ionic liquid as a linker, was developed [211] for large-scale solution processing of CNTs. The water-soluble shell promotes dispersion of the CNT aggregates and brings individual CNTs into aqueous solution while maintaining intact π-conjugation structure and leads to a reduced charge injection barrier and realizes high photovoltaic performance. In addition, the 1-hexadecyl-3-vinylimidazolium bromide (hvimBr), a water-soluble long-chain imidazolium ionic liquid (IL) with surfactant properties, showed the ability to produce stable homogeneous aqueous dispersions of pristine SWCNTs [212]. The improvement of SWCNT dispersing ability was suggested to be due to the effect of different groups in position three of the imidazole ring.

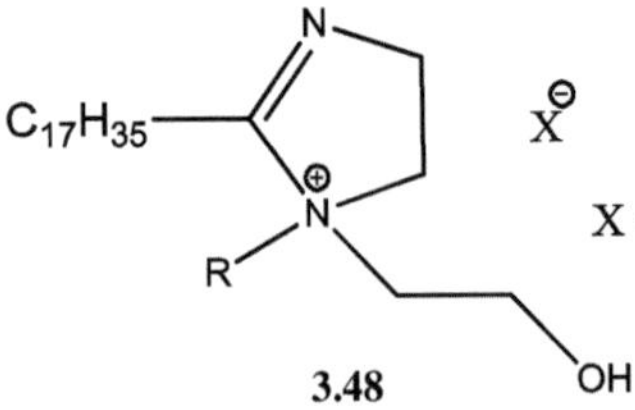

X = Br, I

3.48

Among other CNT composites with N-heteroatom-containing cycles, we note thymine (**3.49**)-modified CNTs (CNT–Thy), which can be dispersed in solution in the presence of diaminotriazine (**3.50**, DAT) end-functionalized polymers, through supramolecular Thy/DAT association [213]. Long-term stable supramolecular CNT dispersions were reversibly aggregated by adding a few droplets of a selective dissociating agent of the Thy/DAT association (DMSO). CNT–Thy, simply recycled by centrifugation or filtration, can be redispersed in another solvent in the presence of a suitable soluble DAT polymer. Dispersion and aggregation can also be switched on and off by choosing a polymer for which a given solvent is close to Θ conditions (a solvent in which polymer coils act like ideal chains), e.g., PS in cyclohexane or PPO/PEO in water. In case of polyvinylpyrrolidone (PVP) **3.51**, homogeneous MWCNT suspensions in water can be easily obtained under ultrasonication [214].

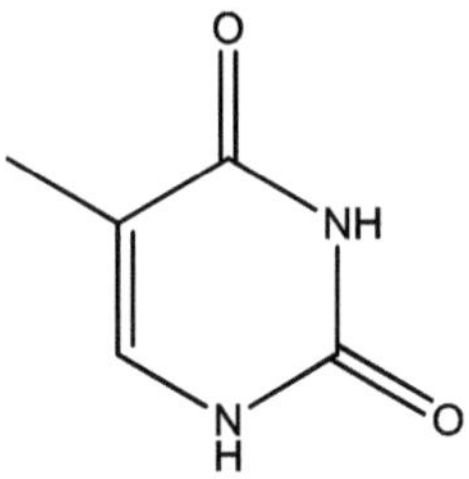

3.49

Thymine

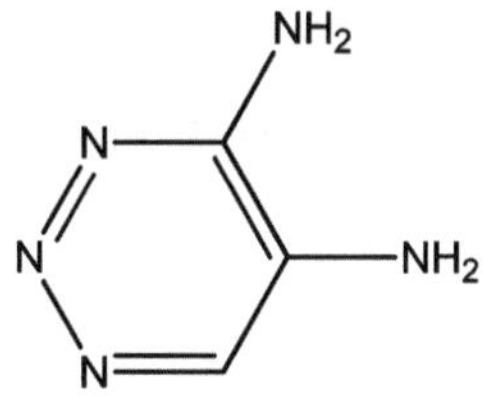

3.50

Diaminotriazine

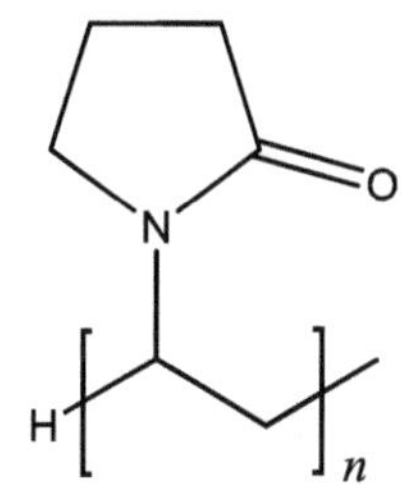

3.51

PVP

Several other N-, N,O-, and N,P-containing organic compounds have been applied for CNT dispersion in solvents, in particular organic hydrazines [215] or polyimides [216]. Thus, a total aromatic polyimide was found [217] to have a high potential to solubilize SWCNTs individually in organic solutions. When SWCNT concentrations increase, the solutions became viscous and then changed to gels. Also, the addition of an azobenzene derivative, Disperse Orange 3 (DO3) **3.52** to dispersions of MWCNTs in THF efficiently helped debundling MWCNT and made dispersions stable for days [218]. The effect of poly(2-ethyl-2-oxazoline) (PEOX, $[-N(COC_2H_5)CH_2CH_2-]_n$) as a surface modifier for MWCNTs in a poly(vinyl alcohol) (PVA) composite was studied [219] through a simple polymer wrapping and melt extrusion processes. MWCNTs were carboxylated in a reaction with HNO_3, followed by mixing with 10 wt.% PEOX in ethanol to produce a MWCNT–PEOX complex. A PEOX layer existing on the surface of MWCNTs was identified. Compared with pristine MWCNTs, MWCNT–PEOX showed an improved dispersion in hydrophilic solvents such as water and ethanol. Also, poly [(4-pyridineoxy)(phenoxy)phosphazene] was prepared [220] by sequential treatment of poly(dichlorophosphazene) $(-PNCl_2-)_n$ with sodium 4-pyridineoxy and sodium phenoxide in THF. MWCNT/poly(organophosphazene) composites (f-MWCNT/PZS) with different feed ratios [Rfeed = 1:1, 1:3, 1:5 and 1:10 (w/w)] were obtained by the treatment of the functionalized f-MWCNT with the protonated poly(organophosphazene) (PZS), achieving excellent dispersions of in water and common organic solvents. Among many useful applications of compounds with N-containing moieties, we emphasize diethylenetriamine pentaacetic acid (DTPA), enabling tight attachment of Gd atoms onto the CNT surface for development of efficient contrast agents for cell labeling [221].

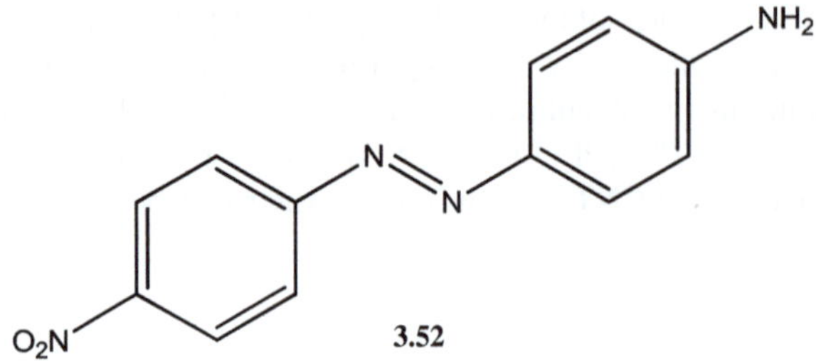

3.52

Disperse Orange 3 (DO3)

3.1.2.4 Compounds with S-Containing Moieties

In a difference with N- or O-containing effective dispersants for CNT solubilization in solvents, the compounds with S-containing moieties are rarer. Among these surfactants or functionalizing agents, we note the surfactants sodium dodecyl sulfate (SDS) $C_{12}H_{25}OSO_3Na$ (and sometimes sodium dodecylbenzene sulfonate), already mentioned above (the section on ionic salts of organic acids), whose presence together with hydroxypropyl methyl cellulose [222] or HNO_3/H_2SO_4 mixture [223], helps to assist the dispersion of CNTs. We note that molecular dynamics (MD) simulation of CNT dispersion in a variety of surfactants, in particular in SDS (Fig. 3.22), was recently reported [224]. Dodecanethiol

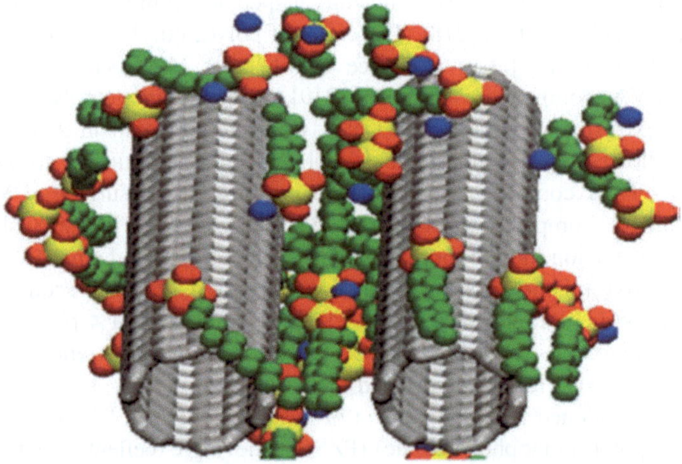

Fig. 3.22 Representative simulation snapshots for SDS aggregates on two approaching SWCNTs at low SDS surface coverage [228]. The SWCNTs are separated by 6.90 Å. *Green*, *red*, and *yellow* spheres represent methyl groups, oxygen, and sulfur atoms of SDS, respectively. *Blue* spheres represent sodium ions. Carbon atoms in nanotubes are connected with bold gray lines. Water is not shown for clarity (Reproduced with permission of the *American Chemical Society*)

$CH_3(CH_2)_{11}SH$ as the reaction agent was used to obtain a stable suspension of thiolated SWCNTs in toluene; the thiolation process is also observed on the exposure of the nanotubes to toluene solutions of dodecanethiol-stabilized Au nanoparticles, using them for labeling or manipulating the location of the chemical reaction sites on the tube wall [225]. Similarly, cyclic disulfides were also used for introduction of sulfur-containing functional groups onto SWCNTs and further treatment with gold nanoparticles [226]. In addition, exhaustive study of thiolated MWCNT dispersion in a wide range of dispersing agents was carried out [227]. It was established that MWCNT–SH can be dispersed in SDS aqueous solution, DMF, colloidal gold, sodium aurothiomalate, and Nafion®/ethanol.

In the previous section on N-containing surfactants, we noted that such simple and classic organic solvents as DMF or DMA could be successfully applied for CNT dispersion. In case of DMSO, polymer-DMSO-CNT systems are discussed, for example, a co-dispersion of cellulose and MWCNTs, which was prepared in a tetrabutylammonium acetate/dimethyl sulfoxide solvent mixture, resulting hydroxide radical functionalized MWCNTs, well dispersed in water [229]. Study of structure and flow behavior of dilute dispersions of CNTs in polyacrylonitrile–DMSO solution [230] revealed an alignment and association of the macromolecules as well as (dis)orientation and (dis)association of CNTs. Other S compounds, used for CNTs dispersion, frequently contain heterocycles, for instance, thiophene group. Thus, as an example of an S heterocycle, the hydrogenation of 2-nitrothiophene gave 2-aminothiophene **3.53** that was used [231] for amidation of SWCNTs functionalized with carboxylic acid groups (SWCNT–COOH). In these modified carbon nanotubes (SWCNT–CONHTh), the thiophenes were covalently attached to the SWCNTs via amide linkages. The modified SWCNTs showed enhanced solubility and thus better dispersion in common organic solvents and were used as dopant in polymer–fullerene photovoltaic cells. Other thiophene derivatives **3.54** and **3.55** were reported in [232], yielding methyl-2-thiophenyl-3-carboxylate- and 2-thiophenyl-3-carbonitrile-functionalized SWCNTs, highly soluble in THF. Highly concentrated CNT dispersions can be prepared [233] using biomolecules on the basis of polythiophene **3.56**, where a dispersion system was developed to investigate the bioinspired interactions between an adenine (A)-terminated poly(3-adeninehexyl thiophene) and CNT–U (uracil). A rich N-"doped" and thiophene-substituted hexaazapentacene molecule, denoted as 4HP (Fig. 3.23) [234], was found to be more favorable for high-purity semiconducting s-SWCNTs with large diameters, selectively entrapping them with high purity and yield. In addition, if small concentrations of some inorganic salts (KBr, KI, $AlCl_3$) are added to THF solutions of poly(3-hexyl thiophene) (P3HT), when a non-dispersed MWCNTs are present, the *salt-in* of the precipitated tubes leads to the formation of a stable MWCNTs (Fig. 3.24) [235]. This effect was explained by a conformational change of polymer chains, induced by salt appearance, whose result is a steric repulsion among polymer-decorated chains.

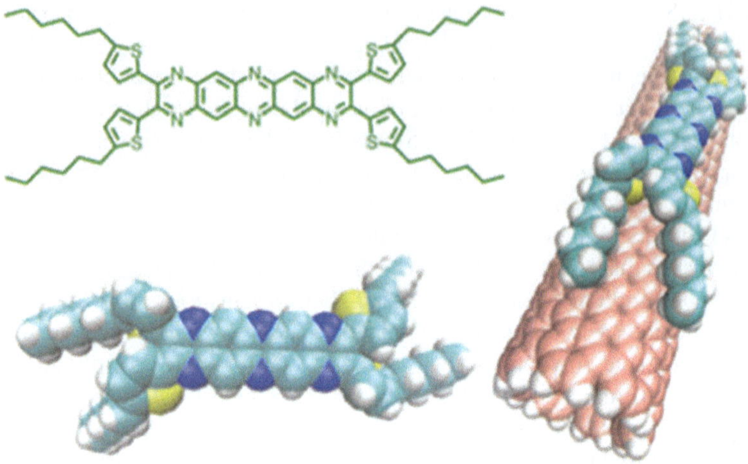

Fig. 3.23 Chemical structure and conformation of a 4HP molecule and the complex of 4HP-SWCNT (Reproduced with permission of the *Royal Society of Chemistry*)

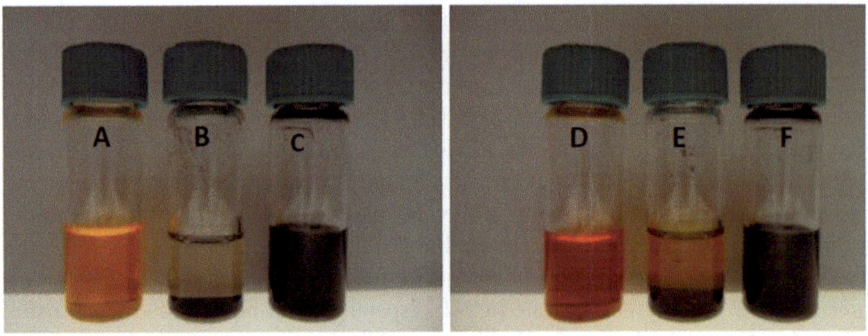

Fig. 3.24 Images of vials containing a THF solution of (**a**) 0.1 mg ml^{-1} of P3HT, (**b**) after addition of 2 mg ml^{-1} of MWCNTs and sonication for 3 h, (**c**) after addition of KBr to a final concentration of 0.5 mM and re-sonication for 1 h, (**d**) 0.25 mg ml^{-1} of P3HT, (**e**) after addition of 2 mg ml^{-1} of MWCNTs and sonication for 3 h, and (**f**) after addition of $AlCl_3$ to a final concentration of 1.2 mM and re-sonication for 1 h (Reproduced with permission of *Royal Society of Chemistry*)

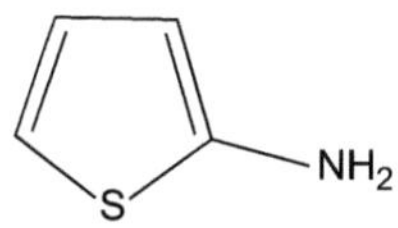

3.53

2-aminothiophene

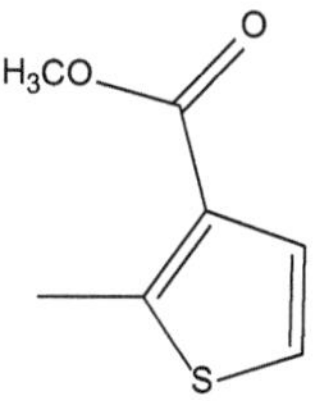

3.54

2-thiophenyl-3-carboxylate-

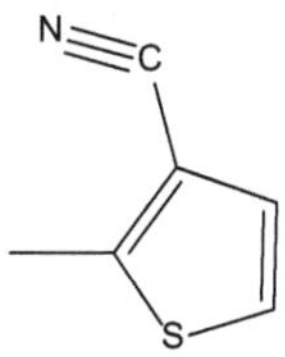

3.55

2-thiophenyl-3-carbonitrile-

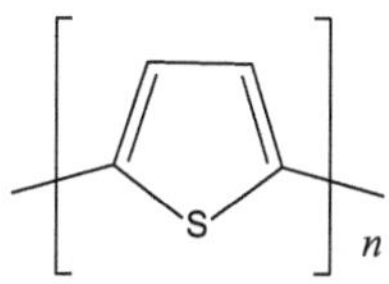

3.56

Polythiophene

Other interesting examples of bifunctional aromatic molecules, containing S, N, or O atoms, are as follow. Large aromatic surfaces of electron donor and acceptor molecules {highly soluble tetrathiafulvalene (TTF) derivatives **3.57** and **3.58** as the electron donors and a coronene bisimide (CBI) derivative **3.59** as the electron acceptor} direct them onto the surface of the graphene/SWCNTs through π–π stacking interactions, whereas the alkyl chains as well as the glycol chains on the aromatic molecules promote the solubility, thereby giving stable dispersions of SWCNT/graphene composites in nonpolar solvents [236]. We note that tetrathiafulvalene copolymers (for instance, with fluorene) are of a permanent interest for the area of CNT composites with S-containing polymers [237].

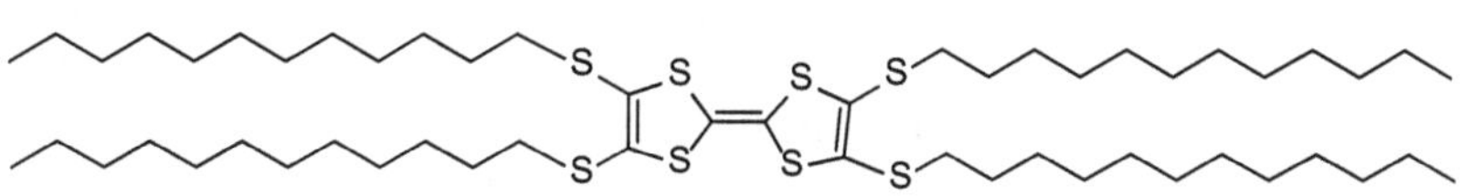

III-57

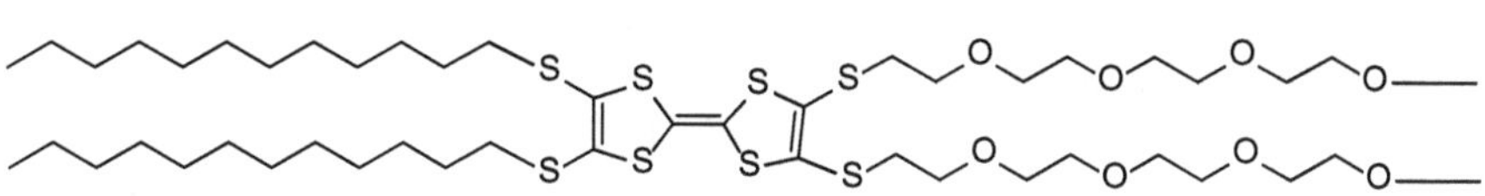

III-58

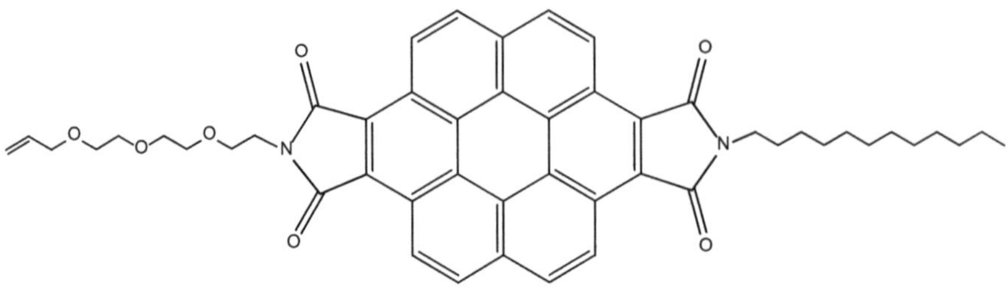

III-59

Molecular structures of electron-donor TTF derivatives and electron-acceptor coronene bisimide (CBI) derivatives.

3.1.2.5 Specific Organic Dispersants

Representative examples of organic agents for CNT dispersion are a series of *dyes*. Thus, fluorophore and dye molecules such as Rhodamine B and fluorescein have both hydrophilic and hydrophobic moieties and also contain π-conjugated systems that can potentially interact with SWCNTs to induce its dispersion. It was shown that SWCNTs can be dispersed [238] in aqueous solution in the presence of various fluorophore or dye molecules. However, the ability of a fluorophore or dye molecule to disperse SWCNTs is not correlated with the stability of the fluorophore/dye–SWCNT complex, suggesting that the on-rate of fluorophore/dye binding to SWCNTs may dominate the efficiency of this process. These results could have potential applications in the delivery of poor cell-penetrating fluorophore molecules. In case of another CNT–dye thermodynamically stable colloidal system [239], by adding sodium chloride electrolyte, SWCNTs flocculated and settled out due to the destabilization of colloidal systems initiated by the increase in ionic strength. The dye molecules can be removed by heat treatment at 300 °C for 5 h following washing with water. In addition, bromocresol green **3.60** (BCG) and bromophenol blue **3.61** (BPB) in acidic and basic forms were used [240] as dispersing agents to suspend MWCNTs into deionized water. Both these dyes exhibited critical concentrations in dispersing MWCNTs; this may result from the formation of dye micelles, which induces osmotic pressure on MWCNTs and aids nanotube reaggregation. An optimum dye concentration for dispersion was proposed to be a competing result between adsorption of dye molecules on nanotube surfaces and formation of micelles in water (Fig. 3.25). This critical value is largely related with the balanced point between hydrophilicity and hydrophobicity of dye molecule and its molecular architecture as well. Discussing diazo dyes, the mixture of SWCNTs and a rigid, planar, and conjugated diazo dye, Congo red (CR), can be dissolved in water with a solubility as high as 3.5 mg/ml for SWCNTs [241]. The authors noted that the π stacking interaction between adsorbed CR and SWCNTs was considered responsible for the high solubility.

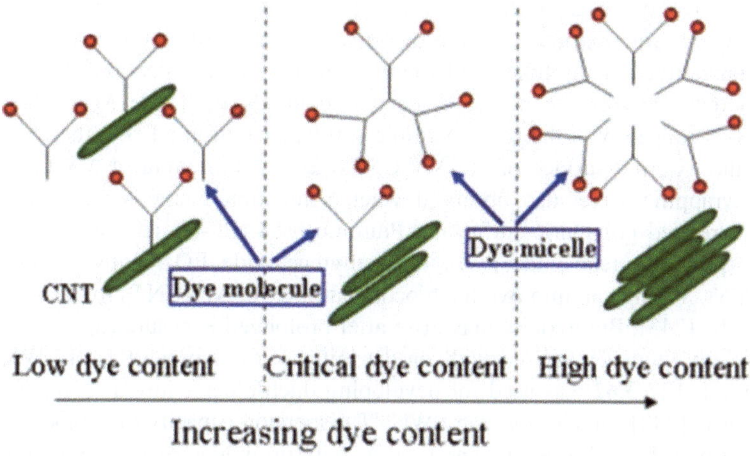

Fig. 3.25 Effect of dye content on the formation of dye micelles and dispersion of CNTs. Red balls represent the hydrophilic group in dye molecule (Reproduced with permission of *Elsevier Science*)

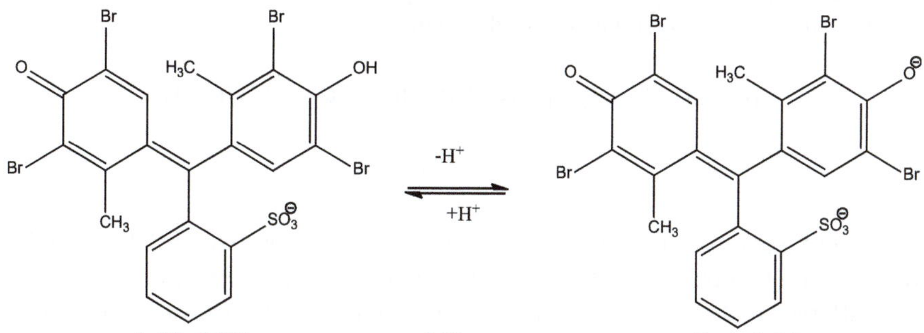

Acidic BCG **3.60** Basic BCG

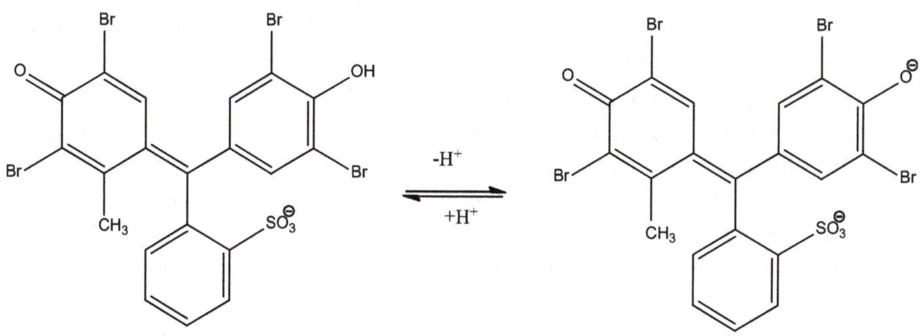

Acidic BPB **3.61** Basic BPB

Chemical structures of BCG and BPB.

MWCNTs were also dispersed [242] in water and in a Pluronic F108 (2-methyloxirane; oxirane) solution by four different dispersion methods (stirring, bath sonication, stirring followed by bath sonication, and sonication probe). It was shown that, irrespective of the dispersion methods used, the MWCNTs were strongly wrapped with the biocompatible surfactant Pluronic F108, thereby modifying the external surface of the MWCNTs. Some shortening of MWCNTs and more wrapping were also observed when sonication methods were used. For another related polymeric surfactant, Pluronic F-68 (PF-68, (polyethylene oxide, $EO)_{75}$-(polypropylene oxide, $PO)_{30}$-(polyethylene oxide, $EO)_{75}$), it was established that PF68 coating can improve the biocompatibility of MWCNTs to mesenchymal stem cells [243]. But toxicity may arise after prolonged exposure (up to 72 h), and the toxicity tendency still depends on the initial surface properties of MWCNTs. Pluronic F-127 **3.62** was used for developing ductile cementitious composites on CNT basis [244]. In addition, the SWCNT dispersion capacity of triblock copolymer Pluronic L121 [poly(ethylene $oxide)_5$-poly(propylene $oxide)_{68}$-poly(ethylene $oxide)_5$] was compared with that of the nonionic surfactant Triton X-100 below, as well as with ionic liquid [BMIM][BF_4] (1-butyl-3-methylimidazolium tetrafluoroborate, see also the section above on imidazolium-containing surfactants) [245]. (See also the section below on polymer-assisted CNT dispersion in solvents, including polyethylene-based polymers.)

$$HO\text{-}[CH_2CH_2O]_{100}\text{-}[CH_2C(CH_3)HO]_{65}\text{-}[CH_2CH_2O]_{100}\text{-}H$$

3.62

Triton X-series surfactants, compared in the section above with SBDS and SDS, with a same hydrophobic functional group {4-(1,1,3,3-tetramethylbutyl)-phenyl} and different hydrophilic polyethoxyl chain lengths were selected to investigate their adsorption onto CNTs and their ability to stabilize CNT suspensions [246]. Adsorption capacities of the surfactants increased with decreasing hydrophilic chain length: Triton-305 < Triton-165 < Triton-114 < Triton-100. Electrostatic interaction and hydrogen bond could be excluded as the main mechanism because adsorption was not significantly affected by pH change. Hydrophobic and π–π interactions between the surfactants and CNTs were the dominant mechanism for their adsorption. Among other applications of this series of surfactants, the effect of MWCNT nanofluid, obtained using Triton X-100, on solar parabolic through collector performance was studied [247]. This surfactant is indeed frequently used for preparation of CNT nanofluids (a liquid suspension of nanometer-sized particles) [248], MWCNT-filled epoxies [249], etc. In addition to CNTs, the boron nitride nanotubes can be also dispersed in Triton X-100 surfactants [250].

Among other numerous functionalities, *calixarenes* [251], have been used for CNT solubilization. Thus, the ability of using *p*-H_2O_3P-calix[8]arene **3.63** to stabilize MWCNTs and a mixture of C_{60} and MWCNTs (C_{60}@MWCNTs) in forming "peapod" structures in water was established [252]. The calixarene-functionalized carbon materials are colloidally stable for weeks without morphology changes and, in addition, can be readily decorated with palladium nanoparticles. We note also *carbenes* [253] (**3.64**), *carboranes* [254] (**3.65**), *phenosafranine* (3,7-diamino-5-phenylphenazinium chloride, **3.66**) [255], and much more compounds [256–259].

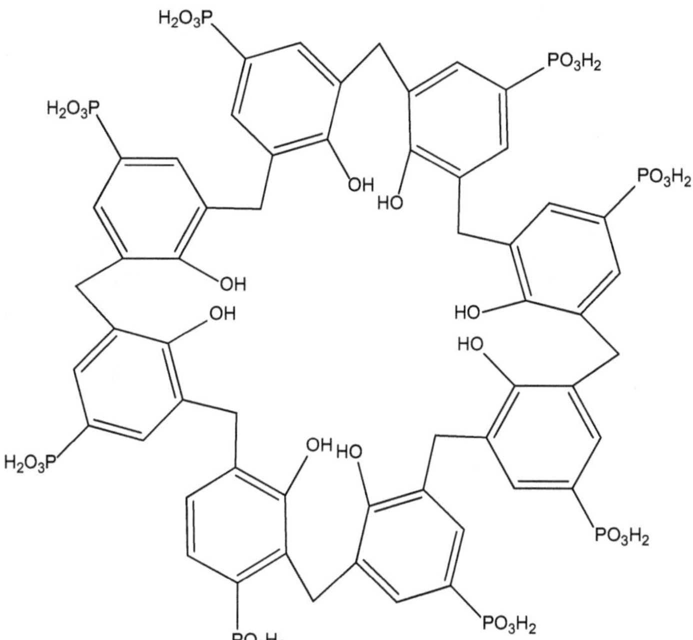

3.63

p-H_2O_3P-calix[8]arene

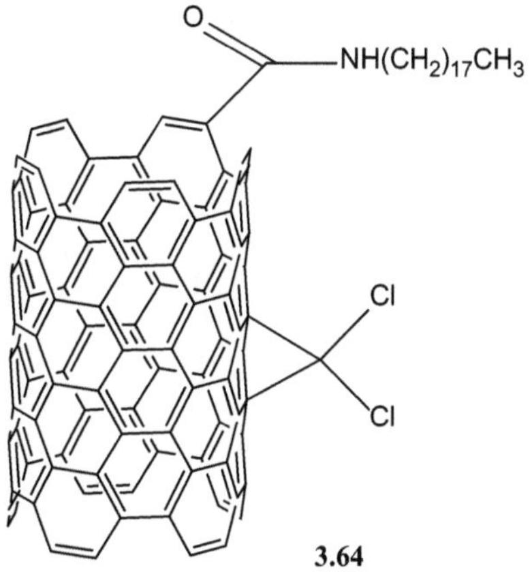

3.64

(s-SWNT)CCl_2

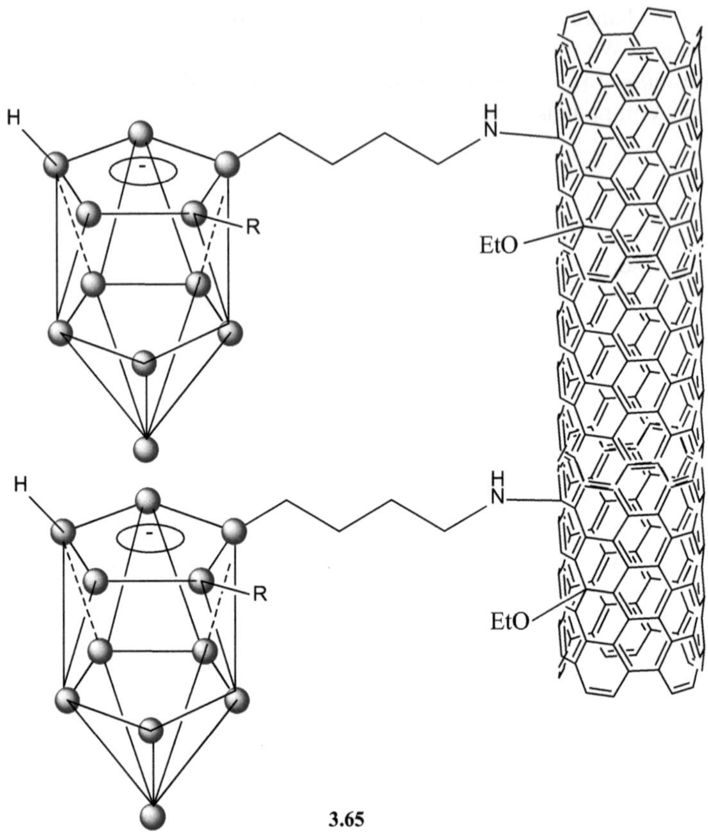

3.65

Carbene-(**3.64**) and carborane-(**3.65**) functionalized SWCNTs.

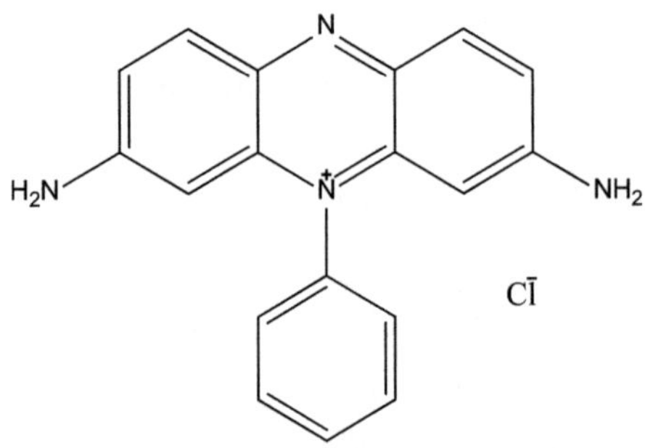

3.66

3.1.3 Coordination and Organometallic Compounds

Metal complexes have a lot of useful applications in organic and organometallic chemistry, catalysis [260], in medicine as anticancer pharmaceutics and for drug delivery [261], in various biological systems [262], polymers [263] and dyes, separation of isotopes [264], and heavy metals [265], among many other uses. Sometimes they are applied for increasing solubility [202, 266] of classic objects, CNTs, which form bundle-like structures with very complex morphologies with a high number of van der Waals interactions, causing extremely poor solubility in water or organic solvents. Metal complexes are also able to serve as precursors to fill CNTs with metals [267] or oxides [268], to decorate CNTs with metal nanoparticles [22, 269], as well as to be encapsulated by CNTs [270].

Various techniques are nowadays applied in order to obtain functionalized CNTs [271–273]. Figure 3.26 shows functionalization possibilities for SWCNTs, some of which will be discussed below for several metal complexes. The simplest functionalization by mineral acids, usually used as a first step in many reports, leads to formation of -OH and -COOH groups, which further can be replaced with more complex organic moieties. In particular, as it will be shown below, a series of

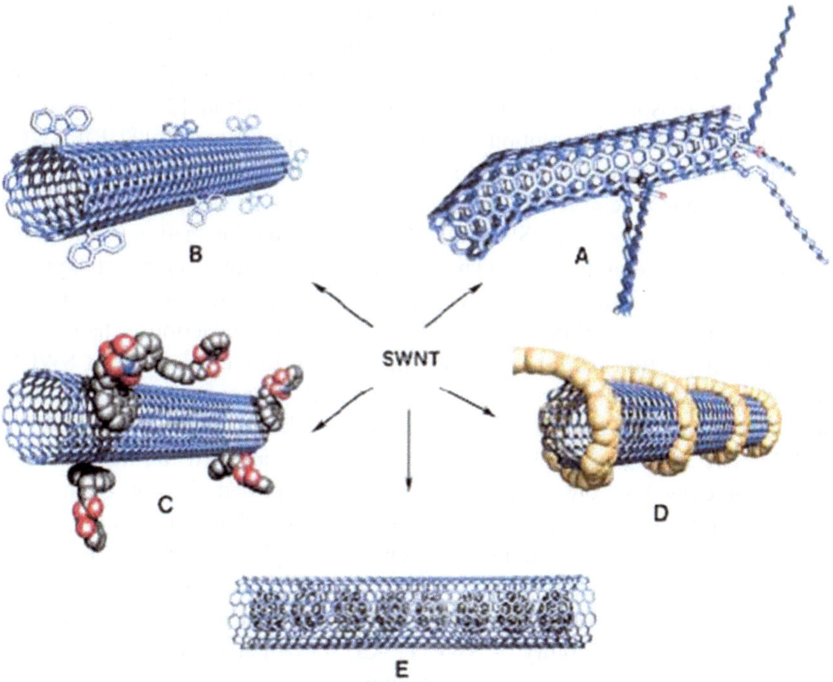

Fig. 3.26 Functionalization possibilities for SWCNTs: (**a**) defect group functionalization, (**b**) covalent sidewall functionalization, (**c**) noncovalent functionalization with surfactants, (**d**) noncovalent exohedral functionalization with polymers, and (**e**) endohedral functionalization [274] (Reproduced with permission of *John Wiley & Sons*)

coordination and organometallic compounds have been successfully anchored onto CNTs by covalent or noncovalent mode. In this chapter, we describe peculiarities of functionalization of CNTs with metal complexes, leading to their dispersion in solvents, paying particular attention to the ligand type (N-, O-, N,O-, N,S-, N, P-containing moieties), bond type inside complexes {coordination bond M–O, M–N, M–S, M–P; σ- and π-metal–carbon bond in organometallics}, and interaction type between CNTs and complex.

3.1.3.1 Solubilization of CNTs Hybrids with Metal Complexes of O-Containing Ligands

A few *crown ethers* without metal ions have been used for CNT functionalization [275–277], showing frequently higher dispersibility of formed hybrids (see also the section above on O-containing compounds). Thus, SWCNTs may be made soluble in a range of organic solvents without sidewall functionalization via their reduction by Na/Hg amalgam in the presence of dibenzo-18-crown-6 [278]. The $[Na(dibenzo\text{-}18\text{-}crown\text{-}6)]_n[SWCNT]$ complex was consistent with no additional sidewall functionalization as compared with raw SWCNTs; the presence of the $[Na(dibenzo\text{-}18\text{-}crown\text{-}6)]^+$ ion was shown. Solubility was found to be greatest in CH_2Cl_2 and DMF being comparable to surfactant dispersed SWCNTs; measurable solubilities were also detected in hexane, toluene, and alcohols. We note that benzo-18-crown-6 covalently linked to multiwalled carbon nanotubes (MWCNTs) can be used as ion sensors, in particular for Pb^{2+} determination [279]. Crown ether (18-crown-6)-assisted dissolution of SWCNT polyelectrolytes $[K(THF)]_nSWCNT$ in DMSO is described in [280] and in a recent patent [281].

Other ligands, containing donor oxygen atom only, are rare. Thus, functionalization of oxidized SWCNTs by a zwitterionic interaction ($COO^-\ NH_3^+$) between protonated amine on crown ether and an oxyanion from a carboxylic acid group on SWCNT was described [282]. The functionalization was achieved by adding 4-aminobenzo-9-crown-3 **3.67** to SWCNTs (reaction scheme 3.17). The ionic interaction led to a considerable increase in the solubility of SWCNTs in both organic and aqueous solvents such as ethanol, dimethyl sulfoxide, dimethylformamide, and H_2O, showing the highest solubility in DMF and DMSO. The ionic-bonded 4-benzo-9-crown-3 ether allowed the hosting of Li^+, and the ionic bond of crown ether to SWCNT was identified. We note the important major differences of ionic functionalization to covalent functionalization made by authors: (a) the acid–base reaction represents the simplest possible route to soluble SWCNTs and can be readily scaled-up at low cost. (b) Unlike the covalent amide bond, it seems that the presence of zwitterions (ionic functionalization) can significantly improve the solubility of SWCNT–CE (crown ether) in aqueous solvents. (c) The cation in crown ether of the ionic bond of $SWCNT\text{–}COO^-NH_3^+$ of SWCNTs can be readily exchanged by other organic and inorganic cations. (d) The authors found that the covalent functionalization approach generally gave a much higher yield (30.4%) of SWCNT–CE than the ionic functionalization approach (26%).

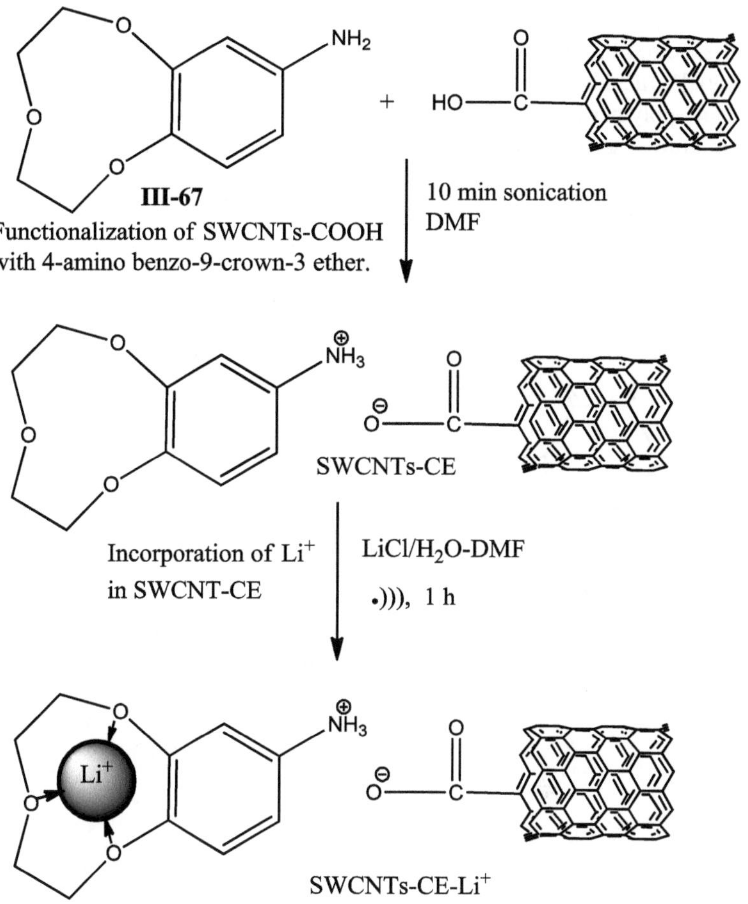

3.1.3.2 Composites of CNTs with Metal Complexes of N-Containing Ligands

Complexes with Bipyridyl Moieties The dispersion of SWCNTs in the presence of water-soluble ruthenium *polypyridyl* complexes [283] of the general formula $[Ru_x(bpy)_yL]^{2+}$ (L = dppz **3.68**, dppn **3.69**, tpphz **3.70**) was achieved [202]. These ligands have extended planar π systems, which aid in the solubilization of SWCNTs via π–π interactions (composites **3.71**, **3.72**, and **3.73**). Another example is a water-soluble ruthenium *tris*(bipyridyl) complex **3.74** linked through peptidic bonds to SWCNTs (Ru–SWCNTs) was prepared by radical addition of thiol-terminated SWCNT to a terminal C = C double bond of a bipyridyl ligand of the ruthenium *tris*(bipyridyl) complex [284]. The resulting macromolecular Ru–SWCNT (≈500 nm, 15.6% ruthenium complex content) was found to be water-soluble. The emission of Ru–SWCNT was 1.6 times weaker than that of a mixture of $[Ru(bpy)_3]^{2+}$ and SWCNT of similar concentration. Time-resolved absorption optical spectroscopy allowed the detection of the $[Ru(bpy)_3]^{2+}$-excited triplet and $[Ru(bpy)_3]^{+}$. In addition, poly(4-vinylpyridine) (P4VP) and P4VP/pentacyanoferrate (II) metallopolymer were used to suspend up to 1 mg mL^{-1} of MWCNTs in ethanol–water mixtures (see also the section below on polymers) [201].

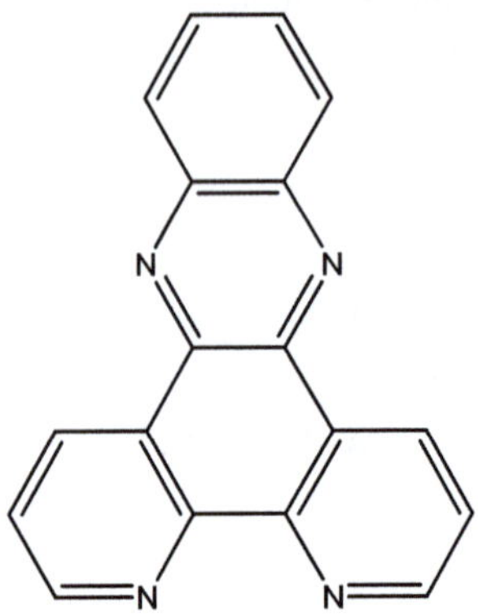

3.68

dppz = dipyrido[3,2-*a*:2',3'-*c*]phenazine

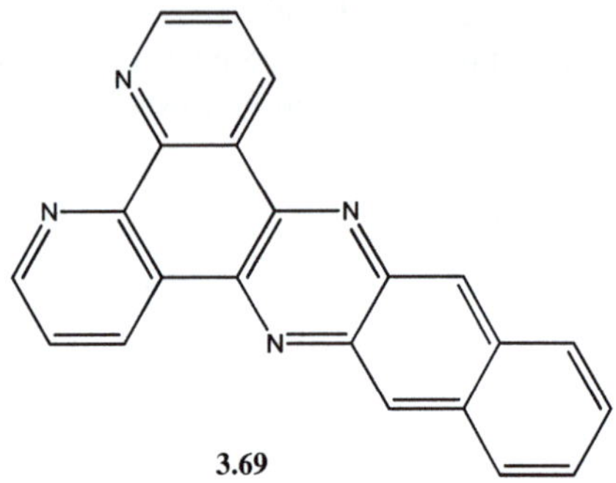

3.69

dppn = 4,5,9,16-tetraaza-dibenzo[a,c]naphthacene, benzo[i]dipyrido[3,2-a:2′,3′-c]phenazine

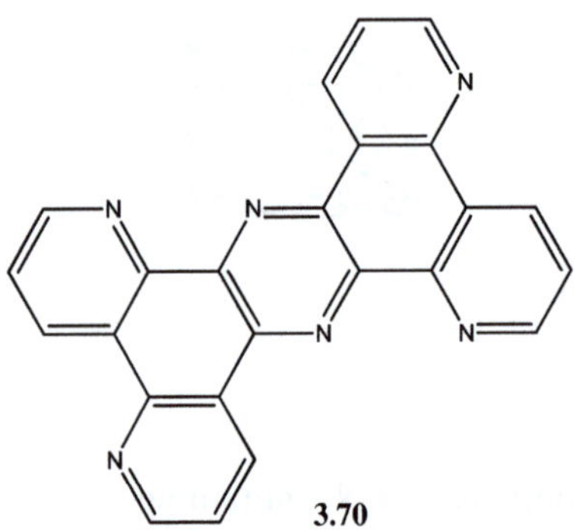

3.70

tpphz = tetrapyridophenazine

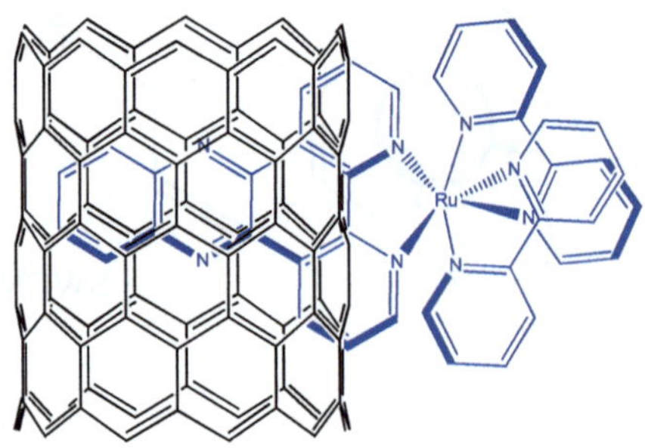

3.71

$[Ru(bpy)_2(dppz)]^{2+}$ composite with SWCNTs

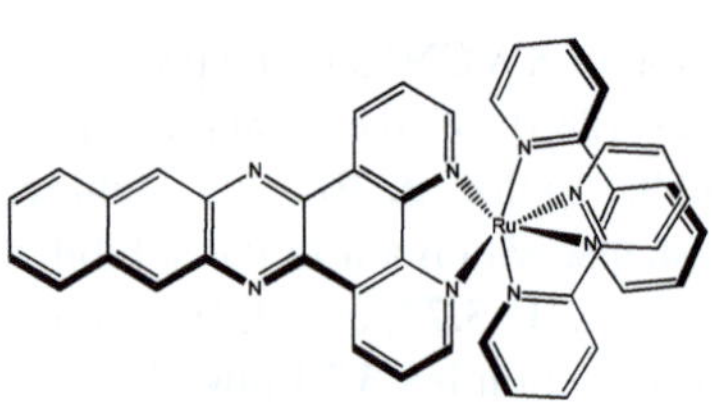

3.72

$[Ru(bpy)_2dppn]^{2+}$

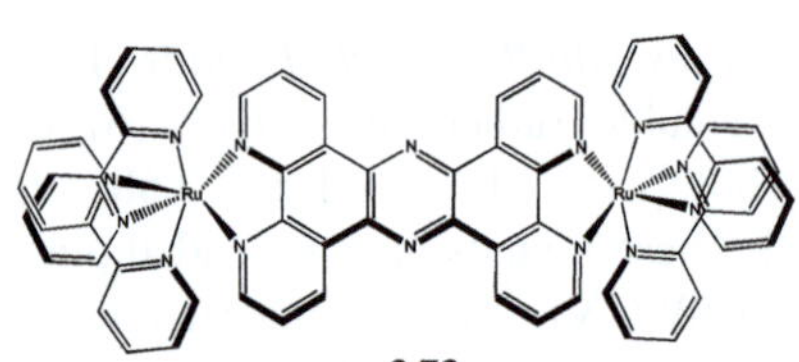

3.73

$[(bpy)_2Ru(tpphz)Ru(bpy)_2]^{4+}$

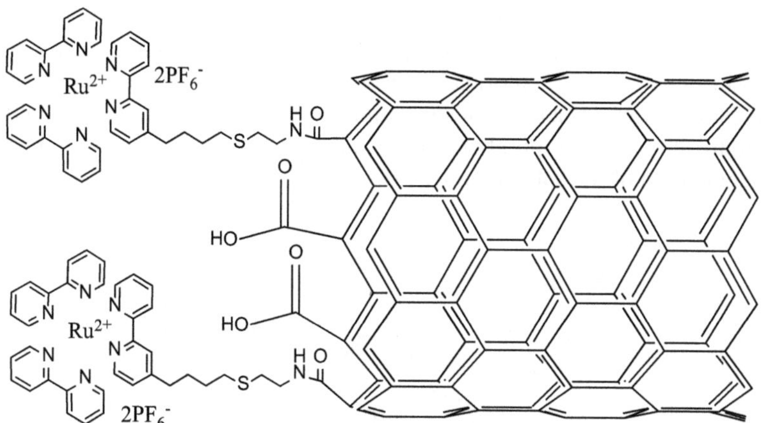

3.74

3.1.3.3 Porphyrin- and Metalloporphyrin-Functionalized Carbon Nanotubes

Porphyrin molecules can dissolve SWCNTs in organic solutions, and the SWCNT–porphyrin hybrid nanomaterials can be separated from the solutions; [285] moreover, both individually dissolved nanotubes and bundled nanotubes coexisted in the solution. The solid purified SWCNT(p-SWCNTs)–porphyrin nanomaterials were readily separated from the p-SWCNT–porphyrin solution, and this nanomaterial was redissolvable in DMF. Both free porphyrins [286, 287] and their metal complexes [288] can be applied for functionalization of CNTs (mainly SWCNTs), frequently in combination with fullerenes [289]. As an example of a non-metalloporphyrin functionalization of CNTs, SWCNTs were modified with porphyrin units [290] with aid of two strategies. In the first approach, the electropolymerization of pyrrole or pyrrole-substituted porphyrin monomers occurred via formation of conjugated positively charged polypyrrole (PPyrr) backbone, while negatively charged carbon nanotubes functionalized with carboxylic groups (SWCNT–COOH) acted as a polymer dopant. In the second case, the SWCNT–porphyrin composites were chemically synthesized via the interaction between glycyl-substituted porphyrin and non-modified CNTs and then entrapped in PPyrr/SWCNT–COOH film.

The formed porphyrin–CNT solutions can be stable for long time; thus, water-soluble porphyrin molecules [*meso*-(tetrakis-4-sulfonatophenyl) porphine dihydrochloride] (**3.75**) were used to solubilize SWCNTs, resulting aqueous solutions stable for several weeks [291]. A suspension, stable more than 1 week, was prepared with the use of an anionic tetra (*p*-carboxyphenyl) porphyrin (TCPP) and

MCNTs, and, as an application, a spectrofluorometric method of DNA hybridization was proposed [292]. Spectroscopic changes of tetraphenylporphyrin CNT composites in a variety of chlorinated solvents such as chloroform, dichloroethane, and dichlorobenzene as a result of sonication were studied [293]. It was established that protonation of the porphyrin core nitrogen atoms occurs as a result of sonodegradation of the solvent molecules.

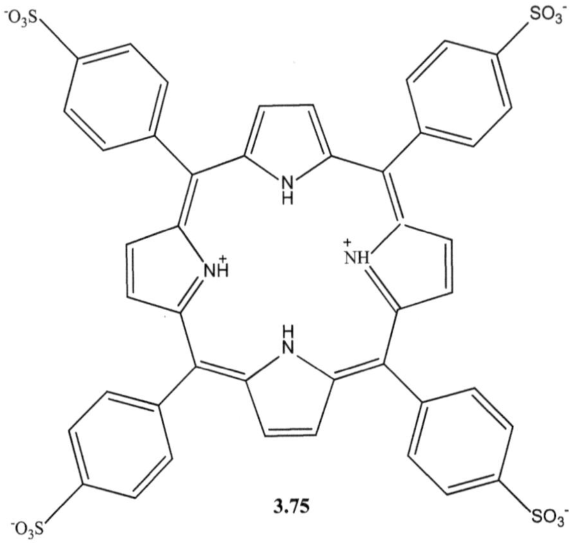

3.75

meso-(Tetrakis-4-sulfonatophenyl) porphine dihydrochloride.

Several *metalloporphyrins*, most frequently those of zinc, iron, and ruthenium, were used for CNT functionalization and solubilization. Thus, dispersible SWCNTs grafted with poly(4-vinylpyridine), SWCNT–PVP, were tested in coordination assays with zinc tetraphenylporphyrin {Zn(TPP)} (TPP = tetraphenylporphyrin **3.76**), showing the formation of a SWCNT–PVP·Zn(TPP) nanohybrid **3.77** (reaction 3.18) [294]. Temperature can be used to control the SWCNT–PVP·Zn(TPP) association; the SWCNT–PVP coordination to ZnP was found to be labile and dynamic. Hence, increasing or decreasing the temperature weakens or strengthens the complex, respectively. In this context, fluorescence emerged as a sensitive temperature probe. Temperature increase, for example, led to a notable reactivation of the ZnP fluorescence, while lower temperatures essentially caused deactivation. Porphyrins and their zinc complexes can be used for the purpose to solubilize CNTs in water. The porphyrins used for the solubilization of CNTs are usually TPP analogues. However, TPP analogues do not favor the formation of π–π interactions because the aryl groups prevent the porphyrin moiety from approaching the CNT surface due to the perpendicular conformation of the porphyrin moiety and its aryl group substituents. To avoid this, non-TPP-type porphyrins **3.78**, with chiral and hydrophilic substituents, which make the porphyrins soluble in water, were synthesized [266]. SWCNTs were effectively dissolved into water by the non-TPP-type chiral porphyrins, and the dissolved chiral porphyrin/SWCNT composites could be easily redissolved. Both the dissolved and redissolved SWCNT solutions were found to be very stable and did not form apparent aggregates even after being kept for 6 months.

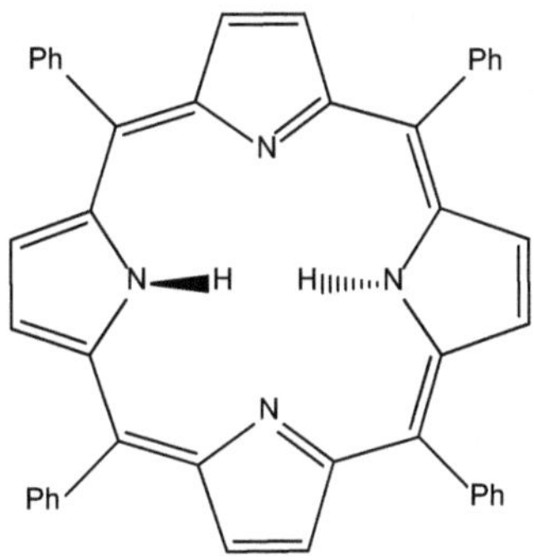

TPP
3.76

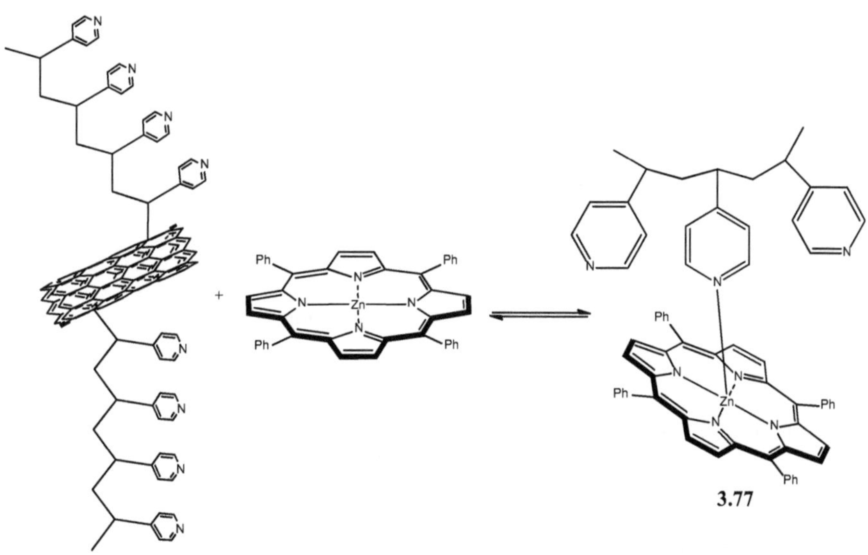

3.77

Partial structure of SWNT-PVP that coordinates axially to Zn(TPP)

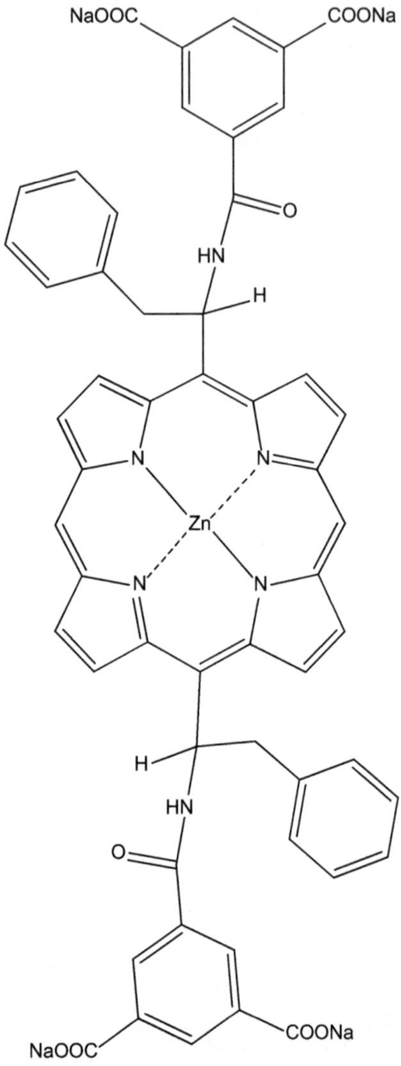

3.78

(R)- and (S)-5,15-bis(1-(1',3'-isophthalic acid disodium salt-5'-carbamoyl)-2-phenyl-ethyl) porphyrin Zn(II).

A highly soluble, conjugated Zn–porphyrin polymer was synthesized in the presence of trifluoroacetic acid in THF and found to strongly interact with the surface of SWCNTs, producing a soluble polymer–nanotube complex, which remains soluble after excess free polymer was removed from solution and could be centrifuged at high speed with no observable sedimentation [295]. The carbon

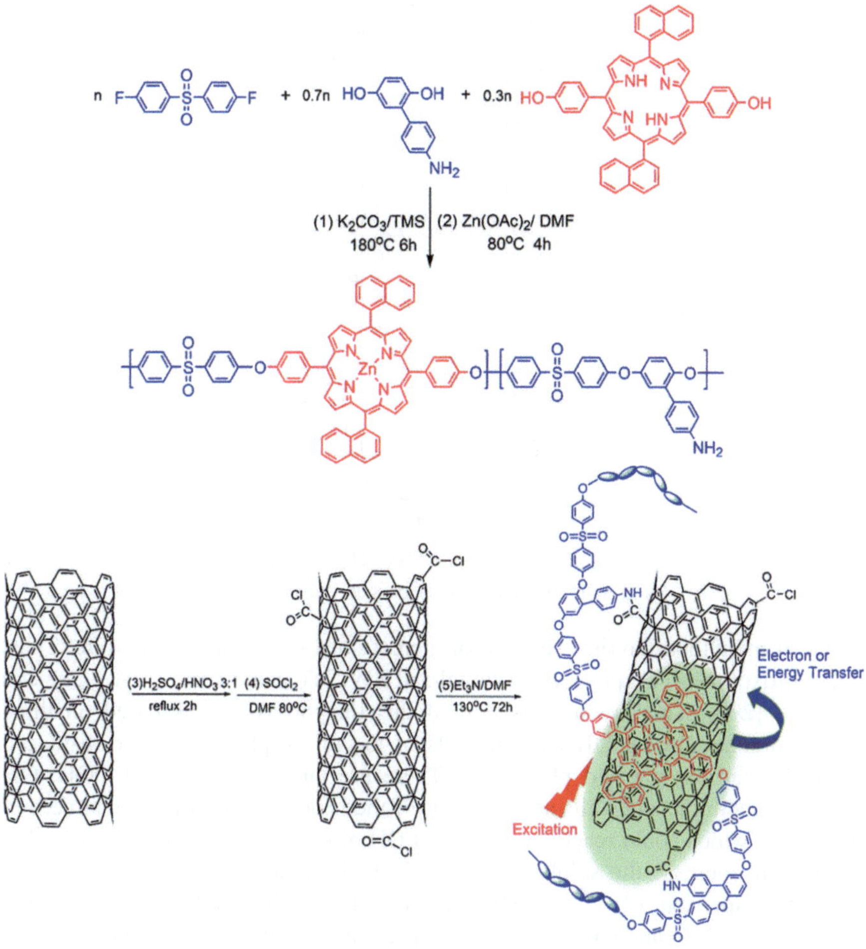

Fig. 3.27 The polymerization of ZnTNP–PAES and the synthetic procedure of MWCNT hybrid. 5,15-bis(4-hydroxyphenyl)-10,20-dinaphthylporphyrin = *trans*-DHTNP (Reproduced with permission of the *Royal Society of Chemistry*; Du et al. [297])

nanotubes can be *purified* from a large quantity of impurities such as carbon nanoparticles by executing a step of adding the carbon nanotubes into a solution (in THF, chloroform, dichloromethane, toluene, benzene, chlorobenzene, dimethylformamide, dimethyl sulfoxide, hexane, acetone, methanol, ethanol, isopropanol, butanol, acetonitrile, or diethyl ether) using Zn–porphyrin complex with their further recovering [296] (**3.79**). Among a variety of other Zn–porphyrin hybrids, a MWCNT hybrid was synthesized (Fig. 3.27) by covalently attaching a porphyrinated poly(arylene ether sulfone, PAES) onto the surface of CNTs [297]. This polymer-functionalized nanotube exhibited excellent solubility in organic solvents, and the weight% of MWCNTs in the resulting hybrid was found to be 36%. Additional details on the use of porphyrin salts in CNT functionalization were reported in [298].

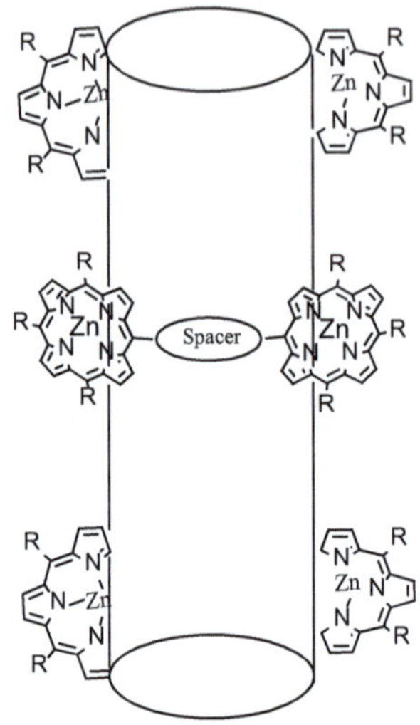

3.79

Zn-porphyrin complex/SWCNT intermediate compound, used for CNTs purification.

Other metalloporphyrins, studied for increase of CNT solubility, are rare. Thus, a dihydroxotin(IV) porphyrin-functionalized SWCNT nanohybrid (Fig. 3.28) was obtained [299]. The structural design of this tin porphyrin was based on three considerations: (1) the OH axial ligands of the tin porphyrins can be displaced easily by carboxylate and phenoxide. There are large numbers of COOH, OH groups on both the sidewalls and the ends of the nanotubes after acid treatment of SWCNTs, which could react with the axial OH of the tin porphyrin; (2) long alkyl chains will increase the solubility of SWCNTs in organic solvents and stabilize SWCNT dispersion; (3) ^{119}Sn NMR can be used to further explore the interaction between SWCNTs and the functional groups, which is difficult in most other cases for SWCNT materials. It was demonstrated that efficient electron transfer occurs within the nanohybrid at the photoexcited state, and the charge-separated state of

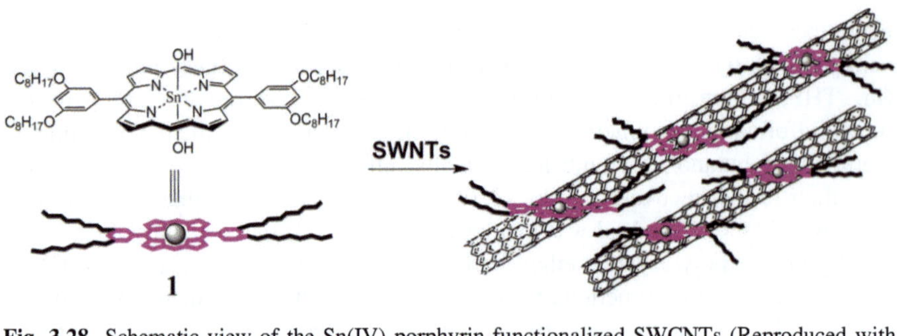

Fig. 3.28 Schematic view of the Sn(IV) porphyrin-functionalized SWCNTs (Reproduced with permission of *MDPI*; Ren et al. [299])

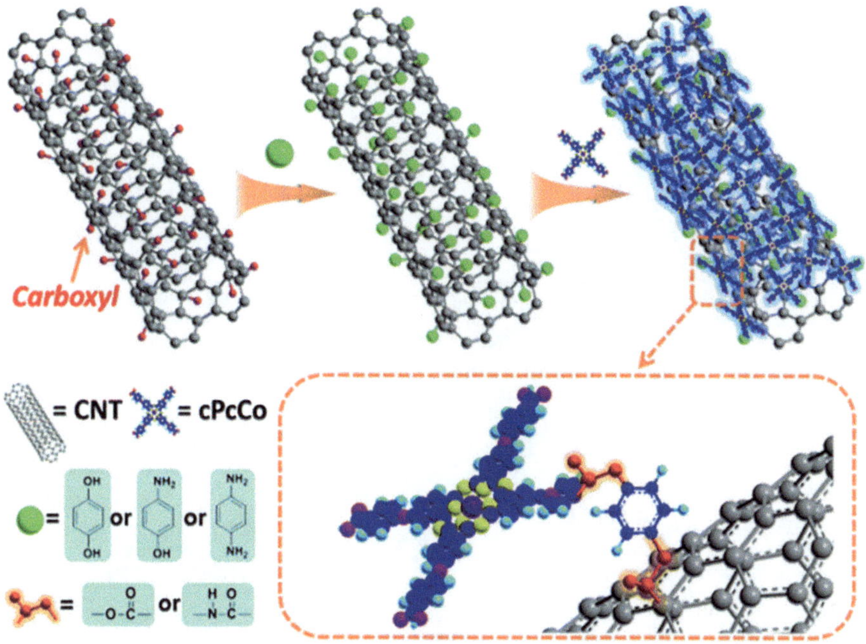

Fig. 3.29 A schematic illustration of the synthesis procedure for cPcCo–B–a-CNT hybrids (Reproduced with permission of the *Royal Society of Chemistry*; Wu et al. [301])

the nanohybrid was observed by transient absorption spectrum. The product possesses certain solubility in a series of organic solvents, for instance (mg/L, r.t.), in 1,2-dichlorobenzene 440, chloroform 358, toluene 237, tetrahydrofuran 209, hexane, ethanol, and methanol – <1 mg L^{-1} under sonication for over 2 h. It was also illustrated that this soluble electron donor–acceptor nanohybrid might be a good candidate as a light-harvesting material in molecular photoelectronic devices.

In comparison with porphyrins, soluble *phthalocyanine*–CNT composites are considerably lesser represented. Thus, a water-soluble composite of oxidized multiwall carbon nanotubes and sulfonic acid sodium salt derivatized copper phthalocyanine for application in bilayer organic solar cells is reported [300]. Tetra-β-carboxyphenyloxyphthalocyanine cobalt(II) (cPcCo) was bonded on the surface of acidified MWCNTs (a-CNTs) by a two-step condensation reaction in the presence of N,N′-dicyclohexylcarbodiimide, using hydroquinone (HQ), *p*-aminophenol (PAP), and *p*-phenylenediamine (PPD) as linking molecules, respectively (Fig. 3.29) [301]. The obtained cPcCo–B–a-CNT (B = HQ, PAP, and PPD) hybrids displayed good dispersibility in ethanol and can be used as H_2S sensors (Fig. 3.30). Water-soluble TSCuPc (tetrasulfonate copper phthalocyanine) was used as a matrix to disperse the SWCNTs and was found to be applied as electrode as a replacement of ITO in organic electronic devices [302]. CNT–molecular semiconductor thin films on the phthalocyanine basis were patented in [303, 304].

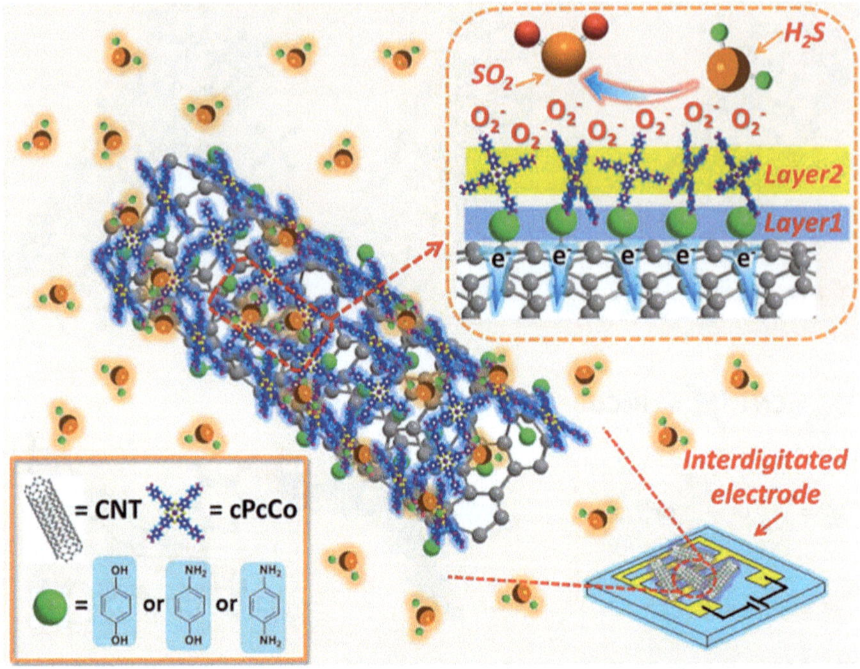

Fig. 3.30 A schematic illustration of the gas sensing mechanism of cPcCo–B–a-CNT sensors upon interaction with H_2S (Reproduced with permission of the *Royal Society of Chemistry*)

3.1.3.4 Composites of CNTs with Complexes of Sulfur-Containing Ligands

The complexes of sulfur-containing ligands, forming "soluble" CNT composites, are practically absent. The only exception is the synthesis, and structural characterization of a stable discrete *bis*-dithiolene complex, $[PPh_4][Cu(DMED)_2]$ (DMED = 1,2-dicarbomethoxy-1,2-dithiolate), involving the reaction between a copper polysulfide precursor with activated acetylene, was reported [305]. This complex, possessing terminal $-COOCH_3$ groups, forms nanospheres by hydrogen bonding in a mixture of solvents containing water as one of the components. These nanospheres further aggregate with water-soluble (carboxylated) carbon nanotubes (wsCNTs). These nanocomposites are assisted by hydrogen bonding between the carboxylic acid groups of the wsCNTs and the peripheral $-COOCH_3$ groups of the coordinated dithiolenes of the nanospheres, which is promoted by water molecules.

3.1.3.5 Functionalization of CNTs with Organometallics and Grignard Reagents

A few organometallics have been used for CNT functionalization having the objective their further dispersion. Thus, SWCNTs and MWCNTs were covalently functionalized with a titanium alkoxide catalyst through a *Diels–Alder* cycloaddition reaction (reaction scheme 3.19) [306] and used for the surface initiated titanium-mediated coordination polymerizations of L-lactide, ε-caprolactone, and *n*-hexyl isocyanate employing the "grafting from" technique. The final polymer-grafted CNTs were readily dissolved in organic solvents as compared to the insoluble pristine and catalyst-functionalized CNTs. The SWCNTs, functionalized by *ferrocene*-grafted poly(*p*-phenylene ethynylene), were found to gelate common organic solvents such as chloroform forming a freestanding CNT organogel that cannot be redispersed in any organic solvents and can be applied for obtaining an insoluble, homogeneous, electroactive SWCNT film [307]. Ferrocene units can be covalently introduced onto SWCNTs also by 1,3-dipolar cycloaddition of azomethine ylides, thus solubilizing them [308]. The use of *Grignard reagents* RC(=S)SMX (wherein R is alkyl, alkenyl, or aryl; wherein any alkyl, alkenyl, or aryl may be optionally substituted; M is Be, Mg, Ca, Sr, or Ba; and X is Cl, Br, or I) in functionalization of CNTs is also known [309].

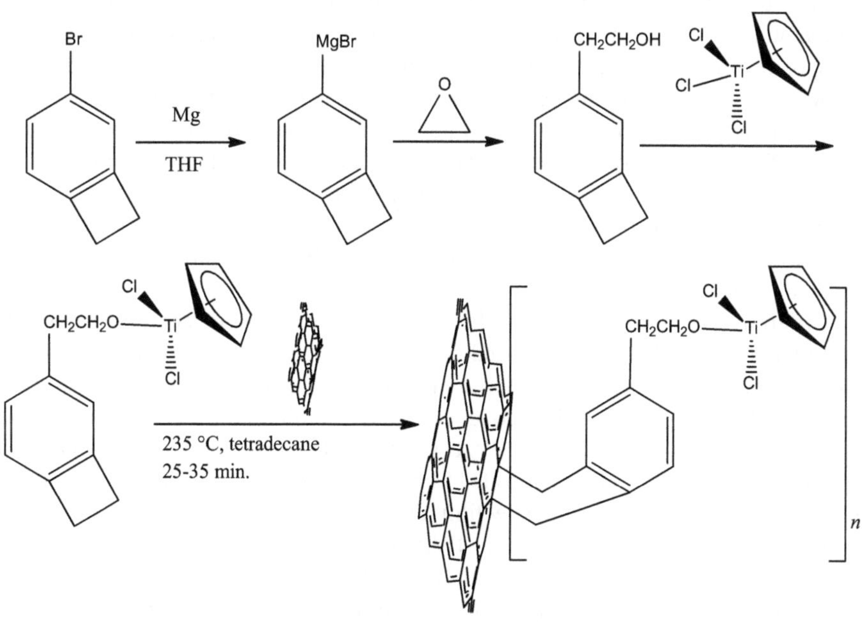

Grignard synthesis of (1-benzocyclobutene - (BCB-EOTiCpCl_2) and covalent functionalization of MWNTs using a [4+2] *Diels-Alder* cycloaddition reaction.

3.1.4 Polymers or Their Precursors

A lot of reports are dedicated to CNT–polymer composites, used, in particular, for polymer reinforcement, drug delivery, or energy storage [310] purposes. Currently, the dispersion processes of CNTs in polymers are the object of serious attention [311–317]. In case, if a polymer can be solubilized in solvents, its CNT composites could also be dispersed; obviously, this is not an only condition for the possibility of polymer-assisted CNT dispersion. In comparison with other functionalizing agents for CNTs, the polymers and their precursors, undoubtedly, have been of the most importance (together with biomolecules) during the last 15 years due to a considerable improvement of mechanical properties and stability of polymer hybrids with CNTs. A host of experimental articles and patents have been published on the basis of results using a broad spectrum of distinct polymers: polystyrene [318–322], polyvinyl pyrrolidone [323–325], polybutadiene [326], poly(*N*-vinylcarbazole) [327], polyaniline [328, 329], polyethylene glycol [330, 331], poly(acrylic acid) and related polymers [332–334], polyurethane [335], and much more [336–345]. The obtained composites (sometimes in form of films) were soluble in water or common organic solvents, depending on the polymer used. Some representative examples of the synthesis of CNT–polymer composites and their applications are represented in Table 3.1.

In case of most simple polymers, such as *poly(vinyl alcohol)* (**3.80**, PVA), a low temperature synthesis method of amorphous carbon nanotubes (a-CNTs) and PVA composite thin films were reported [346]. Initially the a-CNTs were functionalized by acid treatment for getting better dispersion in water. The functionalized a-CNTs showed much enhanced solubility in water. It was revealed that the sheet resistance of the PVA–a-CNT composite monotonically decreases with increasing a-CNT concentrations in the composites. The composite showed a higher crystallinity with increased a-CNT concentration, compared to pure PVA. The swelling test confirmed that due to the introduction of the carboxylic group into a-CNTs' walls, it gradually became hydrophilic.

Different polymers can be used by stepwise preparation of composites. Thus, PEGylated MWCNTs were prepared for the successive fabrication of poly(vinyl alcohol) PVA–MWCNT nanocomposite film by solution casting [347]. The surface modified MWCNTs showed a good colloidal stability in water and improved dispersion stability in aqueous PVA solution. The electrical conductivity of PVA/modified MWCNT composite film was threefold higher than that of PVA/pristine MWCNT composite film due to the much improved distribution uniformity of modified MWCNTs in PVA matrix. PVA dispersant was also used for creation of CNT–water nanofluids [348]. 0.5 to 1.5 wt.% of PVA gave the optimum stability of fluids to the entire range of CNT concentrations. It was also found that the thermal conductivity enhancement of the studied nanofluid stabilized by PVA increased nonlinearly with temperature. The presence of NaDDBS can improve the dispersion of MWCNTs in the PVA solution (Fig. 3.31) [349].

Table 3.1 Representative examples of the composites, prepared on the basis of polymers and carbon nanotubes

Precursors	Product(s)	Properties and possible applications	Ref.
A rigid linear polymer poly(phenylene ethynylene) (PPE) and related compounds, SWCNTs	PPE/SWCNT hybrid	The water solubility of SWCNTs was enhanced to 1.8 mg/ml. Used in a photovoltaic cell with the bulk heterojunction configuration	[397, 398]
Poly(acrylamide) (PAM), SWCNTs (via reversible addition–fragmentation chain transfer (reversible addition–fragmentation chain transfer, RAFT) polymerization)	PAM-g-SWCNT	Relatively uniform polymer coatings present on the surface of individual, debundled nanotubes	[399]
RAFT agents, MWCNTs, cationic polymer (poly(2-(dimethylamino) ethyl methacrylate)), anionic polymer (poly (acrylic acid)), and zwitterionic polymer (poly{3-[*N*-(3- methacrylamidopropyl)-*N*,*N*-dimethyl] ammoniopropane sulfonate})	Poly{3-[*N*-(3-methacrylamidopropyl)-*N*,*N*-dimethyl] ammoniopropane sulfonate} functionalized MWCNTs, poly(acrylic acid) functionalized MWCNTs, and poly (2-(dimethylamino) ethyl methacrylate) functionalized MWCNTs	Good solubility in aqueous solution	[400]
Poly(N-isopropylacrylamide) (PNIPAM), MWCNTs, (via RAFT polymerization)	MWCNT-g-PNIPAM	Good solubility in water, chloroform, and THF Potential applications by grafting other functional polymer chains onto MWCNTs	[401]
Poly-*m*-aminobenzene sulfonic acid (PABS) or polyethylene glycol (PEG), SWCNTs	Graft copolymers SWCNT-PABS (30% loading SWCNTs) and SWCNT-PEG (71% loading SWCNTs) 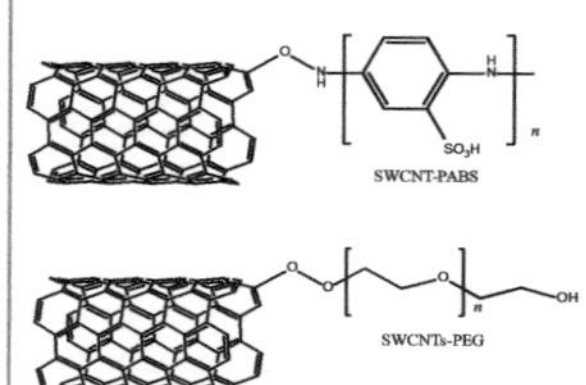	Water soluble (5 mg/L) Fairly uniform length and diameter	[402, 403]

(continued)

Table 3.1 (continued)

Precursors	Product(s)	Properties and possible applications	Ref.
MWCNTs. In situ polymerization of aniline followed by sulfonation with chlorosulfonic acid in an inert solvent and by hydrolysis in water	SPAN/MWCNTs SPAN = sulfonated polyaniline 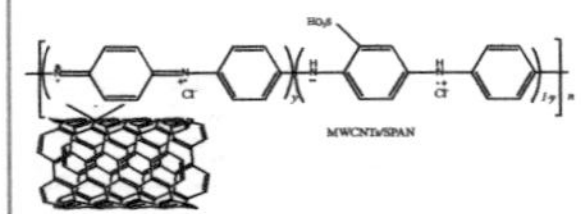	SPAN/MWCNTs are highly dispersible in water. Quinonoid structure of SPAN preferentially interacts with the nanotubes and is stabilized by strong π–π interaction between two components	[404]
3-Aminophenylboronic acid monomers, DNA, SWCNTs	Poly-(anilineboronic acid)/ss-DNA/SWCNT composite (PABA/ss-DNA/SWCNT)	Water soluble; the conductivity of the nanocomposite was much higher than that of the pure self-doped polyaniline in both acidic and neutral solutions	[405]
Poly(1-phenyl-1-alkyne) and poly (diphenylacetylene) derivatives carrying azido functional groups at the ends of their alkyl pendants, SWCNTs	SWCNT–polyacetylene composites: 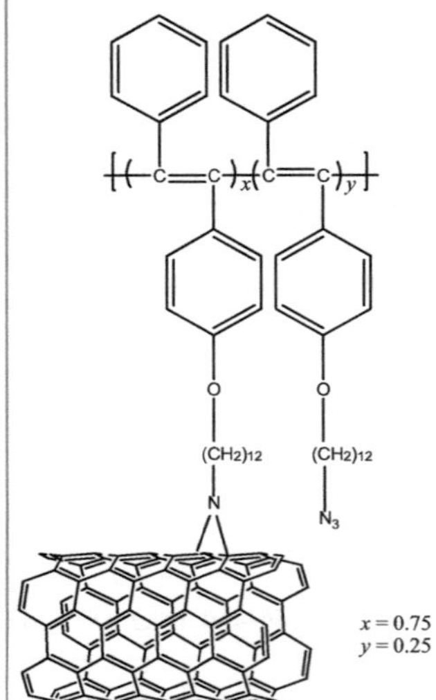	Soluble in common solvents	[406–409]

	or 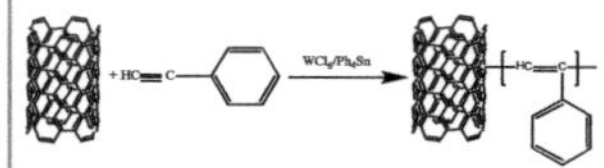		
Double-hydrophilic block copolymer, poly(ethylene oxide)-*b*-poly [2-(*N*,*N*-dimethylamino)ethyl methacrylate] (PEO-*b*-PDMA), MWCNTs	(PEO-*b*-PDMA)/MWCNTs	Yield 26%. Direct evidence for the individual dispersion. Possible applications: Au and Pt nanoparticles were attached on the sidewall of the modified MWCNTs by using the amino groups of PDMA segments	[410]
Poly(ethylene-co-vinyl alcohol) (EVOH) copolymer, SWCNTs, carbodiimide-activated esterification reaction conditions	EVOH–SWCNT	Soluble in highly polar solvent systems such as DMSO and hot ethanol–water mixtures. A fabrication of nanocomposites in which the SWCNTs are homogeneously dispersed in the polymer matrix is possible	[411]
Poly(vinyl alcohol) (PVA), MWCNTs (wet-casting method; esterification reactions)	PVA–MWCNT films	Water soluble. The mechanical properties of the nanocomposite films were significantly improved compared to the neat polymer film. The composite-based electrodes can be used as biosensors for glucose detection	[412–414]
Carboxylic acid-terminated hyperbranched poly(ether-ketone)s (HPEKs), MWCNTs, or SWCNTs	HPEK-*g*-SWCNT and HPEK-*g*-MWCNT nanocomposites	The resultant nanocomposites were homogeneously dispersed in various common polar aprotic solvents as well as in concentrated ammonium hydroxide	[415]

(continued)

Table 3.1 (continued)

Precursors	Product(s)	Properties and possible applications	Ref.
Thionyl chloride, SWCNTs, caprolactam, metallic sodium	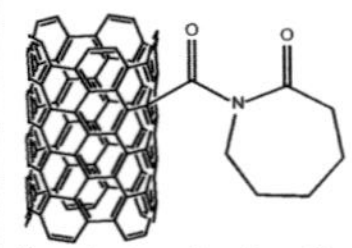 Caprolactam-functionalized SWCNTS 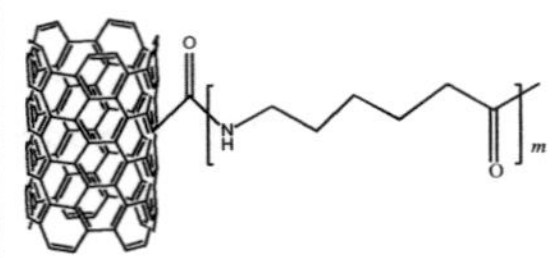 Caprolactam- and nylon-SWCNTs Caprolactam and nylon SWCNTs	Soluble in organic solvents	[416]
Poly(3,4-ethylenedioxythiophene) poly (styrenesulfonate) (PEDOT:PSS), SWCNTs	Thin films were obtained from PEDOT:PSS/SWCNT suspensions	Uniform aqueous suspensions, obtained at final concentrations of up to 0.23 mg/mL (>1% by weight relative to PEDOT:PSS) without use of additional surfactants	[148]
A two-step sequential copolymerization of precursors (polycondensation followed by ring-opening polymerization)	Polycaprolactone-polyimide-polycaprolactone triblock copolymers	Low-boiling organic solvent (THF)	[417]

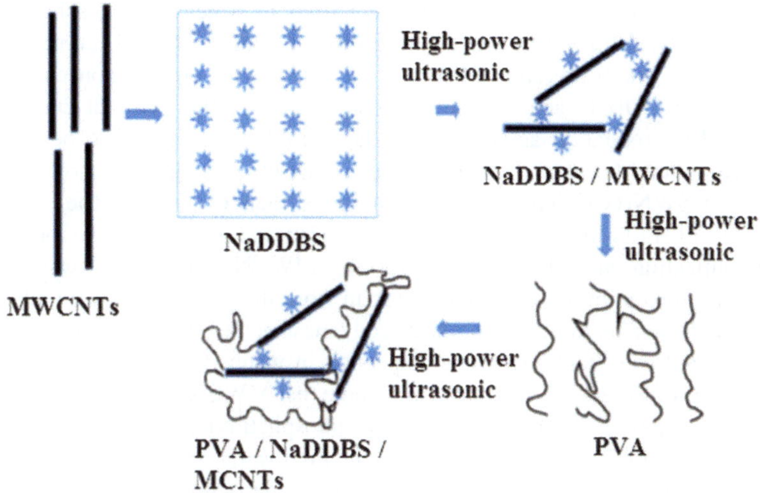

Fig. 3.31 Schematic diagram of NaDDBS in PVA/MWCNT/NaDDBS/water system (Reproduced with permission of the *Elsevier Science*)

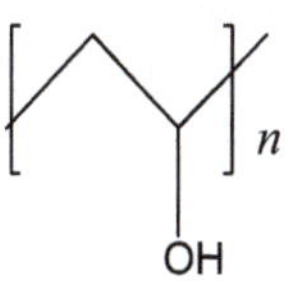

3.80

PVA

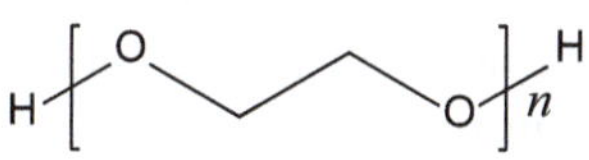

3.81

PEG

Polyethylene glycol **3.81** (PEG), mentioned above, and its derivatives are also very common in CNT functionalization processes. Thus, the CNT–polyethylene glycol graft copolymer was synthesized [350] by covalent functionalization of electric arc-produced SWCNTs with the monofunctional, tetrahydrofurfuryl-terminated polyethylene glycol PEG-THFF (molecular weight ~200), to give a material composed of 80 wt. % SWCNTs. The sequential processing of the resulting material by ultrasonication and high-shear mixing provided a means to disperse the SWCNT-PEG-THFF macromolecules on two different length scales and leads to highly viscous solutions; at a concentration of 10 mg/mL, the kinematic viscosity (v) of an aqueous SWCNT-PEG-THFF dispersion reached a value of v >1000 cSt (for water v ~1 cSt). Analysis of this procedure by means of viscosity measurements and AFM showed that ultrasonication is effective in disrupting the SWCNT bundles, while the high-shear mixing disperses the individual SWCNTs. In order to improve the biocompatibility of SWNHox (carbon nanohorns, new materials that are similar to SWCNTs but have more comparative advantages than SWCNTs), carboxyl polyethylene glycol distearoyl phosphatidylethanolamine (DSPE-PEG-COOH) was chosen [351] to modify them. Different concentrations of DSPE-PEG-COOH were used in water and phosphate buffer solution (PBS) followed by determining the zeta potentials and monitoring the coagulation times. It was shown that 0.25 mg/mL was the optimal concentration of DSPE-PEG-COOH to achieve the best dispersion in PBS. Without DSPE-PEG-COOH, H_2O_2 oxidation assisted by Xe lamp in 1 h was found to be the most effective method because it generates large amount of oxygenated groups on the surface of SWNHs which are of great help for their dispersion. The stabilizing character of PEG-linked butylpyrene (see also the section above on pyrene functionalization and the report [352]) in MWCNT dispersion was studied [353]. The separation of the aromatic group from the polar PEG by an alkyl unit was found to be a requirement for the effectiveness of the surfactant. The use of this surfactant led to a stable nanotube dispersion in water. Discussing drug delivery applications of PEG–CNT composites, we note that MWCNTs were modified [354] with three types of hydrophilic moieties: phosphatidylcholine (PC), PEG, and PC-terminated PEG (PEG-PC). Based on the high water dispersibility of modified CNTs, the extremely high doses of these CNT derivatives were intravenously injected in rats from 12.5 to 200 $mg{\cdot}kg^{-1}$. Based upon pathologic results, most of the injected CNT derivatives accumulated in the lung, whereas CNT-PC dispersed throughout the organism compared to CNT-PEG and CNT-PEG-PC, thereby resulting in a broader distribution in liver and spleen which may be biologically relevant. The modified CNT induced only low acute toxicity, which may contribute to biocompatibility of mammals to these grafted functional groups. In addition, starting from experimental evidence of the goodness of poly(ethylene glycol-bl-propylene sulfide), PEG-PPS, to disperse SWCNTs, atomistic molecular dynamics simulations were performed to study SWCNTs/polymer systems in the presence of water molecules in solution [355]. It was revealed that the hydrophobic nature of PPS systematically ensures a higher SWCNT surface coverage, higher interstitial water depletion, and a much lower degree of water ordering when compared to the PEG homopolymer. In addition, phospholipids with tethered poly(ethylene glycol) chains (PL−PEGs) offer efficient, noncovalent dispersion of CNTs (Fig. 3.32) [356]. The morphological transitions from dense, micelle-like structures to sparser, monolayer-like coatings of PL−PEG upon increasing CNT diameter and/or PEG chain length were observed.

Polyethylene oxide (PEO, **3.82**) and its derivatives are also capable to disperse CNTs [357]; the use of additional surfactants is sometimes needed [358]. For this process, molecular dynamics simulations were carried out to investigate CNT interactions and dispersion in a PEO/water solution [359]. Light-switchable type of "smart" SWCNTs was developed by the reversible host–guest interaction between azobenzene-terminal PEO (AzoPEO) and pyrene-labeled host attached on the sidewalls of nanotubes via π–π stacking [360]. These SWCNT hybrids not only were found to be well dispersed in pure water but also exhibited switchable dispersion/aggregation states upon the alternate irradiation of UV and visible light. Such a reversible host–guest interaction system may open up the possibility to control the dispersion state of SWCNTs by other common polymers. Another representative example of PEO–pyrene combination is "Pluronic F38"

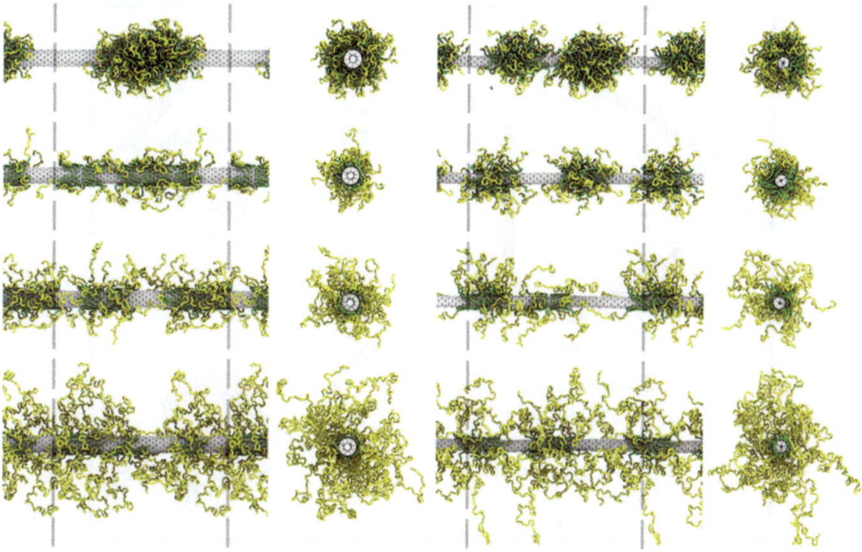

Fig. 3.32 Representative side view and cross-section snapshots of PL–PEG assembly on (18,0) CNT (*left*) and (10,0) CNT (*right*), as determined by CG MD simulation. PEG chain lengths are 12, 24, 48, and 114 monomers (from *top to bottom*). The dashed lines show the borders of the simulation cell. CNTs appear *silver*, aliphatic PL tails are *green*, and PEG chains are *yellow*. Although present in the simulations, water, sodium counterions, and the PL–PEG micelles in solution are omitted for clarity (Reproduced with permission of the *American Chemical Society*)

(PEO-PPO-PEO) micelles with pendent pyrene groups, obtained [361] via UV-induced crosslinking of pentaerythritol tetraacrylate within the micellar core. The pyrene moieties were attached to the periphery of the micelles via reacting OH groups of F38 with pyrenebutyric acid. Also, various effects on dispersion of MWCNTs in organic solvent and in polymer film by means of pyrene-containing polymeric surface modifiers were investigated [362]. Graft surface modifiers of poly(ethylene oxide) and polystyrene with pyrene groups were synthesized by radical copolymerization of corresponding macromonomers and a pyrene-containing methacrylate monomer. The dispersibility of MWCNTs in THF was improved with increasing pyrene content in the graft polymer and was also improved by a branched architecture of the polymeric noncovalent surface modifier.

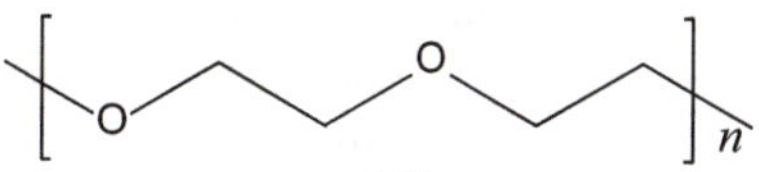

3.82

PEO

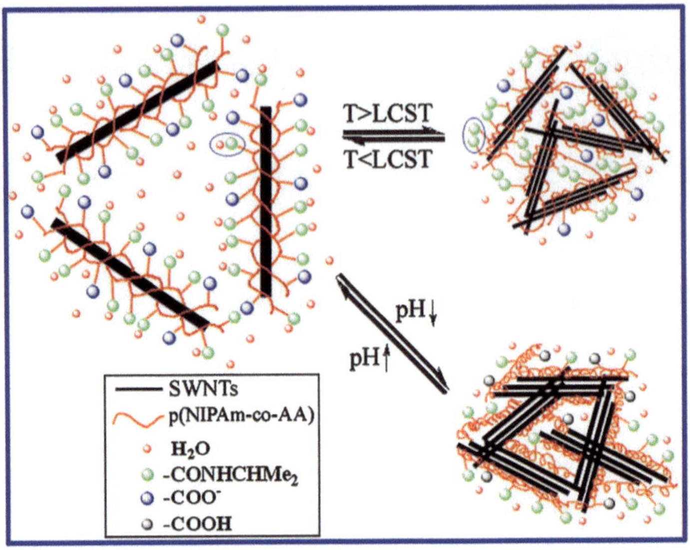

Fig. 3.33 Schematic showing the reversible changes of SWCNTs between the aggregated and well-exfoliated states with pH and temperature of the P(NIPAm-co-AA) aqueous solution. At high pH or low temperature, the polymer was in the extended state, and the SWCNTs were homogeneously dispersed in the aqueous solution. At low pH or high temperature, the polymer was in the coiled state, and the SWCNTs aggregated or precipitated from the aqueous solution (Reproduced with permission of *Elsevier Science*)

Water-soluble MWCNTs were prepared [363] by in situ polymerization of acrylic acid in a poor solvent for *poly(acrylic acid)* (PAA, **3.83**). The solvent type influenced the grafted density and chain length of PAA. MWCNTs with a high grafted density of PAA (22 wt.%) could be well dispersed in water, NaCl aqueous solution (0.9 wt.%), and cell culture media. The biocompatibility of these MWCNTs is sufficient for biological applications; in particular, PAA-g-MWCNTs were utilized for lymph node tracing. In case of PAA derivatives, the SWCNTs were homogeneously dispersed in the aqueous solution of poly (N-isopropylacrylamide-co-acrylic acid) {P(NIPAm-co-AA), **3.84**} with the assistance of sonification [364]. The mixture was sonicated at 25 °C for 3 h to form the P (NIPAm-co-AA)–SWCNT or PNIPAm–SWCNT complex and then centrifuged at 2000 rpm for 30 min. In addition, the SWCNTs were endowed with pH- and thermo-sensitivity at the same time. The SWCNTs switched reversibly between the aggregated and the well-exfoliated states by using pH or temperature as a stimulus (Fig. 3.33). Moreover, the factors including solvent composition while preparing P(NIPAm-co-AA), concentration, and composition of P(NIPAm-co-AA) showed evident influence on the dispersing stability of SWCNTs. In addition, dispersants based on block copolymers of poly(acrylic acid) (PAA) and poly (amide-imide) (PAI) or a homopolymer of PAA were applied for dispersion of carbon nanoparticles [365]. Their dispersibility was found to be increased through the series of copolymers (from most to least dispersed): PAA_2PAI > PAA-*block*-PAI > PAA-*block*-PAI-*block*-PAA > homo-PAA.

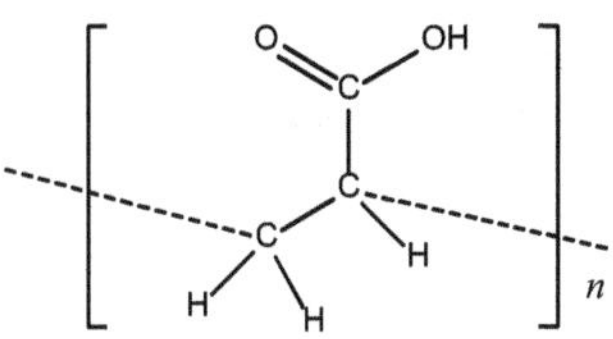

3.83

PAA

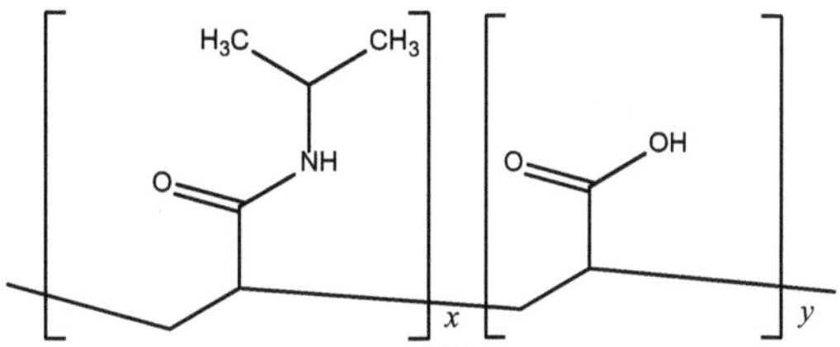

3.84

P(NIPAm-co-AA)

In order to control the level of carbon nanotube exfoliation in water, pH-responsive polymers (i.e., weak polyelectrolytes, PAA, *poly(methacrylic acid*) (PMAA, **3.85**), *poly(allylamine*) (PAAm, **3.86**)), and branched *polyethyleneimine* (BPEI, **3.87**) were used [366] as stabilizers in water. This noncovalent functionalization of SWCNTs resulted in suspensions whose dispersion state can be altered by simply changing pH (Figs. 3.34 and 3.35), similarly to poly (N-isopropylacrylamide) (PNIPAAm, **3.88**, reversible dispersion and aggregation by tuning temperature) [367] and poly-L-lysine (PLL, **3.89**) [368]. The SWCNTs stabilized with these polymers showed a pH tailorable exfoliation and bundling in water. Composite films prepared by drying these aqueous suspensions suggest that nanotube microstructure in the liquid state is largely preserved in the solid composites, with more bundled/networked structures showing higher electrical conductivity. In addition, homogeneous dispersion and functionalization of pristine MWCNTs in various organic solvents was achieved by a simple ultrasonic process in the presence of an azide copolymer, poly(4-azidophenyl methacrylate-co-methyl acrylate) {P(APM-co-MA)} [369]. The copolymers were noncovalently attached to the surface of the MWCNTs via π–π interactions to form MWCNT–P(APM-co-MA) composites. The solution dispersion of the MWCNT–P(APM-co-MA) composites were used to prepare superhydrophobic cotton fabric by a facile dip-coating approach. MWCNTs were covalently attached to the surface of the cotton fabric through the chemical reactions between the azide groups of P(APM-co-MA) with both MWCNTs and cotton fibers. Since MWCNTs were covalently attached on the surface of the cotton fabric, the superhydrophobicity possesses high stability and chemical durability. At last, the ability of poly(acrylonitrile) (PAN) to overcome the van der Waals interaction of MWCNTs was investigated [370], taking into account the premise that the CN group in the PAN molecule interacts with the graphene layers of the carbon crystals, to induce wrapping or coating of the MWCNTs with PAN.

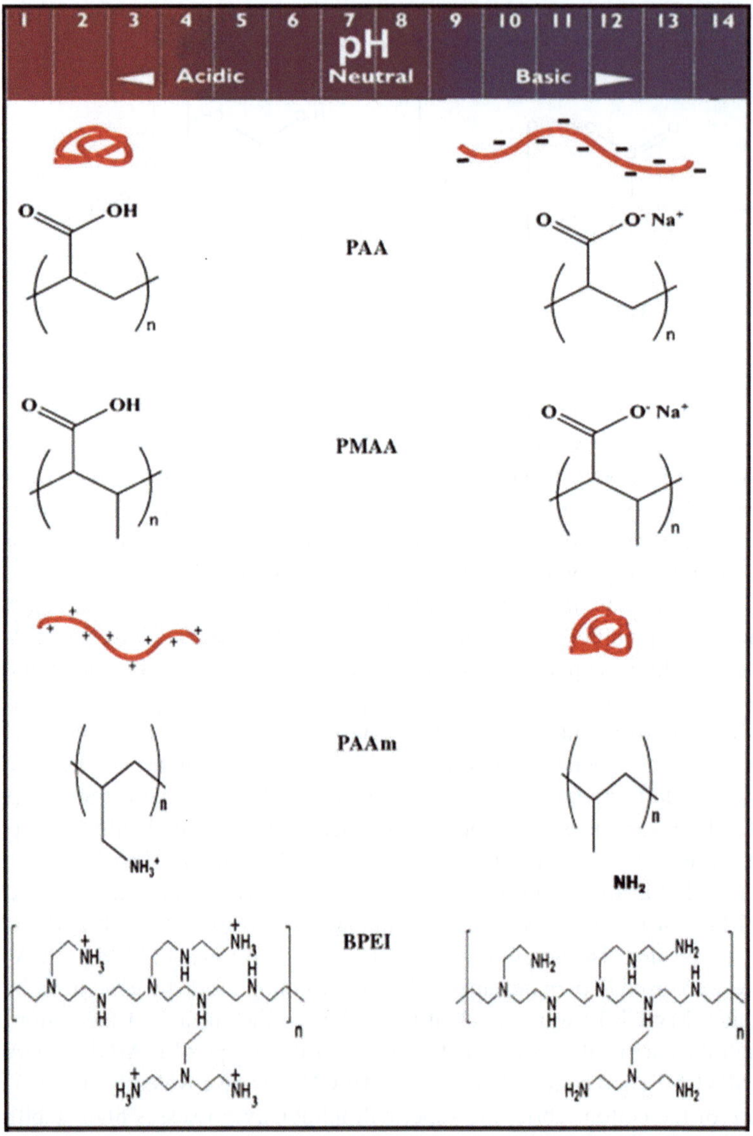

Fig. 3.34 Effect of pH on the chain conformations of *PAA*, *PMAA*, *PAAm*, and *BPEI*. *PAA* and *PMAA* have neutral charge at low pH and become negatively charged at high pH. *PAAm* and *BPEI* are neutral at high pH and attain positive charge at low pH (Reproduced with permission of *Elsevier Science*)

Fig. 3.35 Images of aqueous nanotube suspensions after centrifugation at different pH. All suspensions contained 0.11 wt.% of SWCNT in 1 wt.% of the aqueous polymer solution (Reproduced with permission of *Elsevier Science*)

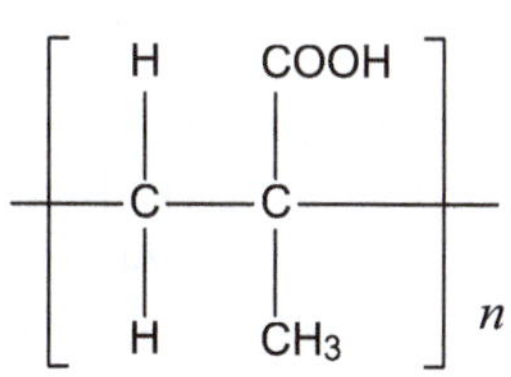

3.85
PMAA

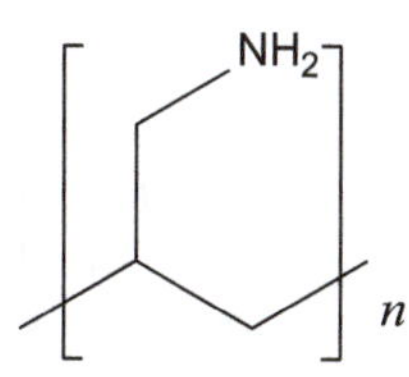

3.86
PAAm

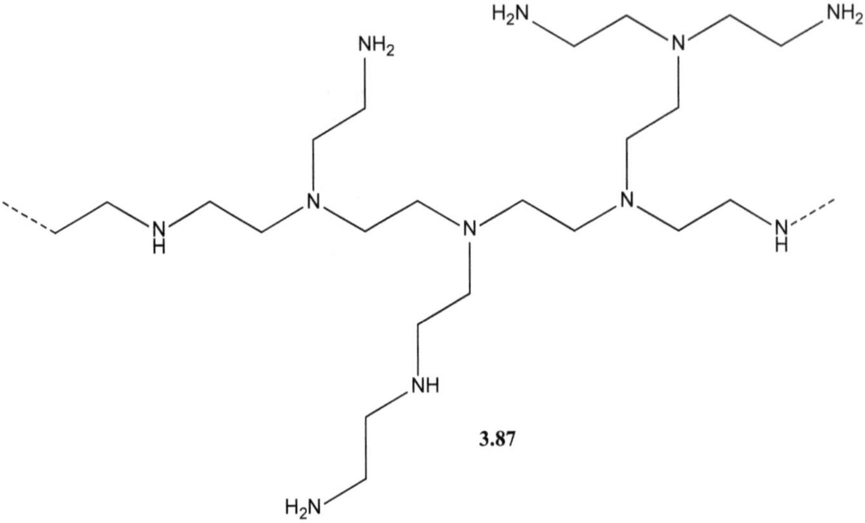

3.87

BPEI

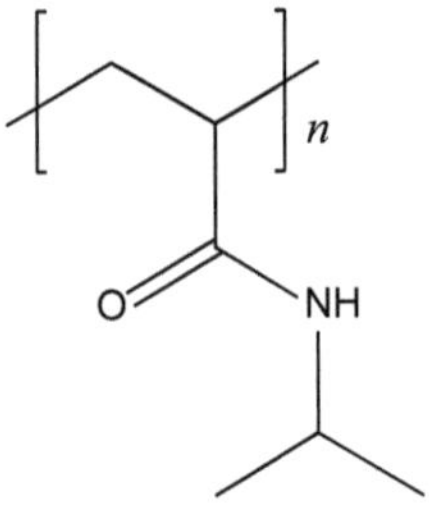

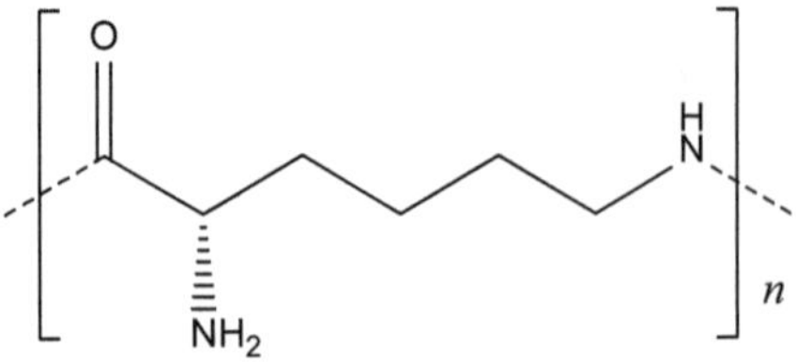

3.88 **3.89**

PNIPAAm Poly-L-lysine

Such classic polymers, frequently used in nanotechnology, as *polyaniline*, *polyurethane*, and *siloxanes*, have been also applied for CNT dispersion. Thus, a straightforward, template-free chemical oxidative polymerization of aniline was used to prepare nanofibrillar polyaniline (nf-PANI, **3.90**) and a set of corresponding composites with MWCNTs. All the products showed [371] remarkable water dispersibility since they are formed by hydrophilic particles of nanometric size. On one hand, the presence of MWCNTs affects neither the chemical structure nor the crystallinity of polyaniline. On the other hand, even small amounts of MWCNTs have a significant effect on the morphology of polyaniline in composites. MWCNT–polyurethane (PU, **3.91**) composites were obtained [372] by an in situ polycondensation approach. The functionalized MWCNTs had more advantages for improving the dispersion and stability in water and N,

N′-dimethylformamide. The tensile strength and elongation at break of the composites exhibited obvious increases with the addition of MWCNT contents below 1 wt.% and then decreases with additions above 1 wt.%.

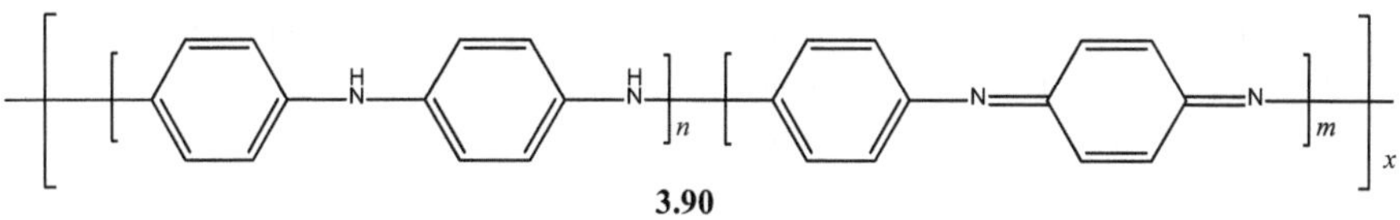

3.90

PANI

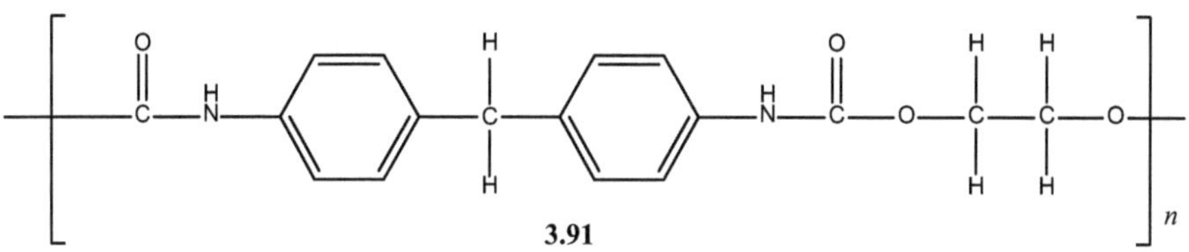

3.91

PU

To endorse sufficient water affinity to MWCNTs, dendritic hyperbranched *poly (ether-ketone)* (HPEK) was first covalently grafted to the surface of a MWCNT via a Friedel–Crafts acylation reaction [373]. The resultant HPEK-grafted MWCNT (HPEK-g-MWCNT) was subsequently sulfonated in chlorosulfonic acid to produce sulfonated HPEK-g-MWCNT (SHPEK-g-MWCNT), which was found to be dispersible well in water showing a zeta potential value of −57.8 mV. The SHPEK-g-MWCNT paper simply formed by filtration of aqueous dispersion has a sheet resistance as low as 63 Ω/sq. Its thin film showed a high electrocatalytic activity for oxygen reduction reaction (ORR). Also, the ability of a mixture of an ethoxy-modified *trisiloxane* (a silicone surfactant, named Ag-64) and a block copolymer F127 to disperse CNTs was investigated, showing [374] that dispersions with large amounts of individual CNTs were obtained. The quantity of dispersed CNTs was obviously larger than each quantity of the dispersions with individual surfactants at the same concentration, even exceeded the sum of them. It can be inferred that Ag-64 and few F127 could wrap onto the surface of CNTs to dispart clusters to individuals, and the other F127 interact with adsorbed Ag-64 and F127 to generate stronger steric stabilization, showing a synergistic effect on dispersing CNTs by the mixture of Ag-64 and F127.

Two kinds of *polyfluorenes* **3.92** bearing two lateral pyrene-terminated alkyl chains and two alkyl chains per repeating unit were synthesized by Suzuki polycondensation and used to disperse SWCNTs in organic solvents [375]. Stable polymer–SWCNT complex can be formed via the multivalent stacking interaction of the lateral pyrene functional groups and the polyfluorene backbone with the outer surface of carbon nanotubes; meanwhile the lateral alkyl chains can impart good solubility to the complex. Polyfluorenes bearing lateral pyrene functional groups and octyl chains exhibited much higher CNT solubility in common organic solvents than the corresponding polyfluorenes bearing only octyl chains. The selective dispersion of SWCNT species (n,m) with conjugated polymers such as poly (9,9-dioctylfluorene) (PFO, **3.93**) and its analogue poly(9,9-dioctylfluorene-co-benzothiadiazole) (F8BT) in organic solvents depends not only on the type of solvent but also on the molecular weight of the polymer [376]. The solution viscosity was found to be one of the factors influencing the apparent selectivity by changing the reaggregation rate of the SWCNTs. In addition, useful information on CNT dispersion in polyfluorenes is given in recent reports [377–380]. Also, *1,10-binaphthyl-conjugated polymers* (**3.94, 3.95**, and **3.96**) appears to disperse SWCNTs well in N-methyl-2-pyrrolidone (NMP) [381]. Related copolymer **3.97** composed of a naphthalene diimide and poly(dimethylsiloxane) efficiently dissolves the SWCNTs in many solvents including methyl ethyl ketone, methyl isobutyl ketone, and toluene (these solvents are widely used sin polymer/resin-related industrial areas) [382].

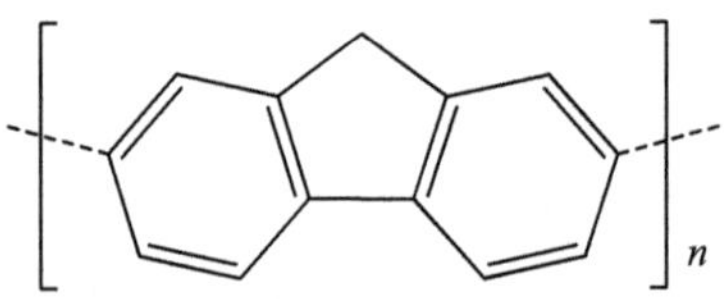

3.92

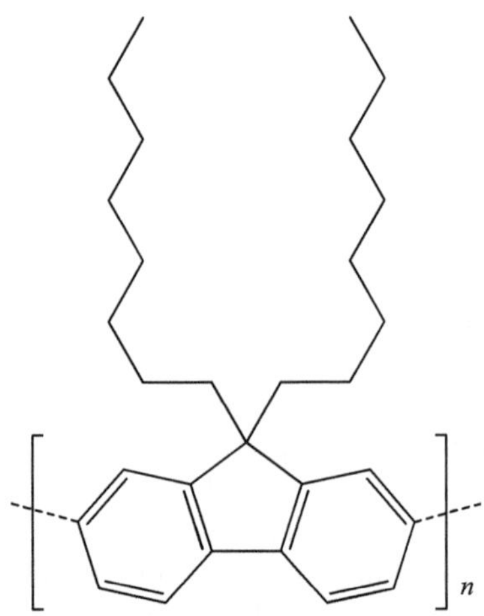

3.93

Polyfluorenes

PFO

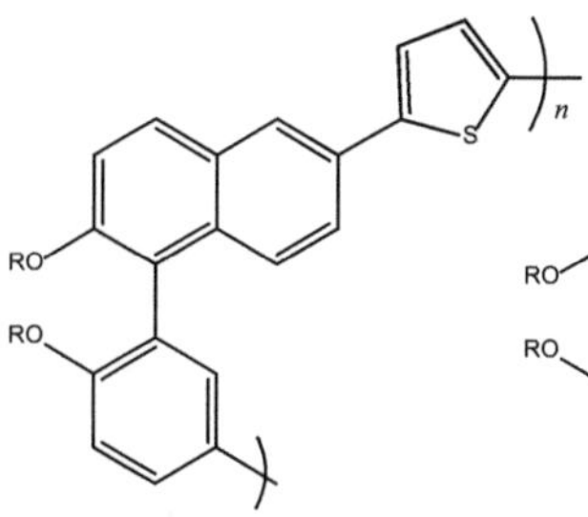

3.94

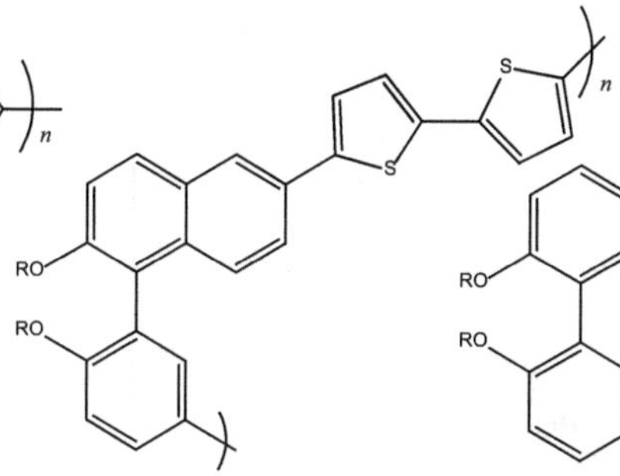

3.95

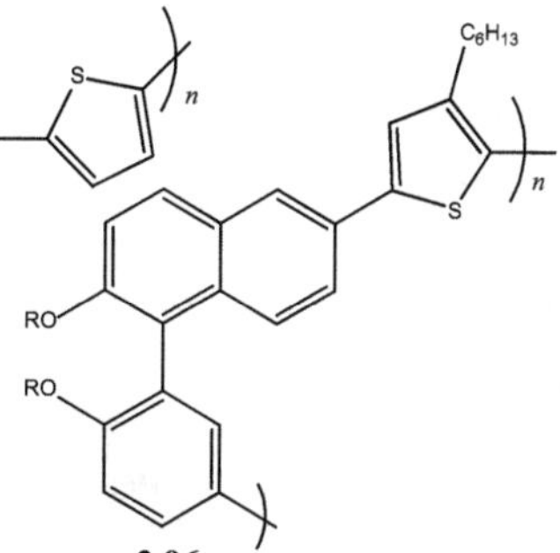

3.96

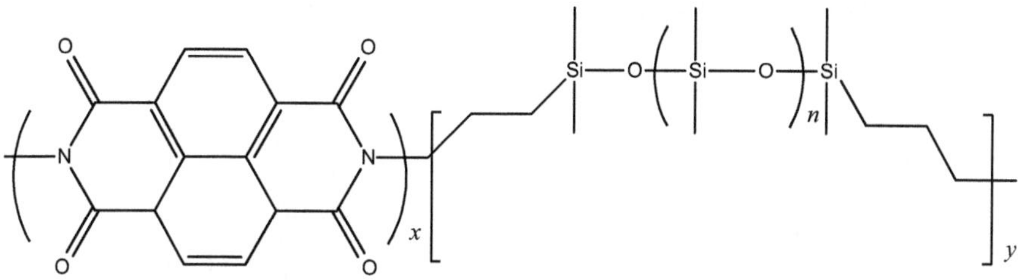

3.97

Poly(4-vinylpyridine) (P4VP **3.98**) is a widely studied polymer for applications in catalysis, humidity sensitive, and antimicrobial materials due to its pyridine group exhibiting coordinative reactivity with transition metals. The noncovalent functionalization of SWCNTs with P4VP in CO_2-expanded liquids was reported [383], showing that P4VP-stabilized SWCNTs revealed good dispersion in both organic solvent and aqueous solution (pH = 2). The ability to manipulate the dispersion state of CNTs in water with P4VP will likely benefit many biological applications, such as drug delivery and optical sensors. Among a series of other polymers, whose CNT composites could be successfully dispersed in solvents, we note poly(vinyl chloride) [384] (PVC) **3.99**, carboxymethylcellulose [385] **3.100**, heteroarm star block terpolymer on polystyrene basis [386], branched block copolymer Tetronic 1107 [387] **3.101**, polyamides [388], phenolic resins [389], phenylene vinylene-conjugated polymers [390], polymers with azobenzene [391], coumarin [392] units, etc. [367, 393–395] Among them, we note *polyethylene-based polymers*, namely, hyperbranched polyethylene (HBPE) featured with distinct highly compact dendritic chain architecture, which were discovered to show outstanding performance in rendering stable nanotube solutions in common low-polarity organic solvents (including tetrahydrofuran (THF) and chloroform, Fig. 3.36) at surprisingly high concentrations [396].

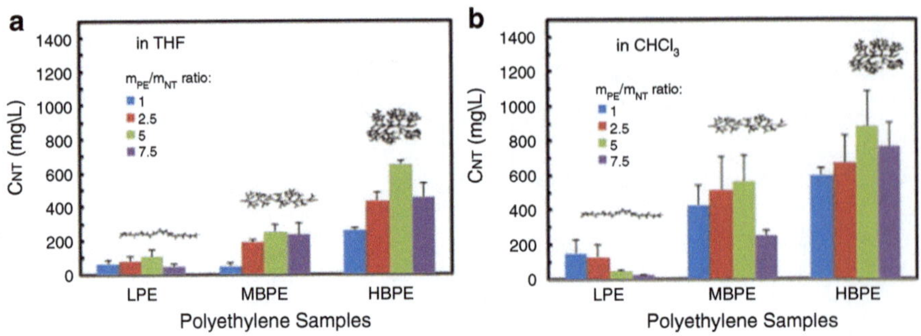

Fig. 3.36 Nanotube solubility achieved with the use of distinct polyethylene-based polymers in THF (**a**) and chloroform (**b**), respectively, at different mPE/mNT ratios and but the same nanotube feed concentration of 2 mg/mL (Reproduced with permission of *Elsevier Science*)

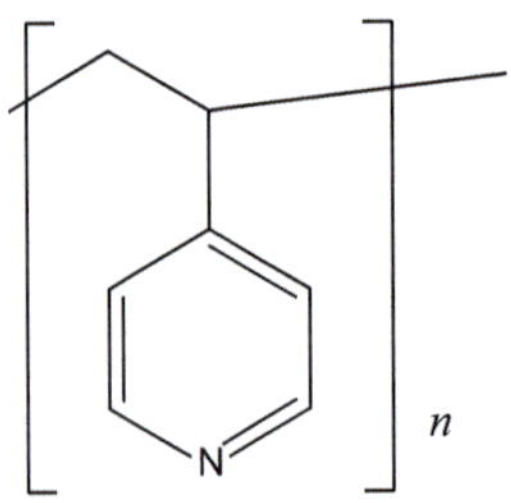

III-98

P4VP

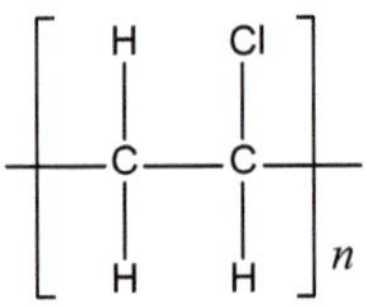

III-99

PVC

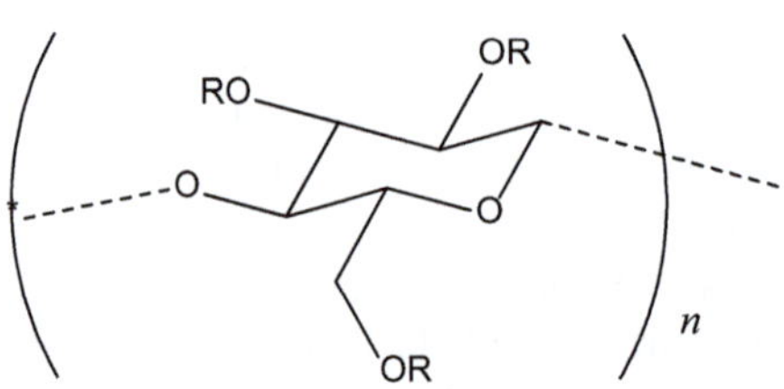

R = H or CH_2COOH

III-100

Carbocymethyl cellulose

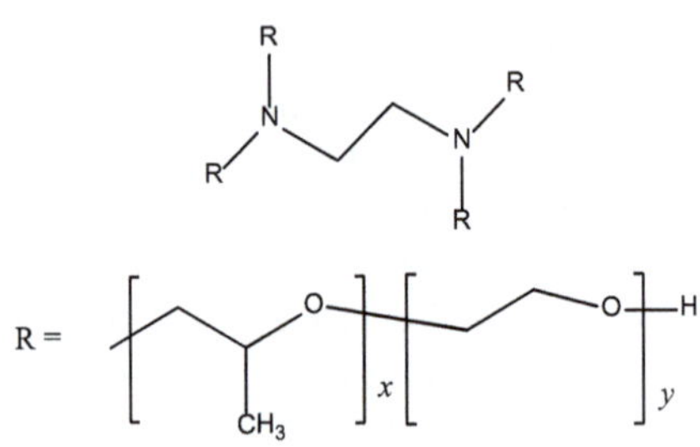

III-101

Tetronic 1107

3.1.5 *Natural Products*

3.1.5.1 Inorganic Natural Products

The ability of *Laponite* (a synthetic-layered magnesium aluminum fluorosilicate modified with an inorganic polyphosphate-dispersing agent) to stabilize aqueous suspensions of MWCNTs was investigated with the help of analytical centrifugation, microscopic image analysis, and measurements of electrical conductivity of hybrid laponite/MWCNT suspensions [418]. It was observed that sonication of MWCNTs at critical minimal concentration of Laponite $X_c \approx 0.25 \pm 0.05$ resulted in efficient dispersion and formation of stabilized suspensions of individual nanotubes. The stabilization of nanotubes in the presence of laponite was explained by adsorption of its particles and formation of a hydrophilic-charged shell on the surface of nanotubes (Fig. 3.37). In another report of the same research group, LapO (organomodified laponite) platelets were found to surround the nanotubes destroying bulky aggregates and splitting the nanotube bundles (Fig. 3.38) [419]. In a related investigation, hybrid nanofillers were produced by simply grinding MWCNTs with *sodium montmorillonite* (MMT ($Na_{0.33}(Al,Mg)_2(Si_4O_{10})(OH)_2 \cdot nH_2O$) [420]. The formed composite possesses a considerably enhanced MWCNT aqueous dispersion. In toluene and CCl_4, this effect was lesser because of strong interaction MMT–water and poor interaction of MMT with toluene and CCl_4.

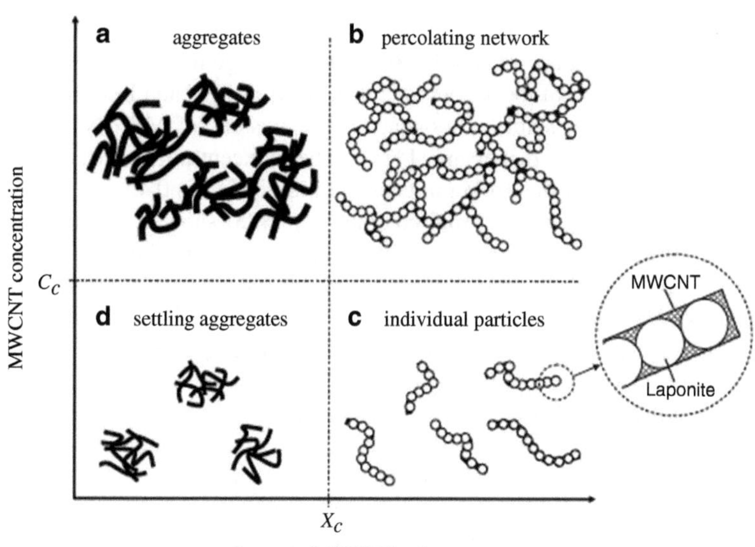

Fig. 3.37 Impact of Laponite content and nanotube concentration on the structure of hybrid laponite + MWCNT suspensions (Reproduced with permission of *Elsevier Science*)

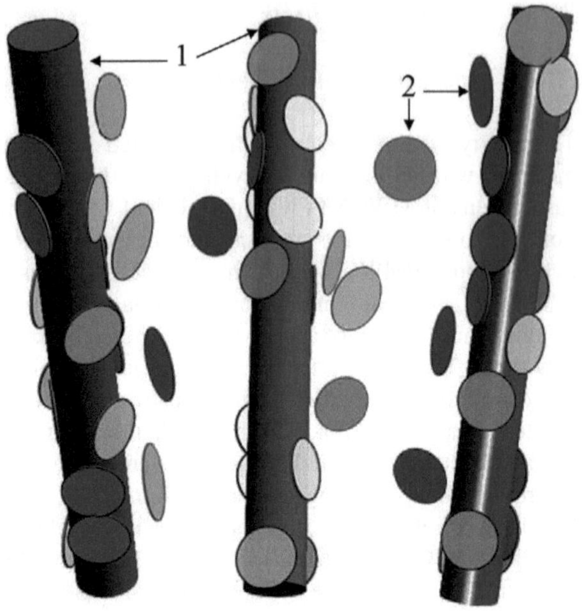

Fig. 3.38 Schematic representation of the structure of CNT (*1*) and LapO (Laponite) organomodified by ion-exchange reactions with the surfactant cetyltrimethylammonium bromide (CTAB, $C_{16}H_{33}$–N$(CH_3)_3Br$) (*2*) particles in LC (liquid crystals) host (Reproduced with permission of *Elsevier Science*)

3.1.5.2 Organic Natural Products

Among a few organic natural products, whose composites with CNTs could be dispersed, the *gum arabic* [421–423] (GA, a natural gum made of hardened sap taken from two species of the acacia tree; *Senegalia senegal* and *Vachellia seyal*) is mostly frequent. Thus, stable homogeneous suspensions of MWCNTs were prepared using GA as a dispersant [424]. The concentration measurement of MWCNTs showed that the optimum concentration of GA is 0.45 $g{\cdot}L^{-1}$. Measurement of their viscosity and thermal conductivity [425] showed that thermal conductivity ratios increased strongly with temperature, especially above 45 °C. Relative viscosities also increased with temperature above 45 °C at relatively higher MWCNT concentration indicating the potential role of Brownian motion. Also, MWCNT/epoxy nanocomposites were fabricated with 0.3 wt.% unmodified, oxidized, and GA-treated MWCNTs [426]. The gum treatment produced better dispersion of the MWCNTs in distilled water and that gum-treated MWCNT/epoxy nanocomposites had a better tensile strength and elastic modulus than did the unmodified and acid-treated MWCNT/epoxy nanocomposites.

Biopolymer dispersant *gellan gum* (a water-soluble polysaccharide produced by *Sphingomonas elodea*, a bacterium) was used to achieve aqueous dispersion of highly concentrated SWCNTs, which can be used to form the SWCNT liquid crystal phase [427]. To achieve alignment of SWCNTs, purification of SWCNTs was found to be very important, and it was achieved by a facile and nondestructive physical method that can prepare large volumes of SWCNTs in high yield for experimental use. Composite membranes of aligned SWCNTs could be obtained by simple evaporation SWCNT liquid crystal. Co-dispersion of native cellulose and SWCNTs in water was demonstrated [428], showing that the pH of the water should be between 6 and 10 for better dispersion. The co-solubility is likely caused through disruption of intramolecular hydrogen bonds in the cellulose by hydroxyl groups present on nanotube surface and the creation of intermolecular hydrogen bonds between cellulose and carbon nanotubes. MWCNTs can be also stably dispersed in water with small amount of *lignin* [429]. One-step dispersion in the 20.0 g/L concentration range was achieved at r.t. with excellent electrical properties of MWCNTs. Lignin was depicted to act as an antistatic agent. In a related report [430], a small amount of softwood kraft lignin (SKL) could facilitate the dispersion of MWCNTs in DMF solutions. Lignin-assisted double sonication represents a major low-cost and renewable alternative to prepare stable and concentrated suspensions of individualized CNTs [431].

In addition, MWCNTs and SWCNTs were surface modified [432] with *humic acids* HA [433] **3.102** {the major organic constituents of soil (humus), peat, coal, many upland streams, dystrophic lakes, and ocean water} from different sources and with surfactants of different ionic types. Both humic acid and surfactant could effectively disperse MWCNTs, but not SWCNTs, into stable suspensions under the studied conditions. The inhibitory effect of peat humic acid was relatively stronger than that of soil humic acid, but the two surfactants had a similar inhibitory effect on atrazine adsorption by the two CNT types. Increases in surfactant concentration resulted in rapid decreases in the adsorption of atrazine by CNTs when the surfactant concentration was less than 0.5 critical micelle concentration. The tribological behavior of CNTs in aqueous HA solutions was studied using a surface forces apparatus and showed promising lubricant additive properties [434]. Adding CNTs to the solution changes the friction forces between two mica surfaces from "adhesion-controlled" to "load-controlled" friction. Lateral sliding promotes a redistribution or accumulation, rather than squeezing out, of nanotubes between the surfaces.

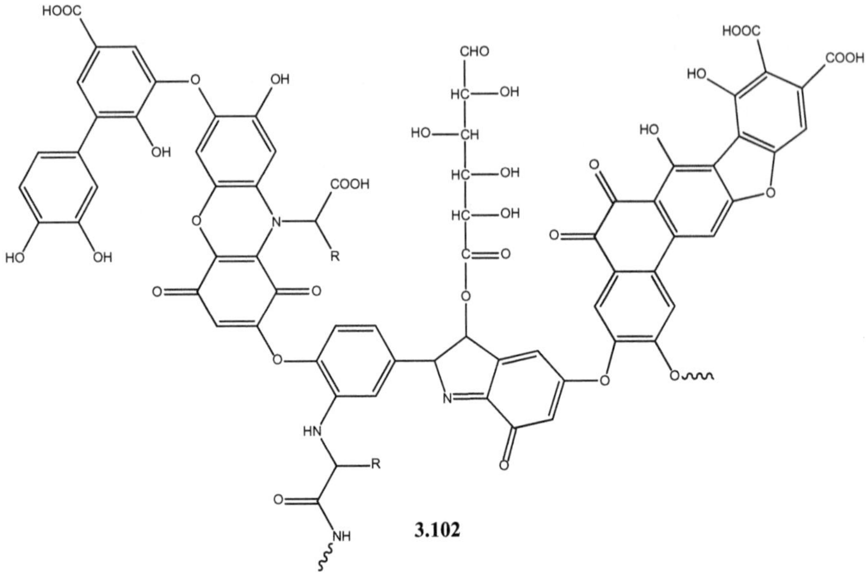

3.102

Typical humic acid

Green tea (tea extracts as sources of polyphenols are widely used in "greener chemistry" approaches for obtaining nanoparticles [435]) was reported [436, 437] to be a good dispersant of SWCNTs in aqueous media and organic solvents. Dimethyl sulfoxide (DMSO) was found to be a good solvent of green tea extract for dispersing SWCNTs. A combination of green tea (dispersant)/DMSO (solvent)/ polyvinyl alcohol (PVA) (nanotube wrap) was obtained that resulted in the dispersion of SWCNTs almost to individual nanotubes or to very thin nanotube bundles. Dispersions of SWCNTs in various surfactant solutions were also evaluated using natural products catechins **3.103**, phenolic acids, and flavonoids (a class of plant secondary metabolites) [438]. In addition, the polyphenol *curcumin* (**3.104**) was loaded by PEG-functionalized SWCNTs through $\pi-\pi$ stacking (loading capacity of 235–327 mg/g (curcumin/f-SWCNTs)), and the resulting composite was found to form good dispersion stability in water [439]. Its excellent biocompatibility as a drug carrier was confirmed.

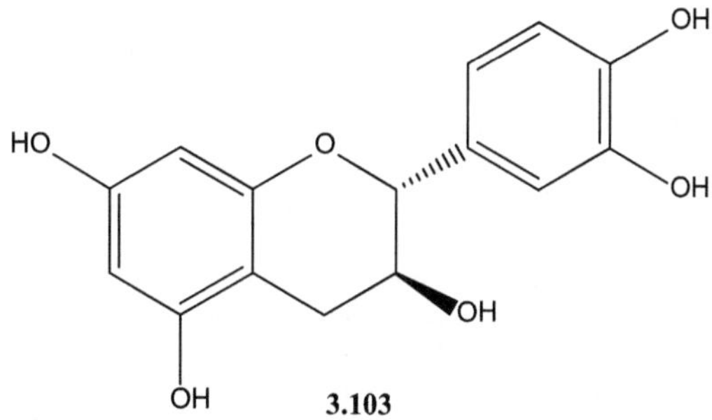

3.103

Catechin

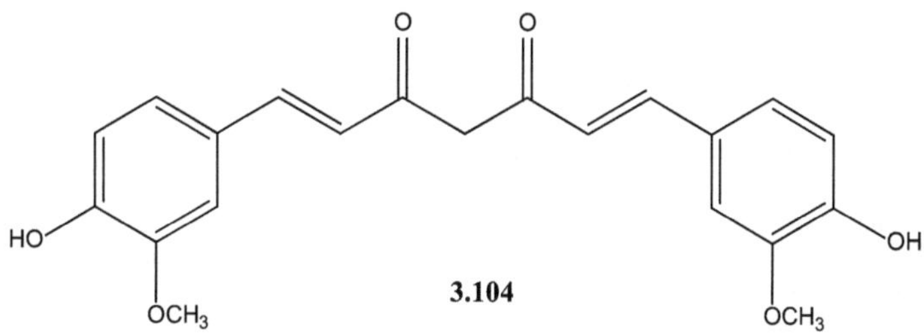

3.104

References

1. S. Kim, Y.-I. Lee, D.-H. Kim, K.-J. Lee, B.-S. Kim, M. Hussain, Y.-H. Choa, Estimation of dispersion stability of UV/ozone treated multi-walled carbon nanotubes and their electrical properties. Carbon **51**(1), 346–354 (2013)
2. Y.-I. Lee, S. Kim, K.-J. Lee, N.V. Myung, Y.-H. Choa, Inkjet printed transparent conductive films using water-dispersible single-walled carbon nanotubes treated by UV/ozone irradiation. Thin Solid Films **536**, 160–165 (2013)
3. E. Najafi, J.-Y. Kim, S.-H. Han, K. Shin, UV-ozone treatment of multi-walled carbon nanotubes for enhanced organic solvent dispersion. Colloids and Surfaces A: Physicochem. Eng. Aspects **284–285**, 373–378 (2006)
4. Z. Zaw Lin, M. Yoshida, Y. Uesugi, et al., Dispersion of relatively long multi-walled carbon nanotubes in water using ozone generated by dielectric barrier discharge. Journal of the Institute of Electrical Engineers of Japan A (Basic, Materials) **136**(4), 180–185 (2016.) https://www.jstage.jst.go.jp/article/ieejfms/136/4/136_180/_article/-char/ja/
5. J.-H. Kim, B. Gil Min, Functionalization of multi-walled carbon nanotube by treatment with dry ozone gas for the Enhanced Dispersion and Adhesion in polymeric composites. Carbon Lett. **11**(4), 298–303 (2010)
6. O. Byl, J. Jie Liu, J.T. Yates Jr., Etching of carbon nanotubes by OzonesA surface area study. Langmuir **21**, 4200–4204 (2005)

7. J.M. Simmons, B.M. Nichols, S.E. Baker, et al., Effect of ozone oxidation on single-walled carbon nanotubes. J. Phys. Chem. B **110**(14), 7113–7118 (2006)
8. Z. Zhang, K. Peng, Y. Chen, Mechanical performance of ozone functionalized MWCNTs/PC nanocomposites. Express Polym Lett **5**(6), 516–525 (2011)
9. A.S. Ayesh, S. Sobhy, S.S. Ibrahim, A.A. Aljaafari, Electrical, optical, and rheological properties of ozone-treated multiwalled carbon nanotubes–polystyrene nanocomposites. J. Reinf. Plast. Compos. **32**(6), 359–370 (2013)
10. N.V. Qui, P. Scholz, T.F. Keller, K. Pollok, B. Ondruschka, Ozonated Multiwalled carbon nanotubes as highly active and selective catalyst in the oxidative dehydrogenation of ethyl benzene to styrene. Chem. Eng. Technol. **36**(2), 300–306 (2013)
11. S. Fogden, C.A. Howard, R.K. Heenan, N.T. Skipper, M.S.P. Shaffer, Scalable method for the reductive dissolution, purification, and separation of single-walled carbon nanotubes. ACS Nano **6**(1), 54–62 (2012)
12. T. Shirafuji, Y. Noguchi, T. Yamamoto, et al., Functionalization of Multiwalled carbon nanotubes by solution plasma processing in ammonia aqueous solution and preparation of composite material with polyamide 6. Jpn. J. Appl. Phys. **12R**, 52 (2013)
13. Y. Feng, H. Zhang, Y. Hou, et al., Room temperature purification of few-walled carbon nanotubes with high yield. ACS Nano **2**(8), 1634–1638 (2008)
14. M.S. Han, Y.K. Lee, W.N. Kim, et al., Effect of multi-walled carbon nanotube dispersion on the electrical, morphological and rheological properties of polycarbonate/multi-walled carbon nanotube composites. Macromol. Res. **17**(11), 863–869 (2009)
15. Y. Peng, H. Liu, Effects of oxidation by hydrogen peroxide on the structures of Multiwalled carbon nanotubes. Ind. Eng. Chem. Res. **45**(19), 6483–6488 (2006)
16. V. Datsyuk, M. Kalyva, K. Papagelis, Chemical oxidation of multiwalled carbon nanotubes. Carbon **46**, 833–840 (2008)
17. P. Liu, T. Wang, Concise route to water-soluble multi-walled carbon nanotubes. Curr. Nanosci. **6**(1), 54–58 (2010)
18. L. Zhang, Y. Hashimoto, T. Taishi, Q.-Q. Ni, Mild hydrothermal treatment to prepare highly dispersed multi-walled carbon nanotubes. Appl. Surf. Sci. **257**(6), 1845–1849 (2011)
19. B.I. Kharisov, O.V. Kharissova, U. Ortiz Mendez, in *Microwave Hydrothermal and Solvothermal Processing of Materials and Compounds*, ed. by W. Cao. Microwave Heating, (INTECH, Rijeka, Croatia, 2012)
20. S. Huang, C. Liang, Y.-J. Chen, Persulfate chemical functionalization of carbon nanotubes and associated adsorption behavior in aqueous phase. Ind. Eng. Chem. Res. **55**, 6060−6068 (2016)
21. O.-k. Park, N. Hoon, K. Kin-tak, L.J.H. Lee, Effect of surface treatment with potassium persulfate on dispersion stability of multi-walled carbon nanotubes. Mater. Lett. **64**, 718–721 (2010)
22. B.I. Kharisov, O.V. Kharissova, U. Ortiz-Mendez, Decoration of carbon nanotubes with metal nanoparticles: Recent trends. Synthesis and Reactivity in Inorganic, Metal-Organic, and Nano-Metal Chemistry **46**(1), 55–76 (2016)
23. N. Jha, S. Ramaprabhu, Thermal conductivity studies of metal dispersed multiwalled carbon nanotubes in water and ethylene glycol based nanofluids. J. Appl. Phys. **106**, 084317 (2009)
24. R. Zhou, M. Shi, X. Chen, M. Wang, Y. Yang, X. Zhang, H. Chen, Water-soluble and highly fluorescent hybrids of multi-walled carbon nanotubes with uniformly arranged gold nanoparticles. Nanotechnology **18**(48), 485603 (2007)
25. N. Kocharova, T. Ääritalo, J. Leiro, J. Kankare, J. Lukkari, Aqueous dispersion, surface thiolation, and direct self-assembly of carbon nanotubes on gold. Langmuir **23**(6), 3363–3371 (2007)
26. Y.L. Hsin, J.Y. Lai, K.C. Hwang, S.C. Lo, F.R. Chen, J.J. Kai, Rapid surface functionalization of iron-filled multi-walled carbon nanotubes. Carbon **44**(15), 3328–3335 (2006)

27. V. Georgakilas, V. Tzitzios, D. Gournis, D. Petridis, Attachment of magnetic nanoparticles on carbon nanotubes and their soluble derivatives. Chem. Mater. **17**(7), 1613–1617 (2005)
28. F. Liang, L. Alemany, J.M. Beach, E.W. Billups, Structure analyses of dodecylated single-walled carbon nanotubes. J. Amer. Chem. Soc. **127**(40), 13941–13948 (2005)
29. J.J. Stephenson, A.K. Sadana, A.L. Higginbotham, J.M. Tour, Highly functionalized and soluble Multiwalled carbon nanotubes by reductive alkylation and Arylation: The Billups reaction. Chem. Mater. **18**, 4658–4661 (2006)
30. Z. Gu, F. Liang, Z. Chen, A. Sadana, C. Kittrell, W.E. Billups, R.H. Hauge, R.E. Smalley, In situ Raman studies on lithiated single-wall carbon nanotubes in liquid ammonia. Chem. Phys. Lett. **410**(4), 467–470 (2005)
31. E. W. Billups, A. Sadana, J. Chattopadhyay, Two-step method of functionalizing carbon allotropes and pegylated carbon allotropes made by such methods. WO07051071, 2007
32. J. Chattopadhyay, F.d.J. Cortez, S. Chakraborty, N.K.H. Slater, W.E. Billups, Synthesis of water-soluble PEGylated single-walled carbon nanotubes. Chem. Mater. **18**, 5864–5868 (2006)
33. Y. Liu, Y. Shen, L. Sun, Elemental superdoping of graphene and carbon nanotubes. Nat. Commun. **7**, 10921 (2016)
34. H. Song, Y. Ishii, A. Al-Zubaidi, T. Sakai, S. Kawasaki, Temperature-dependent water solubility of iodine-doped single-walled carbon nanotubes prepared using an electrochemical method. Phys. Chem. Chem. Phys. **15**(16), 5767–5770 (2013)
35. E. Joo Park, J. Lee, D. Jung, S. Eun Shim, Dispersion stability of fluorinated multi-walled carbon nanotubes in FC-27 refrigerant. *Journal*. J. Dispers. Sci. Technol. **32**(10), 1485–1492 (2011)
36. E.T. Mickelson, I.W. Chiang, J.L. Zimmerman, et al., Solvation of fluorinated Single-Wall carbon nanotubes in alcohol solvents. J. Phys. Chem. B **103**(21), 4318–4322 (1999)
37. S. Gupta, J. Farmer, Multiwalled carbon nanotubes and dispersed nanodiamond novel hybrids: Microscopic structure evolution, physical properties, and radiation resilience. J. Appl. Phys. **109**., 014314, 15 pp (2011)
38. S.S.J. Jyothirmayee Aravind, S. Ramaprabhu, Graphene wrapped multiwalled carbon nanotubes dispersed nanofluids for heat transfer applications. J. Appl. Phys. **112**, 124304 (2012)
39. S. Ciftan Hens, G. Cunningham, G. McGuire, O. Shenderova, Nanodiamond-assisted dispersion of carbon nanotubes and hybrid nanocarbon-based composites. Nanosci. Nanotechnol. Lett. **3**(1), 75–82 (2011)
40. C. Zhang, L. Ren, X. Wang, T. Liu, Graphene oxide-assisted dispersion of pristine multiwalled carbon nanotubes in aqueous media. J. Phys. Chem. C **114**(26), 11435–11440 (2010)
41. L. Tian, M.J. Meziani, F. Lu, C.Y. Kong, L. Cao, T.J. Thorne, Y.-P. Sun, Graphene oxides for homogeneous dispersion of carbon nanotubes. ACS Appl. Mater. Interfaces **2**(11), 3217–3222 (2010)
42. Y.-T. Yeong-Tarng Shieh, B.-R. Huang, M.-L. Tsai, Graphene oxide-assisted dispersion of carbon nanotubes in Sulfonated chitosan-modified electrode for simultaneous detections of sodium nitrite, hydroquinone, and catechol. Int. J. Electrochem. Sci. **10**, 3867–3884 (2015)
43. X. Chen, M. Qiu, H. Ding, K. Fu, Y. Fan, A reduced graphene oxide nanofiltration membrane intercalated by well-dispersed carbon nanotubes for drinking water purification. Nanoscale **8**, 5696–5705 (2016)
44. D. Yu, L. Dai, Self-assembled graphene/carbon nanotube hybrid films for Supercapacitors. J. Phys. Chem. Lett. **1**, 467–470 (2010)
45. G. Lasarte-Aragonés, R. Lucena, S. Cárdenas, M. Valcárcel, Effervescence-assisted carbon nanotubes dispersion for the micro-solid-phase extraction of triazine herbicides from environmental waters. Anal. Bioanal. Chem. **405**(10), 3269–3277 (2013)
46. Z. Li, H. Guan, N. Yu, Q. Xu, I. Imae, J. Wei, Conjugated polymer poly(2-methoxy-5-(3′,7′-dimethyloctyloxy)-1,4- phenylenevinylene) modification on carbon nanotubes with

assistance of supercritical carbon dioxide: Chemical interaction, solubility, and light emission. J. Phys. Chem. C **114**(22), 10119–10125 (2010)
47. L. He, X. Zheng, Q. Xu, Modification of carbon nanotubes using poly(vinylidene fluoride) with assistance of supercritical carbon dioxide: The impact of solvent. (2010). J. Phys. Chem. B **114**(16), 5257–5262 (2010)
48. C. Chen, M. Bortner, J.P. Quigley, D.G. Baird, Using supercritical carbon dioxide in preparing carbon nanotube Nanocomposite: Improved dispersion and mechanical properties. Polym. Compos. **33**(6), 1033–1043 (2012)
49. C. Sun, K. Sun, Preparation of multi-walled carbon nanotubes/SiO_2 core–shell nanocomposites by a two-step Stöber process. IET Micro & Nano Letters **11**(2), 67–70 (2016)
50. P. Stynoski, P. Mondal, E. Wotring, C. Marsh, Characterization of silica-functionalized carbon nanotubes dispersed in water. J. Nanopart. Res. **15**(1), 1396 (2013)
51. T. Matsuda, D. Minami, F. Khoerunnisa, et al., Aqueous Nanosilica dispersants for carbon nanotube. Langmuir **31**, 3194–3202 (2015)
52. Y.C. Tsai, C.C. Chiu, M.C. Tsai, J.Y. Wu, T.F. Tseng, T.M. Wu, S.F. Hsu, Dispersion of carbon nanotubes in low pH aqueous solutions by means of alumina-coated silica nanoparticles. Carbon **45**(14), 2823–2827 (2007)
53. S. Ou, S. Patel, B.A. Bauer, Free energetics of carbon nanotube association in pure and aqueous ionic solutions. J. Phys. Chem. B **116**(28), 8154–8168 (2012)
54. S. Niyogi, S. Boukhalfa, S.B. Chikkannanavar, et al., Selective aggregation of single-walled carbon nanotubes via salt addition. J. Am. Chem. Soc. **129**, 1898–1899 (2007)
55. A.I. Frolov, R.N. Arif, M. Kolar, A.O. Romanova, M.V. Fedorov, A.G. Rozhin, Molecular mechanisms of salt effects on carbon nanotube dispersions in an organic solvent (N-methyl-2-pyrrolidone). Chem. Sci. **3**(2), 541–548 (2012)
56. N.M. Uddin, F. Capaldi, B. Farouk, Molecular dynamics simulations of carbon nanotube interactions in water/surfactant systems. *Journal of Engineering Materials and Technology, Transactions of the*. ASME **132**(2), 0210121–0210125 (2010)
57. Z.-X. Cai, X.-P. Yan, In situ electrostatic assembly of CdS nanoparticles onto aligned multiwalled carbon nanotubes in aqueous solution. Nanotechnology **17**(16), 4212–4216 (2006)
58. B.J. Landi, C.M. Evans, J.J. Worman, S.L. Castro, S.G. Bailey, R.P. Raffaelle, Noncovalent attachment of CdSe quantum dots to single wall carbon nanotubes. Mater. Lett. **60**(29), 3502–3506 (2006)
59. S.T. Lyu, L.D. Troshkina, E.G. Rakov, Sorption of inorganic salts on carbon nanomaterials and magnetite. Russ. J. Phys. Chem. A **90**(11), 2275–2279 (2016)
60. J. Singh, N.C. Kothiyal, D. Pathania, Synthesis of highly dispersed single walled carbon nanotubes from furnace oil and light diesel oil by modified chemical vapour deposition method. *International Journal of Theoretical and Applied*. Science **3**(2), 15–20 (2011)
61. A. Yu, C.-C.L. Su, I. Roes, B. Fan, R.C. Haddon, Gram-scale preparation of surfactant-free, carboxylic acid groups functionalized, individual single-walled carbon nanotubes in aqueous solution. Langmuir **26**(2), 1221–1225 (2010)
62. S. Sahebian, S.M. Zebarjad, J. vahdati Khaki, A. Lazzeri, A study on the dependence of structure of multi-walled carbon nanotubes on acid treatment. Journal of Nanostructure in Chemistry **5**(3), 287–293 (2015)
63. N.A. Zawawi, Z.A. Majid, N. Aini, A. Rashid, Effect of acid oxidation methods on multiwalled carbon nanotubes (MWCNT) for drug delivery application. International Journal of Advanced Scientific Research and Management **1**(11), 14–22 (2016)
64. A. Tomova, G. Gentile, A. Grozdanov, Functionalization and characterization of MWCNT produced by different methods. Acta Phys. Pol. A **129**(3), 405–408 (2016)
65. A.M. Dimiev, A. Gizzatov, L.J. Wilson, J.M. Tour, Stable aqueous colloidal solutions of intact surfactant-free graphene nanoribbons and related graphitic nanostructures. Chem. Commun. **49**(26), 2613–2615 (2013)

66. J. M. Tour, J. L. Hudson, C. R. Dyke, J. J. Stephenson, Functionalization of carbon nanotubes in acidic media. WO05113434, 2005
67. J.J. Stephenson, J.L. Hudson, A.D. Leonard, B.K. Price, J.M. Tour, Repetitive functionalization of water-soluble single-walled carbon nanotubes. Addition of acid-sensitive addends. Chem. Mater. **19**, 3491–3498 (2007)
68. F. Liang, E. W. Billups, Water-soluble single-wall carbon nanotubes as a platform technology for biomedical applications. US20070110658, 2007
69. F. Liang, J.M. Beach, P.K. Rai, W. Guo, R.H. Hauge, M. Pasquali, R.E. Smalley, W.E. Billups, Highly exfoliated water-soluble single-walled carbon nanotubes. Chem. Mater. **18**, 1520–1524 (2006)
70. J. M. Tour, J. L. Bahr, J. Yang, Derivatization of carbon nanotubes using diazonium species. GB2413123, 2005
71. J.L. Hudson, M. Casavant, J.M. Tour, Water-soluble, exfoliated, nonroping single-wall carbon nanotubes. J. Amer. Chem. Soc. **126**(36), 11158–11159 (2004)
72. M. Farbod, S.K. Tadavani, A. Kiasat, Surface oxidation and effect of electric field on dispersion and colloids stability of multiwalled carbon nanotubes. Colloids Surf. A Physicochem. Eng. Asp. **384**(1–3), 685–690 (2011)
73. A.G. Osorio, I.C.L. Silveira, V.L. Bueno, C.P. Bergmann, $H_2SO_4/HNO_3/HCl$ — Functionalization and its effect on dispersion of carbon nanotubes in aqueous media. Appl. Surf. Sci. **255**, 2485–2489 (2008)
74. R.A. Moraes, C.F. Matos, E.G. Castro, W.H. Schreiner, M.M. Oliveira, A.J.G. Zarbin, The effect of different chemical treatments on the structure and stability of aqueous dispersion of iron- and iron oxide-filled multi-walled carbon nanotubes. J. Braz. Chem. Soc. **22**(11), 2191–2201 (2011)
75. S. Zhang, J.-Y. Wu, C.-T. Tse, J. Niu, Effective dispersion of multi-wall carbon nano-tubes in hexadecane through physiochemical modification and decrease of supercooling. Sol. Energy Mater. Sol. Cells **96**(1), 124–130 (2012)
76. V.T. Le, C.L. Ngo, Q.T. Le, et al., Surface modification and functionalization of carbon nanotube with some organic compounds. Adv. Nat. Sci. Nanosci. Nanotechnol. **035017** (5 pp), 4 (2013)
77. H. Sadegh, R. Shahryari-Ghoshekandi, Functionalization of carbon nanotubes and its application in nanomedicine: A review. Nanomed. J. **2**(4), 231–248 (2015)
78. A. Di Crescenzo, V. Ettorre, A. Fontana, Non-covalent and reversible functionalization of carbon nanotubes. Beilstein J. Nanotechnol. **5**, 1675–1690 (2014)
79. G. Tobias, E. Mendoza, B. Ballesteros, in *Encyclopedia of Nanotechnology*, ed. by B. Bhushan. Functionalization of Carbon Nanotubes, (2017), pp. 1–12. Continuously updated edition, Springer, Luxemburg, New York
80. N.M. Mubara, J.N. Sahu, J.R. Wong, et al., in *Chemistry and Applications* ed. by V.K. Thakur, M.K. Thakur. Chemical Functionalization of Carbon Nanomaterials, (CRC Press, 2015), pp. 82–101
81. X. Li, J. Shi, Y. Qin, Q. Wang, H. Luo, P. Zhang, Z.X. Guo, D.K. Park, Alkylation and arylation of single-walled carbon nanotubes by mechanochemical method. Chem. Phys. Lett. **444**(4), 258–262 (2007)
82. P. Lin, Y. Cong, B. Zhang, Fluorinated liquid crystalline surfactants for dispersion and alignment of carbon nanotubes. J. Mater. Sci. **50**, 4187–4199 (2015)
83. C. Backes, C.D. Schmidt, K. Rosenlehner, F. Hauke, J.N. Coleman, A. Hirsch, Nanotube surfactant design: The versatility of water-soluble perylene bisimides. Adv. Mater. **22**(7), 788–802 (2010)
84. R. Filipe Araújo, C.J.R. Silva, M. Conceição Paiva, M. Franco, M.F. Proença, Efficient dispersion of multi-walled carbon nanotubes in aqueous solution by non-covalent interaction with perylene bisimides. RSC Adv. **3**, 24535–24542 (2013)

85. Y. Tsarfati, V. Strauss, S. Kuhri, et al., Dispersing Perylene Diimide/SWCNT hybrids: Structural insights at the molecular level and fabricating Advanced materials. J. Am. Chem. Soc. **137**(23), 7429–7440 (2015)
86. Y. Matsuzawa, H. Kato, H. Ohyama, D. Nishide, H. Kataura, M. Yoshida, Photoinduced dispersibility tuning of carbon nanotubes by a water-soluble stilbene as a dispersant. Adv. Mater. **23**(34), 3922–3925 (2011)
87. P. Gerstel, S. Klumpp, F. Hennrich, et al., Highly selective dispersion of single-walled carbon nanotubes via polymer wrapping: A combinatorial study via modular conjugation. ACS Macro Lett. **3**(1), 10–15 (2014)
88. X. Zhang, L. Hou, P. Samorì, Coupling carbon nanomaterials with photochromic molecules for the generation of optically responsive materials. Nat. Commun. **7**, 11118 (2016)
89. T. Yamamoto, J. Motoyanagi, Y. Murakami, Y. Miyauchi, S. Maruyama, M. Kato, Surfactant-stabilized single-walled carbon nanotubes using triphenylene derivatives remain individually dispersion in both liquid and dried solid states. Appl. Phys. Express **2**(5), 055501 (2009)
90. J. Jay Lee, A. Yamaguchi, A. Alam, et al., Discotic ionic liquid crystals of Triphenylene as dispersants for orienting single-walled carbon nanotubes. Angew. Chem. **51**(34), 8490–8494 (2012)
91. H. Li, Q. Chen, B.-H. Han, Sugar-functionalized triptycenes used for dispersion of single-walled carbon nanotubes in aqueous solution by supramolecular interaction. New J. Chem. **40**, 3300–3307 (2016)
92. P. Hammershøj, P.H.H. Bomans, R. Lakshminarayanan, J. Fock, S.H. Jensen, T.S. Jespersen, T. Brock-Nannestad, T. Hassenkam, J. Nygård, N.A.J.M. Sommerdijk, K. Kilså, T. Bjørnholm, J.B. Christensen, A triptycene-based approach to solubilising carbon nanotubes and C_{60}. Chem. Eur. J. **18**(28), 8716–8723 (2012)
93. S.-Z. Zu, X.-X. Sun, D. Zhou, B.-H. Han, Supramolecular modification of single-walled carbon nanotubes with a water-soluble triptycene derivative. Carbon **49**(15), 5339–5347 (2011)
94. S. Majeed, V. Filiz, S. Shishatskiy, J. Wind, C. Abetz, V. Abetz, Pyrene-POSS nanohybrid as a dispersant for carbon nanotubes in solvents of various polarities: Its synthesis and application in the preparation of a composite membrane. Nanoscale Res. Lett. **7**, 296 (2012.) 11 pp
95. X. Kou, S.E. Morgan, The effect of polyhedral oligomeric silsesquioxane (POSS) structure on dispersion of multi-walled carbon nanotubes. *International SAMPE Technical Conference*, 2012, 9 p. http://www.sampe.org/events/SAMPETechStudentPosterContest.aspx
96. C.H. Li, A.M.C. Ng, C.S.K. Mak, A.B. Djurišić, W.K. Chan, Ruthenium complex containing block copolymer for the enhancement of carbon nanotube photoconductivity. ACS Appl. Mater. Interfaces **4**(1), 74–80 (2012)
97. X. Zhang, M. Kah, M.T.O. Jonker, T. Hofmann, Dispersion state and humic acids concentration-dependent sorption of Pyrene to carbon nanotubes. Environ. Sci. Technol. **46** (13), 7166–7173 (2012)
98. Y. Yan, S. Yang, J. Cui, et al., Synthesis of pyrene-capped polystyrene for dispersion of pristine single-walled carbon nanotubes. Polymer Int. **60**(10), 1425–1433 (2011)
99. L. Wang, Y. Zhao, in *Carbon Nanotubes – Current Progress of their Polymer Composites*, eds. by M.R. Berber, I.H. Hafez. Functionalization of Carbon Nanotubes with Stimuli-Responsive Molecules and Polymers, (INTECH, Rijeka, Croatia, 2016)
100. CO(2) -responsive “smart” single-walled carbon nanotubes. http://www.pubpdf.com/pub/23132767/CO2-responsive-smart-single-walled-carbon-nanotubes. Accessed on 19 Jan 2017
101. Y. Ding, S. Chen, H. Xu, et al., Reversible dispersion of single-walled carbon nanotubes based on a CO_2-responsive dispersant. Langmuir **26**(22), 16667–16671 (2010)
102. A. Feng, L. Peng, B. Liu, et al., Electrochemical redox switchable dispersion of single-walled carbon nanotubes in water. ACS Appl. Mater. Interfaces **8**(17), 11024–11030 (2016)

103. Y. Gao, M. Minmin Shi, R. Zhou, et al., Solvent-dependent fluorescence property of multi-walled carbon nanotubes noncovalently functionalized by pyrene-derivatized polymer. Nanotechnology **20**(13), 135705 (2009)
104. A.P. Periasamy, Y.-H. Ho, S.-M. Chen, Multiwalled carbon nanotubes dispersed in carminic acid for the development of catalase based biosensor for selective amperometric determination of H_2O_2 and iodate. Biosens. Bioelectron. **29**, 151–158 (2011)
105. A. Heidari, M.H. Beheshty, H. Rahimi, Functionalization of multi-walled carbon nanotubes via direct friedel-crafts acylation in an optimized PPA/P_2O_5 medium. Fullerenes Nanotubes and Carbon Nanostructures **21**(6), 516–524 (2013)
106. N. Li, H. Liu, X. Zhang, Functionalized Multiwalled carbon nanotubes in mild Polyphosphoric acid/phosphorous Pentoxide/phosphoric acid and their Composites with epoxy resin. Polym. Compos. **35**(7), 1275–1284 (2014)
107. Y. Liu, I. Zhitomirsky, Aqueous electrostatic dispersion and heterocoagulation of multiwalled carbon nanotubes and manganese dioxide for the fabrication of supercapacitor electrodes and devices. RSC Adv. **4**, 45481–45489 (2014)
108. V.N. Vijayakumar, M.L.N. Madhu Mohan, Dispersion of multi walled carbon nanotubes in a hydrogen bonded liquid crystal. Phys. B Condens. Matter **405**(21), 4418–4423 (2010)
109. S.A. Kumar, S.-F. Wang, T.C.-K. Yang, C.-T. Yeh, Acid yellow 9 as a dispersing agent for carbon nanotubes: Preparation of redox polymer-carbon nanotube composite film and its sensing application towards ascorbic acid and dopamine. Biosens. Bioelectron. **25**(12), 2592–2597 (2010)
110. M. Alvaro, P. Atienzar, P.d.l. Cruz, J.L. Delgado, H. Garcia, F. Langa, Synthesis and photochemistry of soluble, pentyl ester-modified single wall carbon nanotube. Chem. Phys. Lett. **386**(4), 342–245 (2004)
111. S. Li, Y. Qin, J. Shi, Z.-X. Guo, Y. Li, D. Zhu, Electrical properties of soluble carbon nanotube/polymer composite films. Chem. Mater. **17**, 130–135 (2005)
112. Y.-P. Sun, W. Huang, Y. Lin, K. Fu, A. Kitaygorodskiy, L.A. Riddle, Y.J. Yu, D.L. Carroll, Soluble Dendron-functionalized carbon nanotubes: Preparation, characterization, and properties. Chem. Mater. **13**, 2864–2869 (2001)
113. L. Feng, H. Li, F. Li, Z. Shi, Z. Gu, Functionalization of carbon nanotubes with amphiphilic molecules and their Langmuir-Blodgett films. Carbon **41**(12), 2385–2391 (2003)
114. V.A. Basiuk, L.V. Henao-Holguın, V. Meza-Laguna, E.V. Basiuk, Solvent-free derivatization of oxidized single-walled carbon nanotubes and nanodiamond with aminobenzo-crown ethers. Fullerenes, Nanotubes and Carbon Nanostructures **24**(10), 653–661 (2016)
115. M. Massaro, S. Riela, G. Cavallaro, et al., Properties and structural studies of Multi-Wall carbon nanotubes-phosphate Ester hybrids. International Journal of Organic Chemistry **3**, 26–34 (2013)
116. E. Bertels, K. Bruyninckx, M. Kurttepeli, M. Smet, S. Bals, B. Goderis, Highly efficient Hyperbranched CNT surfactants: Influence of molar mass and functionalization. Langmuir **30**(41), 12200–12209 (2014)
117. G. Konesky, Cross-linked carbon nanotubes, in *US20060275956*, (2006)
118. N. Nakashima, T. Tonomari, H. Murakami, Water-soluble single-walled carbon nanotubes via noncovalent sidewall-functionalization with a pyrene-carrying ammonium ion. Chem. Lett. **31**(6), 638–639 (2002)
119. Y. Tonomari, H. Murakami, N, N., Solubilization of single-walled carbon nanotubes using polyciclic aromatic ammonium amphiphiles in water – strategy for the design of solubilizers with high perfomance. Chem. Eur. J. **12**, 4027–4034 (2006)
120. K. Wusiman, H. Jeong, K. Tulugan, H. Afrianto, H. Chung, Thermal performance of multi-walled carbon nanotubes (MWCNTs) in aqueous suspensions with surfactants SDBS and SDS. International Communications in Heat and Mass Transfer **41**, 28–33 (2013)
121. J. Gao, W.-Y. Wang, L.-T. Chen, L.-J. Cui, X.-Y. Hu, H.-Z. Geng, Optimizing processes of dispersant concentration and post-treatments for fabricating single-walled carbon nanotube transparent conducting films. Appl. Surf. Sci. **277**(15), 128–133 (2013)

122. F. De Nicola, M. Salvato, C. Cirillo, et al., Record efficiency of air-stable multi-walled carbon nanotube/silicon solar cells. Carbon **101**, 226–234 (2016)
123. E.S. Kastrisianaki-Guyton, L. Chen, S.E. Rogers, T. Cosgrove, J.S. van Duijneveldt, Adsorption of sodium dodecylsulfate on single-walled carbon nanotubes characterised using small-angle neutron scattering. J. Colloid Interface Sci. **472**, 1–7 (2016)
124. M. Khazaee, D. Ye, A. Majumder, L. Baraban, J. Opitz, G. Cuniberti, Non-covalent modified multi-walled carbon nanotubes: Dispersion capabilities and interactions with bacteria. Biomedical Physics & Engineering Express **2**(5.), 055008), 8 pp (2016)
125. P. Lutsyk, R. Arif, J. Hruby, A sensing mechanism for the detection of carbon nanotubes using selective photoluminescent probes based on ionic complexes with organic dyes. Light: Science & Applications **5**, e16028 (2016)
126. A.Y. Vlasov, A.V. Venediktova, D.A. Videnichev, I.M. Kislyakov, E.D. Obraztsova, E.P. Sokolova, Effects of antifreezes and bundled material on the stability and optical limiting in aqueous suspensions of carbon nanotubes. *Physica Status Solidi (B) Basic*. Research **249**(12), 2341–2344 (2012)
127. A.V. Venediktova, A.Y. Vlasov, E.D. Obraztsova, D.A. Videnichev, I.M. Kislyakov, E.P. Sokolova, Stability and optical limiting properties of a single wall carbon nanotubes dispersion in a binary water-glycerol solvent. Appl. Phys. Lett. **100**(25), 251903 (2012)
128. H.-K. Fu, W.-J. Ye, M.-Y. Liu, J. Hu, M. Du, Q. Zheng, Preparation of layered double hydroxide/carbon nanotubes composite and its dispersion in organic solvents. Gaodeng *Xuexiao Huaxue Xuebao*. Chem. J. Chin. Univ. **34**(3), 538–544 (2013)
129. L. Ju, W. Zhang, X. Wang, J. Hu, Y. Zhang, Aggregation kinetics of SDBS-dispersed carbon nanotubes in different aqueous suspensions. Colloids Surf. A Physicochem. Eng. Asp. **409**, 159–166 (2012)
130. Y.F. Hsieh, B.Y. Huang, C.P. Liu, C.Y. Liu, NaDDBS as a dispersion agent for multiwalled carbon nanotubes in capillary EKC separation of nucleotides. Electrophoresis **31**(19), 3288–3295 (2010)
131. O. Matarredona, H. Rhoads, Z. Li, J.H. Harwell, L. Balzano, D.E. Resasco, Dispersion of single-walled carbon nanotubes in aqueous solutions of the anionic surfactant NaDDBS. J. Phys. Chem. B **107**, 13357–13367 (2003)
132. F. Tardani, L. Gentile, G.A. Ranieri, C. La Mesa, Shear orientation in nematic carbon nanotube dispersions: A combined NMR investigation. J. Phys. Chem. C **117**(16), 8556–8562 (2013)
133. F. Tardani, C. La Mesa, Elasticity of dispersions based on carbon nanotubes dissolved in a lyotropic nematic solvent. J. Phys. Chem. C **115**(19), 9424–9431 (2011)
134. M. Suttipong, N.R. Tummala, A. Striolo, C.S. Batista, J. Fagan, Salt-specific effects in aqueous dispersions of carbon nanotubes. Soft Matter **9**(14), 3712–3719 (2013)
135. V. Sa, K.G. Kornev, Analysis of stability of nanotube dispersions using surface tension isotherms. Langmuir **27**(22), 13451–13460 (2011)
136. N.J. O'Driscoll, T. Messier, M.D. Robertson, J. Murimboh, Suspension of multi-walled carbon nanotubes (CNTs) in freshwaters: Examining the effect of CNT size. Water Air Soil Pollut. **208**(1–4), 235–241 (2010)
137. T. Hiroi, S. Seisuke Ata, M. Shibayama, Transitions of aggregation states for concentrated carbon nanotube dispersion. J. Phys. Chem. C **120**, 5776−5782 (2016)
138. J.C. Goak, S.H. Lee, J.H. Han, S.H. Jang, K.B. Kim, Y. Seo, Y.-S. Seo, N. Lee, Spectroscopic studies and electrical properties of transparent conductive films fabricated by using surfactant-stabilized single-walled carbon nanotube suspensions. Carbon **49**(13), 4301–4313 (2011)
139. N.M. Uddin, F.M. Capaldi, B. Farouk, Molecular dynamics simulations of carbon nanotube dispersions in water: Effects of nanotube length, diameter, chirality and surfactant structures. Comput. Mater. Sci. **53**(1), 133–144 (2012)

140. R.M.F. Fernandes, B. Abreu, B. Claro, M. Buzaglo, O. Regev, I. Furó, E.F. Marques, Dispersing carbon nanotubes with ionic surfactants under controlled conditions: Comparisons and insight. Langmuir **31**(40), 10955–10965 (2015)
141. P. Alafogianni, K. Dassios, S. Farmaki, et al., On the efficiency of UV–vis spectroscopy in assessing the dispersionquality in sonicated aqueous suspensions of carbon nanotubes. Colloids and Surfaces A: Physicochem. Eng. Aspects **495**, 118–124 (2016)
142. W. Zhong, J.P. Claverie, Probing the carbon nanotube-surfactant interaction for the preparation of composites. Carbon **51**(1), 72–84 (2013)
143. I.R. Da Silva, P.L.M. Barreto, I.C. Bellettini, Study of aqueous dispersions of carbon nanotubes using different surfactants [Estudo das dispersões aquosas de nanotubos de carbono utilizando diferentes surfactantes]. Quim. Nova **36**(1), 5–9 (2013)
144. B. Wu, L. Bai, Q.-M. Gong, J. Liang, Effect of non-ionic surfactants on the dispersion of multiwalled carbon nanotubes at high loading in ethanol. Wuli Huaxue Xuebao/ Acta Physico – Chimica Sinica **25**(6), 1065–1069 (2009)
145. H. Kato, K. Mizuno, M. Shimada, A. Nakamura, K. Takahashi, K. Hata, S. Kinugasa, Observations of bound Tween80 surfactant molecules on single-walled carbon nanotubes in an aqueous solution. Carbon **47**(15), 3434–3440 (2009)
146. M. Liu, Z. Jia, C. Zhou, Dispersion of single-walled carbon nanotubes in water by a conjugated surfactant. Adv. Mater. Res. **415-417**, 562–565 (2012)
147. E. Detsri, S.T. Dubas, Dispersion of multiwalled carbon nanotubes with water-soluble polyaniline blend poly(sodium 4-Styrenesulfonate). Appl. Mech. Mater. **229-231**, 223–227 (2012)
148. D.J. Beesley, B.K. Price, S. Hunter, M.S.P. Shaffer, J.C. de Mello, Direct dispersion of SWCNTs in highly conductive solvent-enhanced PEDOT:PSS films. Nano **2**(3), 135–140 (2016)
149. N. Naohiro Terasawa, K. Asaka, High-performance PEDOT:PSS/single-walled carbon nanotube/ionic liquid Actuators combining electrostatic double-layer and faradaic capacitors. Langmuir **32**(28), 7210–7218 (2016)
150. D.-J. Yun, Y. Jin Jeong, H. Ra, et al., Effective way to enhance the electrode performance of multiwall carbon nanotube and poly(3,4-ethylenedioxythiophene): Poly(styrene sulfonate) composite using HCl–methanol treatment. J. Phys. Chem. C **120**(20), 10919–10926 (2016)
151. J.-Z. Zheng, J. Hu, F.-P. Du, Modification of carbon nanotubes with sodium p-aminobenzenesulfonate and its effect on Cu^{2+} adsorption. *Xinxing Tan Cailiao*. New Carbon Materials **28**(1), 14–19 (2013)
152. Y. Jeong, J. Kim, G.-W. Lee, Optimizing functionalization of multiwalled carbon nanotubes using sodium lignosulfonate. Colloid Polym. Sci. **288**(1), 1–6 (2010)
153. C. Li, X. Lv, J. Dai, J. Cui, Y. Yan, Synthesis of water-soluble single-walled carbon nanotubes and its application in poly(vinyl alcohol) composites. Polym. Adv. Technol. **24** (4), 376–382 (2013)
154. I. Madni, C.-Y. Hwang, S.-D. Park, Y.-H. Choa, H.-T. Kim, Mixed surfactant system for stable suspension of multiwalled carbon nanotubes. Colloids Surf. A Physicochem. Eng. Asp. **358**(1–3), 101–107 (2010)
155. T. Wu, Y. Pan, L. Li, Study on superhydrophobic hybrids fabricated from multiwalled carbon nanotubes and stearic acid. J. Colloid Interface Sci. **348**(1), 265–270 (2010)
156. V.O. Almeida, N.M. Balzaretti, T.M.H. Costa, G.B. Machado, M.R. Gallas, Surfactants for CNTs dispersion in zirconia-based ceramic matrix by sol-gel method. J. Sol-Gel Sci. Technol. **65**(2), 143–149 (2013)
157. S. Lin, D. Blankschtein, Role of the bile salt surfactant sodium cholate in enhancing the aqueous dispersion stability of single-walled carbon nanotubes: A molecular dynamics simulation study. J. Phys. Chem. B **114**(47), 15616–15625 (2010)
158. P. Lukaszczuk, E. Borowiak-Palen, M.H. Rümmeli, R.J. Kalenczuk, On the efficiency of bile salt for stable suspension and isolation of single-walled carbon nanotubes-spectroscopic and microscopic investigations. Appl. Phys. Mater. Sci. Process. **100**(2), 505–510 (2010)

159. M. Jang, S. Kim, H. Jeong, S.-Y. Ju, Affinity-mediated sorting order reversal of single-walled carbon nanotubes in density gradient ultracentrifugation. Nanotechnology **27**,41LT01, 6 pp (2016)
160. F. Li, B. Tang, J. Xiu, S. Zhang, Hydrophilic modification of multi-walled carbon nanotube for building photonic crystals with enhanced color visibility and mechanical strength. Molecules **21**,547, 9 pp (2016)
161. B. Sohrabi, N. Poorgholami-Bejarpasi, N. Nayeri, Dispersion of carbon nanotubes using mixed surfactants: Experimental and molecular dynamics simulation studies. J. Phys. Chem. B **118**, 3094–3310 (2014)
162. H. Paloniemi, T. Ääritalo, T. Laiho, H. Liuke, N. Kocharova, K. Haapakka, F. Terzi, J. Lukkari, Water-soluble full-length single-wall carbon nanotube polyelectrolytes: Preparation and characterization. J. Phys. Chem. B **109**(18), 8634–8642 (2005)
163. H. Shirai, M. Kimura, N. Miki, Complex of amphiphilic compound with soluble carbon nanotube. JP2006265151, 2006
164. D.M. Guldi, G.M.A. Rahman, N. Jux, D. Balbinot, U. Hartnagel, N. Tagmatarchis, M. Prato, Functional single-wall carbon nanotube nanohybrids–associating SWCNTs with water-soluble enzyme model systems. J. Amer. Chem. Soc. **127**(27), 9830–9838 (2005)
165. G. Yu, M. Xue, Z. Zhang, J. Li, C. Han, F. Huang, A water-soluble pillar[6]arene: Synthesis, host-guest chemistry, and its application in dispersion of multiwalled carbon nanotubes in water. J. Am. Chem. Soc. **134**(32), 13248–13251 (2012)
166. J. Yang, G. Yu, D. Xia, F. Huang, A pillar[6]arene-based UV-responsive supra-amphiphile: Synthesis, self-assembly, and application in dispersion of multiwalled carbon nanotubes in water. Chem. Commun. **50**, 3993–3995 (2014)
167. A. Di Crescenzo, P. Di Profio, G. Siani, R. Zappacosta, A. Fontana, Optimizing the interactions of surfactants with graphitic surfaces and Clathrate hydrates. Langmuir **32**, 6559−6570 (2016)
168. L. Wan, X. Wang, S. Li, Q. Li, R. Tian, M. Li, J. Cheng, Microwave-assisted chemical functionalization of single-walled carbon nanotubes with organic peroxides. Chin. J. Chem. **27**(2), 359–364 (2009)
169. Thi Thuy Nguyen, Sy Uan Nguyen, Dinh Tam Phuong et al, Dispersion of denatured carbon nanotubes by using a dimethylformamide solution. Adv. Nat. Sci. Nanosci. Nanotechnol. **2** (035015), 4 (2011)
170. P. Mishra, P. Sehrawat, N. Tripathi, S.S. Islam, Purification and dispersion of chemically functionalized SWCNTs by using Dimethylformamide solution. Adv. Sci. Lett. **20**(7–9), 1545–1547 (2014)
171. M. Lebron-Colon, M.A. Meador, Methods of functionalization of carbon nanotubes by photooxidation. US 9,486 **772**, B1 (2016)
172. A.R. Mohamed, R. Bahru, W.-M. Yeoh, K.A. Yaacob, Dimethyl formamide as dispersing agent for Electrophoretically deposited of multi-walled carbon nanotubes. *Int. J. Petrochem.* Sci. Eng. **1**(1), 00004 (2016)
173. H. Pourgheysari, H. Pourzamani, A. Ebrahimi, G. Bonyadinezhad, Comparison between distilled water and dimethylformamid as solvent to fabricate electrodes coated with single wall carbon nanotubes. Int. J. Env. Health. Eng. **5**(3), 6 (2016)
174. A.J. Clancy, J. Melbourne, M.S.P. Shaffer, A one-step route to solubilised, purified or functionalised single-walled carbon nanotubes. *J. Mater. Chem. A, Mater*. Energy Sustain. **3**(32), 16708–16715 (2015)
175. A.A. Dyshin, O.V. Eliseeva, G.V. Bondarenko, A.M. Kolker, M.G. Kiselev, Dispersion of single-walled carbon nanotubes in Dimethylacetamide and a Dimethylacetamide–Cholic acid mixture. Russ. J. Phys. Chem. A **90**(12), 2434–2439 (2016)
176. G. Gabriel, G. Sauthier, J. Fraxedas, M. Moreno-Manas, M.T. Martinez, C. Miravitlles, J. Casabo, Preparation and characterisation of single-walled carbon nanotubes functionalised with amines. Carbon **44**(10), 1891–1897 (2006)
177. Research news, Bonded to change – nanotechnology. Mater. Today **6**(11), 12 (2003)

178. K.J. Donovan, K. Scott, Anomalous effective hydrodynamic radius of octadecylamine functionalised single walled carbon nanotubes. Carbon **50**, 3807–3815 (2012)
179. M. Abdel Salam, R. Burk, Synthesis and characterization of multi-walled carbon nanotubes modified with octadecylamine and polyethylene glycol. Arab. J. Chem. **10**(1), S921–S927 (2017)
180. Y. Chen, J. Tao, A. Ezzeddine, Superior performance nanocomposites from uniformly dispersed octadecylamine functionalized multi-walled carbon nanotubes. Journal of Carbon Research (C) **1**(1), 58–76 (2015)
181. M. Xu, Q. Huang, Q. Chen, P. Guo, Z. Sun, Synthesis and characterization of octadecylamine grafted multi-walled carbon nanotubes. Chem. Phys. Lett. **375**(5), 598–604 (2003)
182. M.A. Hamon, H. Hu, P. Bhowmik, S. Niyogi, B. Zhao, M.E. Itkis, R.C. Haddon, End-group and defect analysis of soluble single-walled carbon nanotubes. Chem. Phys. Lett. **347**(1), 8–12 (2001)
183. Y. Lian, Y. Maeda, T. Wakahara, T. Akasaka, S. Kazaoui, N. Minami, N. Choi, H. Tokumoto, Assignment of the fine structure in the optical absorption spectra of soluble single-walled carbon nanotubes. J. Phys. Chem. B **107**, 12082–12087 (2003)
184. C. Desai, S. Addo Ntim, S. Mitra, Antisolvent precipitation of hydrophobic functionalized multiwall carbon nanotubes in an aqueous environment. J. Colloid Interface Sci. **368**(1), 115–120 (2012)
185. Y. Tian, L. Zhang, P. Duan, F. Liu, B. Zhang, C. Liu, M. Liu, Fabrication of organogels composed from carbon nanotubes through a supramolecular approach. New J. Chem. **34**(12), 2847–2852 (2010)
186. H. Oh, B.M. Jung, H.P. Lee, J.Y. Chang, Dispersion of single walled carbon nanotubes in organogels by incorporation into organogel fibers. J. Colloid Interface Sci. **352**(1), 121–127 (2010)
187. C. Spinato, D. Giust, I.A. Vacchi, et al., Different chemical strategies to aminate oxidized multi-walled carbon nanotubes for siRNA complexation and delivery. J. Mater. Chem. B **4**, 431–441 (2016)
188. J. Shen, W. Huang, L. Wu, Y. Hu, M. Ye, Study on amino-functionalized multiwalled carbon nanotubes. Mater. Sci. Eng. A **464**(1), 151–156 (2007)
189. D. Pramanik, P.K. Maiti, Dendrimer assisted dispersion of carbon nanotubes: A molecular dynamics study. Soft Matter **12**, 8512–8520 (2016)
190. C.-H. Hsu, H.-Y. Liao, P.-L. Kuo, Aniline as a dispersant and stabilizer for the preparation of Pt nanoparticles deposited on carbon nanotubes. J. Phys. Chem. C **114**(17), 7933–7939 (2010)
191. S. Zhu, H. Zhang, R. Bai, Microwave-accelerated dissolution of MWCNT in aniline. Mater. Lett. **61**(1), 16–18 (2007)
192. Y. Sun, S. Wilson, Method for dissolving carbon nanotubes. WO02088025, 2002
193. R. C. Haddon, M. A. Hamon, Method of solubilizing carbon nanotubes in organic solutions. US6531513, 2003
194. R. C. Haddon, J. Chen, M. A. Hamon, Method of solubilizing unshortened carbon nanotubes in organic solutions. US6368569, 2002
195. R. Ramasubramaniam, J. Chen, H. Liu, Homogeneous carbon nanotube/polymer composites for electrical applications. Appl. Phys. Lett. **83**, 2928 (2003)
196. M. Saoudi, R. Ajjel, B. Zaidi, Experimental and theoretical study on the charge transfer between polyaniline and single walled carbon nanotubes. J. Mater. Environ. Sci. **7**(12), 4435–4447 (2016)
197. R.Y. Suckeveriene, E. Zelikman, G. Mechrez, M. Narkis, Literature review: Conducting carbon nanotube/polyaniline nanocomposites. Rev. Chem. Eng. **27**, 15–21 (2011)
198. N.A. Ogurtsov, Y.V. Noskov, V.N. Bliznyuk, et al., Evolution and interdependence of structure and properties of Nanocomposites of multiwall carbon nanotubes with Polyaniline. J. Phys. Chem. C **120**(1), 230–242 (2016)

199. A. Kausar, S. Meer, T. Iqbal, Structure, morphology, thermal, and electro-magnetic shielding properties of polystyrene microsphere/polyaniline/multi-walled carbon nanotube nanocomposite. *J. Plast. Film Sheeting*. 2016, in press. http://journals.sagepub.com/doi/full/10.1177/8756087916663813
200. M. Alvaro, P. Atienzar, P.d.l. Cruz, J.L. Delgado, V. Troiani, H. Garcia, F. Langa, L. Echegoyen, Synthesis, photochemistry, and electrochemistry of single-wall carbon nanotubes with pendent pyridyl groups and of their metal complexes with zinc porphyrin. Comparison with pyridyl-bearing fullerenes. J. Amer. Chem. Soc. **128**(20), 6626–6635 (2006)
201. S.A.V. Jannuzzi, B. Martins, L.E.S.C. Huamaní, A.L.B. Formiga, Supramolecular approach to decorate multi-walled carbon nanotubes with negatively charged iron(II) complexes. J. Braz. Chem. Soc. **28**(1), 2–10 (2017)
202. D. Jain, A. Saha, A.A. Martí, Non-covalent ruthenium polypyridyl complexes-carbon nanotubes composites: An alternative for functional dissolution of carbon nanotubes in solution. Chem. Commun. **47**(8), 2246–2248 (2011)
203. H. Wei, X.-S. Wu, G.-Y. Wen, Y. Qiao, Imidazolium ionic liquid functionalized carbon nanotubes for improved interfacial charge transfer and simultaneous determination of Dihydroxybenzene isomers. Molecules **21**, 617 (2016.) 10 pp
204. Y. Lin, Q. Liu, J. Fan, et al., Highly dispersed palladium nanoparticles on poly(N1,N3-dimethylbenzimidazolium)iodide-functionalized multiwalled carbon nanotubes for ethanol oxidation in alkaline solution. RSC Adv. **6**, 102582–102594 (2016)
205. Y. Liu, L. Yu, S. Zhang, J. Yuan, L. Shi, L. Zheng, Dispersion of multiwalled carbon nanotubes by ionic liquid-type Gemini imidazolium surfactants in aqueous solution. Colloids and Surfaces A: Physicochem. Eng. Aspects **359**, 66–70 (2010)
206. A. Di Crescenzo, S. Cambré, R. Germani, P. Di Profio, A. Fontana, Dispersion of SWCNTs with Imidazolium-rich surfactants. Langmuir **30**(14), 3979–3987 (2014)
207. N. Nishar Hameed, J.S. Church, N.V. Salim, et al., Dispersing single-walled carbon nanotubes in ionic liquids: A quantitative analysis. RSC Adv. **3**, 20034–20039 (2013)
208. J. Janine Carvalho Padilha, J.-M. Noël, J.-F. Bergamini, J. Rault-Berthelot, C. Corinne Lagrost, Functionalization of carbon materials by reduction of Diazonium Cations produced in situ in a Brønstedt acidic ionic liquid. ChemElectroChem **3**(4), 572–580 (2016)
209. A. Zafour, D. Abdelkader, E. Flahaut, N. Rebizi, A. Skender, Dispersion of the carbon nanotubes (DWNTC) by the cationic surfactants imidazolinium type. IOP Conf. Ser. Mater. Sci. Eng. **28**(1), 012010 (2012)
210. R. Mincheva, F. Meyer, P. Verge, J.-M. Raquez, L. Billiet, F. Du Prez, P. Dubois, Synthesis of clicked imidazolium-containing biosourced copolymers and application in carbon nanotube dispersion. Macromol. Rapid Commun. **32**(24), 1960–1964 (2011)
211. L. Zhao, Y. Li, Z. Liu, H. Shimizu, Carbon nanotube-conducting polymer core-shell hybrid using an imidazolium-salt-based ionic liquid as a linker: Designed as a potential platinum electrode alternative material for large-scale solution processing. Chem. Mater. **22**(21), 5949–5956 (2010)
212. A.D. Crescenzo, M. Aschi, E.D. Canto, S. Giordani, D. Demurtas, A. Fontana, Structural modifications of ionic liquid surfactants for improving the water dispersibility of carbon nanotubes: An experimental and theoretical study. Phys. Chem. Chem. Phys. **13**(23), 11373–11383 (2011)
213. A. Prevoteau, C. Soulié-Ziakovic, L. Leibler, Universally dispersible carbon nanotubes. J. Am. Chem. Soc. **134**(49), 19961–19964 (2012)
214. C. Qu, F. Cheng, H. Su, Y. Zhao, Dispergation and modification of multi-walled carbon nanotubes in aqueous solution. Russ. J. Phys. Chem. A **90**(11), 2230–2236 (2016)
215. T. Yokoi, S.I. Iwamatsu, S.I. Komai, T. Hattori, S. Murata, Chemical modification of carbon nanotubes with organic hydrazines. Carbon **43**(14), 2869–2874 (2005)

216. F. Yang, Y. Li, S. Zhang, M. Tao, J. Zhao, C. Hang, Functionalization of multiwalled carbon nanotubes and related polyimide/carbon nanotubes composites. Synth. Met. **160**(15–16), 1805–1808 (2010)
217. M. Shigeta, M. Komatsu, N. Nakashima, Individual solubilization of single-walled carbon nanotubes using totally aromatic polyimide. Chem. Phys. Lett. **418**(1), 115–118 (2006)
218. G.D. Costanzo, S. Goyanes, S. Ledesma, Using photosensitive dye to improve multi walled carbon nanotubes dispersion. J. Phys. Conf. Ser. **274**(1), 012117 (2011)
219. S.H. Tan, K.L. Suk, J.C. Goak, S.C. Hong, J.-Y. Kim, S.-W. Lee, S. Kim, Y. Seo, N. Lee, Effect of poly(2-ethyl-2-oxazoline) on multi-walled carbon nanotubes reinforced poly(vinyl alcohol) composites. Polym. Polym. Compos. **18**(5), 251–256 (2010)
220. E. Okutan, G.O. Aydn, F. Hacvelioğlu, A. Kl, S.K. Beyaz, S. Yeilot, Synthesis and characterization of soluble multi-walled carbon nanotube/poly(organophosphazene) composites. Polymer **52**(5), 1241–1248 (2011)
221. A. Servant, I. Jacobs, C. Bussy, et al., Gadolinium-functionalised multi-walled carbon nanotubes as a T_1 contrast agent for MRI cell labelling and tracking. Carbon **97**, 126–133 (2016)
222. B. Suarez, B.M. Simonet, S. Cardenas, M. Valcarcel, Separation of carbon nanotubes in aqueous medium by capillary electrophoresis. J. Chromatogr. A **1128**(1), 282–289 (2006)
223. L. Licea-Jimenez, P.Y. Henrio, A. Lund, T.M. Laurie, S.A. Perez-Garcia, L. Nyborg, H. Hassander, R.W. Rychwalski, r.m.-f.c.a.-c. MWCNT, Compos. Sci. Technol. **67**(5), 844–854 (2007)
224. S.M. Fatemi, M. Foroutan, Recent developments concerning the dispersion of carbon nanotubes in surfactant/polymer systems by MD simulation. J. Nanostruct. Chem. **6**, 29–40 (2016)
225. J.B. Cui, C.P. Daghlian, U.J. Gibson, Solubility and electrical transport properties of thiolated single-walled carbon nanotubes. J. Appl. Phys. **98**, 044320 (2005)
226. T. Nakamura, T. Ohana, M. Ishihara, Hasegawa; Koga, Y., chemical modification of single-walled carbon nanotubes with sulfur-containing functionalities. Diamond & Related Materials **16**(4), 1091–1094 (2007)
227. R. García-González, A. Fernández-La Villa, A. Costa-García, M.T. Fernández-Abedu, Dispersion studies of carboxyl, amine and thiol-functionalized carbon nanotubes for improving the electrochemical behavior of screen printed electrodes. Sensors Actuators B **181**, 353–360 (2013)
228. N.R. Tummala, B.H. Morrow, D.E. Resasco, A. Striolo, Stabilization of aqueous carbon nanotube dispersions using surfactants: Insights from molecular dynamics simulations. ACS Nano **4**, 7193–7204 (2010)
229. Z. Jiang, D. Chen, Y. Yu, J. Miao, Y. Liu, L. Zhang, Composite fibers prepared from multi-walled carbon nanotubes/cellulose dispersed/dissolved in ammonium/dimethyl sulfoxide mixed solvent. RSC Adv. **7**, 2186–2192 (2017)
230. E. Karpushkin, N. Gvozdik, M. Klimenko, et al., Structure and flow behavior of dilute dispersions of carbon nanotubes in polyacrylonitrile–dimethylsulfoxide solution. Colloid Polym. Sci. **294**, 1187–1195 (2016)
231. M.M. Stylianakis, J.A. Mikroyannidis, E. Kymakis, A facile, covalent modification of single-wall carbon nanotubes by thiophene for use in organic photovoltaic cells. Sol. Energy Mater. Sol. Cells **94**(2), 267–274 (2010)
232. H. Leinonen, M. Lajunen, Direct functionalization of pristine single-walled carbon nanotubes by diazonium-based method with various five-membered S- or N- heteroaromatic amines. J. Nanopart. Res. **14**. 1064, 12 pp (2012)
233. Y.-T. Lin, R. Singh, S.-W. Kuo, F.-H. Ko, Bio-inspired Supramolecular chemistry provides highly concentrated dispersions of carbon nanotubes in Polythiophene. Materials **9**(6), 438 (2016)
234. J. Han, Q. Ji, H. Li, A photodegradable hexaaza-pentacene molecule for selective dispersion of large-diameter semiconducting carbon nanotubes. Chem. Commun. **52**, 7683–7686 (2016)

235. C. Bounioux, A. Bar-Hena, R. Yerushalmi-Rozen, Salting-in effect in organic dispersions of poly(3-hexyl thiophene)-carbon-nanotubes. Chem. Commun. **51**, 6343–6345 (2015)
236. A. Ghosh, K.V. Rao, R. Voggu, S.J. George, Non-covalent functionalization, solubilization of graphene and single-walled carbon nanotubes with aromatic donor and acceptor molecules. Chem. Phys. Lett. **488**(4–6), 198–201 (2010)
237. S. Liang, Y. Zhao, A. Adronov, Selective and reversible Noncovalent functionalization of single-walled carbon nanotubes by a pH-responsive Vinylogous Tetrathiafulvalene–Fluorene copolymer. J. Am. Chem. Soc. **136**, 970–977 (2014)
238. B. Koh, G. Kim, H.K. Yoon, J.B. Park, R. Kopelman, W. Cheng, Fluorophore and dye-assisted dispersion of carbon nanotubes in aqueous solution. Langmuir **28**(32), 11676–11686 (2012)
239. W. Zhang, S.R.P. Silva, Raman and FT-IR studies on dye-assisted dispersion and flocculation of single walled carbon nanotubes. Spectrochimica Acta – Part A: Molecular and Biomolecular Spectroscopy **77**(1), 175–178 (2010)
240. W. Zhang, S.R.P. Silva, The effects of phenolic hydrogens and methyl substitute groups in organic dyes on their dispersion of multiple-walled carbon nanotubes. Carbon **48**(7), 2063–2071 (2010)
241. C. Hu, Z. Chen, A. Shen, X. Shen, J. Li, S. Hu, Water-soluble single-walled carbon nanotubes via noncovalent functionalization by a rigid, planar and conjugated diazo dye. Carbon **44**(3), 428–434 (2006)
242. J. Mejia, F. Tichelaar, C. Saout, O. Toussaint, B. Masereel, Z. Mekhalif, S. Lucas, J. Delhalle, Effects of the dispersion methods in Pluronic F108 on the size and the surface composition of MWCNTs and their implications in toxicology assessment. J. Nanopart. Res. **13**(2), 655–667 (2011)
243. M.-Z. Yao, Y.-L. Hu, X.-X. Sheng, et al., Toxicity analysis of various Pluronic F-68-coated carbon nanotubes on mesenchymal stem cells. Chem. Biol. Interact. **250**, 47–58 (2016)
244. S. Parveen, S. Rana, R. Fangueiro, M.C. Paiva, Characterizing dispersion and long term stability of concentrated carbon nanotube aqueous suspensions for fabricating ductile cementitious composites. Powder Technol. **307**, 1–9 (2017)
245. C. Gasbarri, F. Croce, I. Meschini, et al., Single-walled carbon nanotubes in highly viscous media: A comparison between the dispersive agents [BMIM][BF_4], L121, and triton X-100. Chem. **22**(2), 546–549 (2016)
246. Y. Bai, D. Lin, F. Wu, Z. Wang, B. Xing, Adsorption of triton X-series surfactants and its role in stabilizing multi-walled carbon nanotube suspensions. Chemosphere **79**(4), 362–367 (2010)
247. H. Singh, P. Singh, An experimental appraisal on the efficacy of MWCNT-H_2O Nanofluid on the performance of solar parabolic trough collector. J. Fundam. Renewable Energy Appl. **6** (1), 6 (2016)
248. H. Younes, A. Al Ghaferi, I. Saadat, H. Hong, in *Advances in Carbon Nanostructures,* eds. by A.M.T. Silva, S.A.C. Carabineiro. Nanofluids based on carbon nanostructures, (INTECH, 2016), pp. 177–196
249. C.L. Poha, M. Mariatti, M.N. Ahmad Fauzi, et al., Dielectric and thermal properties of treated and untreated MWCNT filled DER 332 epoxy and OP392 epoxy. Procedia Chemistry **19**, 865–870 (2016)
250. S. Mahmood Fatemi, M. Foroutan, Study of dispersion of boron nitride nanotubes by triton X-100 surfactant using molecular dynamics simulations. J. Theor. Comput. Chem. **13**, 1450063 (2014.) 16 pp
251. T. Ogoshii, T.-A. Yamagishi, Y. Nakamoto, Supramolecular single-walled carbon nanotubes (SWCNTs) network polymer made by hybrids of SWCNTs and water-soluble calix[8]arenes. *Chem. Commun.* (45), 4776–4778 (2007).
252. X. Chen, C.T. Gibson, J. Britton, et al., *p*-Phosphonic acid calix[8]arene assisted dispersion and stabilisation of pea-pod C_{60}@multi-walled carbon nanotubes in water. Chem. Commun. **51**(12), 2399–2402 (2015)

253. H. Hu, B. Zhao, M.A. Hamon, K. Kamaras, M.E. Itkis, R.C. Haddon, Sidewall functionalization of single-walled carbon nanotubes by addition of dichlorocarbene. J. Amer. Chem. Soc. **125**(48), 14893–14900 (2003)
254. Z. Yinghuai, A.T. Peng, K. Carpenter, J.A. Maguire, N.S. Hosmane, M. Takagaki, Substituted carborane-appended water-soluble single-wall carbon nanotubes: New approach to boron neutron capture therapy drug delivery. J. Amer. Chem. Soc. **127**(27), 9875–9880 (2005)
255. S.A. Curran, A.V. Ellis, A. Vijayaraghavan, P.M. Ajayan, Functionalization of carbon nanotubes using phenosafranin. J. Chem. Phys. **120**, 4886 (2004)
256. H. Isobe, T. Tanaka, R. Maeda, E. Noiri, N. Solin, M. Yudasaka, S. Iijima, E. Eiichi Nakamura, Preparation, purification, characterization, and cytotoxicity assessment of water-soluble, transition-metal-free carbon nanotube aggregates. Angew. Chem. Int. Ed. Engl. **45**(40), 6676–6680 (2006)
257. Y.T. Liu, W. Zhao, Z.Y. Huang, Y.F. Gao, X.M. Xie, X.H. Wang, X.Y. Ye, Noncovalent surface modification of carbon nanotubes for solubility in organic solvents. Carbon **44**(8), 1613–1616 (2006)
258. A. Arrais, E. Diana, D. Pezzini, R. Rossetti, E. Boccaleri, A fast effective route to pH-dependent water-dispersion of oxidized single-walled carbon nanotubes. Carbon **44**(3), 587–590 (2006)
259. D. Kumar, E. Paul, Kavita, Singh, K., Bhatti, H.S. Synthesis and characterization of water soluble carbon nanotubes. *Journal of Nanoscience and.* Technology **2**(1), 64–65 (2016)
260. P. Pérez, *Alkane C-H Activation by Single-Site Metal Catalysis (Catalysis by Metal Complexes)* (Springer, New York, 2012), 200 pp
261. W. Rehman, N. Bashir. *Transition Metal Complexes: The Future Medicines: Synthetic Route and Bioassay of Transition Metal Complexes*, (VDM Verlag Dr. Müller, 2010), 64 pp
262. N. Hadjiliadis, E. Sletten (eds.), *Metal Complex – DNA Interactions* (Wiley-Blackwell, Hoboken, NJ, Saarbrücken, Germany, 2009), 544 pp
263. A.D. Pomogailo, *Catalysis by Polymer-Immobilized Metal Complexes* (CRC Press, Taylor & Francis Group, Abingdon, 1999), 424 pp
264. B.M. Andreev, *Separation of Isotopes of Biogenic Elements in Two-Phase Systems* (Elsevier Science, Amsterdam, 2007), 316 pp
265. H. Bradl, *Heavy Metals in the Environment: Origin, Interaction and Remediation*, vol 6. (Interface Science and Technology) (Elsevier Academic Press, Amsterdam, 2005), 282 pp
266. X. Peng, H. Qin, L. Li, Y. Huang, J. Peng, Y. Cao, N. Komatsu, Water redissoluble chiral porphyrin-carbon nanotube composites. J. Mater. Chem. **22**(12), 5764–5769 (2012)
267. J. Cheng, X.P. Zou, G. Zhu, M.F. Wang, Y. Su, G.Q. Yang, X.M. Lu, Synthesis of iron-filled carbon nanotubes with a great excess of ferrocene and their magnetic properties. Solid State Commun. **149**(39–40), 1619–1622 (2009)
268. M.C. Schnitzler, M.M. Oliveira, D. Ugarte, A.J.G. Zarbin, One-step route to iron oxide-filled carbon nanotubes and bucky-onions based on the pyrolysis of organometallic precursors. Chem. Phys. Lett. **381**(5), 541–548 (2003)
269. V. Georgakilas, D. Gournis, V. Tzitzios, L. Pasquato, D.M. Guldi, M. Prato, Decorating carbon nanotubes with metal or semiconductor nanoparticles. J. Mater. Chem. **17**, 2679–2694 (2007)
270. D. Kocsis, D. Kaptas, A. Botos, A. Pekker, K. Kamaras, Ferrocene encapsulation in carbon nanotubes: Various methods of filling and investigation. Phys. Status Solidi B **248**(11), 2512–2515 (2011)
271. C. Backes, *Noncovalent Functionalization of Carbon Nanotubes: Fundamental Aspects of Dispersion and Separation in Water* (Springer, Berlin, Heidelberg, 2012), 260 pp
272. P.J.F. Harris, *Carbon Nanotube Science: Synthesis, Properties and Applications*, 2nd edn. (Cambridge University Press, Cambridge, 2011), 314 pp
273. L. Meng, C. Fu, Q. Lu, Advanced technology for functionalization of carbon nanotubes. Prog. Nat. Sci. **19**, 801–810 (2009)

274. S. Banerjee, T. Hemraj-Benny, S.S. Wong, Covalent surface chemistry of single-walled carbon nanotubes. Adv. Mater. **17**(1), 17–29 (2005)
275. L. Henao-Holguín, V. Meza-Laguna, T.Y. Gromovoy, E. Basiuk, M. Rivera, V.A. Basiuk, Solvent-free covalent functionalization of fullerene C60 and pristine multi-walled carbon nanotubes with crown ethers. J. Nanosci. Nanotechnol. **16**(6), 6173–6184 (2016)
276. Khaled, E., Hassan, H.N.A., Ahmed, M.A., El-Attar, R.O. Crown ether/carbon nanotubes based Biperiden disposable potentiometric sensor. Electroanalysis. (2016), in press. http://onlinelibrary.wiley.com/doi/10.1002/elan.201600635/abstract
277. E. Khaled, M.S. Kamel, H.N.A. Hassan, Novel multi walled carbon nanotubes /crown ether based disposable sensors for determination of lead in water samples. Analytical Chemistry Letters **5**(6), 329–337 (2015)
278. R.E. Anderson, A.R. Barron, Solubilization of single-wall carbon nanotubes in organic solvents without sidewall functionalization. J. Nanosci. Nanotechn. **7**(10), 3646–3640 (2007)
279. G. Kerric, E.J. Parra, G.A. Crespo, F.X. Riusa, P. Blondeau, Nanostructured assemblies for ion-sensors: Functionalization of multi-wall carbon nanotubes with benzo-18-crown-6 for Pb $^{2+}$ determination. J. Mater. Chem. **22**, 16611–16617 (2012)
280. C. Jiang, A. Saha, C. Xiang, et al., Increased solubility, liquid-crystalline phase, and selective functionalization of single-walled carbon nanotube polyelectrolyte dispersions. ACS Nano **7** (5), 4503–4510 (2013)
281. A.A. Marti-Arbona, C. Jiang, A. Saha, M. Pasquali, C. Young, Liquid crystals from single-walled carbon nanotube polyelectrolytes and their use for making various materials. US 9,249 **023**, B2 (2016)
282. A. Khazaei, M.K. Borazjani, K.M. Moradian, Functionalization of oxidized single-walled carbon nanotubes with 4-benzo-9-crown-3 ether. J. Chem. Sci. **124**(5), 1127–1135 (2012)
283. K. Huang, A. Saha, K. Dirian, C. Jiang, P.-L. Chu, J.M. Tour, D.M. Guldi, A.A. Martí, Carbon nanotubes dispersed in aqueous solution by ruthenium(II) polypyridyl complexes. Nanoscale **8**, 13488–13497 (2016)
284. R. Martín, L. Jiménez, M. Alvaro, J.C. Scaiano, H. Garcia, Two-photon chemistry in ruthenium 2,2′-Bipyridyl-functionalized Single-Wall carbon nanotubes. Chem. Eur. J. **16** (24), 7282–7292 (2010)
285. H. Murakami, T. Nomura, N. Nakashima, Noncovalent porphyrin-functionalized single-walled carbon nanotubes in solution and the formation of porphyrin-nanotube nanocomposites. Chem. Phys. Lett. **378**(5), 481–485 (2003)
286. E.M.N. Mhuircheartaigh, S. Giordani, D. MacKernan, S.M. King, D. Rickard, L.M. Val Verde, M.O. Senge, W.J. Blau, Molecular engineering of Nonplanar Porphyrin and carbon nanotube assemblies: A linear and nonlinear spectroscopic and modeling study. Journal of Nanotechnology **2011**,745202, 12 pp (2011)
287. Z. Guo, J. Mao, Q. Ouyang, Y. Zhu, L. He, X. Lv, L. Liang, D. Ren, Y. Chen, J. Zheng, Noncovalent functionalization of single-walled carbon nanotube by porphyrin: Dispersion of carbon nanotubes in water and formation of self-assembly donor-acceptor nanoensemble. J. Dispers. Sci. Technol. **2010, 31**(1), 57–61
288. Y. Kim, S.O. Kim, W. Lee, D. Lee, W. Lee. Metal-porphyrin carbon nanotubes for use in fuel cell electrodes. US Patent 20130030175, 2013
289. O. Ito, F. D'Souza, Recent advances in Photoinduced electron transfer processes of fullerene-based molecular assemblies and Nanocomposites. Molecules **17**, 5816–5835 (2012)
290. L. Lvova, M. Mastroianni, G. Pomarico, M. Santonico, G. Pennazza, C. Di Natale, R. Paolesse, A. D'Amico, Carbon nanotubes modified with porphyrin units for gaseous phase chemical sensing. Sensors Actuators B **170**, 163–171 (2012)
291. J. Chen, P.C. Collier, Noncovalent functionalization of single-walled carbon nanotubes with water-soluble porphyrins. J. Phys. Chem. B **109**(16), 7605–7609 (2005)
292. C.Z. Huang, Q.G. Liao, Y.F. Li, Non-covalent anionic porphyrin functionalized multi-walled carbon nanotubes as an optical probe for specific DNA detection. Talanta **75**(1), 163–166 (2008)

293. E.M.N. Mhuircheartaigh, W.J. Blau, M. Prato, S. Giordani, Spectroscopic changes induced by sonication of porphyrin-carbon nanotube composites in chlorinated solvents. Carbon **45** (13), 2665–2671 (2007)
294. D.M. Guldi, G.M. Aminur Rahman, S. Qin, M. Tchoul, W.T. Ford, M. Marcaccio, D. Paolucci, F. Paolucci, S. Campidelli, M. Prato, Versatile coordination chemistry towards multifunctional carbon nanotube Nanohybrids. Chem. Eur. J. **12**, 2152–2161 (2006)
295. F. Cheng, A. Adronov, Noncovalent functionalization and solubilization of carbon nanotubes by using a conjugated Zn-porphyrin polymer. Chemistry **12**(19), 5053–5059 (2006)
296. N. Komatsu, A. Osuka, S. Isoda, N. Nakashima, H. MurakamiCarbon nanotube and method of purifying the same. EP1702885, 2006
297. Y. Du, N. Dong, M. Zhang, Porphyrin–poly(arylene ether sulfone) covalently functionalized multi-walled carbon nanotubes: Synthesis and enhanced broadband nonlinear optical properties. RSC Adv. **6**, 75530–75540 (2016)
298. D.M. Guldi, G.N.A. Rahman, J. Ramey, M. Marcaccio, D. Paolucci, F. Paolucci, S. Qin, M. Prato, Donor-acceptor nanoensembles of soluble carbon nanotubes. . Chemical communications, (18), p.2034-2035. Chem. Commun. **2004**(18), 2034–2035 (2004)
299. D.-M. Ren, Z. Guo, F. Du, Z.-F. Liu, Z.-C. Zhou, X.-Y. Shi, Y.-S. Chen, J.-Y. Zheng, A novel soluble tin(IV) Porphyrin modified single-walled carbon nanotube Nanohybrid with light harvesting properties. Int. J. Mol. Sci. **9**, 45–55 (2008)
300. R.A. Hatton, N.P. Blanchard, A.J. Miller, S.R.P. Silva, A multi-wall carbon nanotube-molecular semiconductor composite for bi-layer organic solar cells. Physica E **37**(1), 124–127 (2007)
301. H. Wu, Z. Chen, J. Zhang, et al., Stably dispersed carbon nanotubes covalently bonded to phthalocyanine cobalt(II) for ppb-level H_2S sensing at room temperature. J. Mater. Chem. A **4**, 1096–1104 (2016)
302. M. Raïssi, L. Vignau, E. Cloutet, B. Ratier, Soluble carbon nanotubes/phthalocyanines transparent electrode and interconnection layers for flexible inverted polymer tandem solar cells. Org. Electron. **21**, 86–91 (2015)
303. A. H. Ross, S. S. Ravi, Production of carbon nanotube-molecular semiconductor thin film. GB2428135, 2007
304. R. A. Hatton, S. R. Silva, Improvements in thin film production. WO07007061, 2007
305. K. Malika Tripathi, A. Begum, S. Kumar Sonkar, S. Sarkar, Nanospheres of copper(III) 1,2-dicarbomethoxy-1,2-dithiolate and its composite with water soluble carbon nanotubes. New J. Chem. **37**, 2708–2715 (2013)
306. D. Priftis, N. Petzetakis, G. Sakellariou, M. Pitsikalis, D. Baskaran, J.W. Mays, N. Hadjichristidis, Surface-initiated titanium-mediated coordination polymerization from catalyst-functionalized single and Multiwalled CarbonNanotubes. Macromolecules **42**, 3340–3346 (2009)
307. J. Chen, C. Xue, A new method for the preparation of stable carbon nanotube organogels. Carbon **44**(11), 2142–2146 (2006)
308. N. Tagmatarchis, M. Prato, D.M. Guldi, Soluble carbon nanotube ensembles for light-induced electron transfer interactions. Physica E **29**(3), 546–550 (2005)
309. A. V. Ellis, Functionalised carbon nanotubes and methods of preparation. WO07067079, 2007
310. D. Yuan, Y.G. Zhu, C. Jia, in *Carbon Nanotubes – Current Progress of their Polymer Composites*, ed. by M.R. Berber, I.H.Hafez. Carbon Nanotube-Polymer Composites for Energy Storage Applications, (2016), pp. 439–457
311. R. Atif, F. Inam, Reasons and remedies for the agglomeration of multilayered graphene and carbon nanotubes in polymers. Beilstein J. Nanotechnol. **7**, 1174–1196 (2016)
312. H.-C. Chang, J.-T. Wang, D.-H. Li, et al., Conjugated fluorene-moiety-containing pendant polymers for the dispersion of single-wall carbon nanotubes: Polymer wrapping abilities and electrical properties. Polym. J. **48**, 421–429 (2016)

313. N.A. Rice, A.V. Subrahmanyam, B.R. Coleman, A. Adronov, Effect of induction on the dispersion of semiconducting and metallic single-walled carbon nanotubes using conjugated polymers. Macromolecules **48**(15), 5155–5161 (2015)
314. K.E. Wise, C. Park, J.H. Kang, E.J. Siochi, J.S. Harrison, Nanocomposites from stable dispersions of carbon nanotubes in polymeric matrices using dispersion interaction. US Patent-9 **493**, 635 (2016)
315. J. Ma, R.M. Larsen, Comparative study on dispersion and interfacial properties of single walled carbon nanotube/polymer composites using Hansen solubility parameters. ACS Appl. Mater. Interfaces **5**(4), 1287–1293 (2013)
316. M. Giulianini, N. Motta, in *Self-Assembly of Nanostructures*, ed. by S. Bellucci. Polymer Self-assembly on Carbon Nanotubes, Vol 12 of the series Lecture Notes in Nanoscale Science and Technology, (Springer, Luxemburg, New York, 2012), pp.1–72
317. C.-X. Chao-Xuan Liu, J.-W. Choi, Improved dispersion of carbon nanotubes in polymers at high concentrations. Nano **2**, 329–347 (2012)
318. Z. Zhang, J. Zhang, P. Chen, B. Zhang, J. He, G.H. Hu, Enhanced interactions between multi-walled carbon nanotubes and polystyrene induced by melt mixing. Carbon **44**(4), 692–698 (2006)
319. D.Q. Fan, J.P. He, W. Tang, J.T. Xu, Y.L. Yang, Synthesis of polymer grafted carbon nanotubes by nitroxide mediated radical polymerization in the presence of spin-labeled carbon nanotubes. Eur. Polym. J. **43**(1), 26–34 (2007)
320. N. Sluzarenko, B. Heurtefeu, M. Maugey, C. Zakri, P. Poulin, S. Lecommandoux, Diblock copolymer stabilization of multi-wall carbon nanotubes in organic solvents and their use in composites. Carbon **44**(15), 3207–3212 (2006)
321. R. Tong, H.-X. Wu, X.-Q. Qiu, S.-X. Qian, Y.-H. Lin, R.-F. Cai, Optical performance and nonlinear scattering of soluble polystyrene grafted multi-walled carbon nanotubes. Chinese Physics Lett. **23**(8), 2105–2108 (2006)
322. H. Li, A. Adronov, Water-soluble SWCNTs from sulfonation of nanotube-bound polystyrene. Carbon **45**(5), 984–990 (2007)
323. M.J. O'Connell, P. Boul, L.M. Ericson, C. Huffman, Y. Wang, E. Haroz, C. Kuper, R.E. Smalley, Reversible water-solubilization of single-walled carbon nanotubes by polymer wrapping. Chem. Phys. Lett. **342**(3), 265–271 (2001)
324. Y. Sakakibara, M. Tokumoto, H. Kataura, Saturable absorber of polyimide containing dispersed carbon nanotubes. EP1772770, 2007
325. L.-J. Li, R.J. Nicholas, C.-Y. Chen, R.C. Darton, S.C. Baker, Comparative study of photoluminescence of single-walled carbon nanotubes wrapped with sodium dodecyl sulfate, surfactin and polyvinylpyrrolidone. Nanotechnology **16**(5), S202–S205 (2005)
326. C. Li, C. Liu, F. Li, Q. Gong, Optical limiting performance of two soluble multi-walled carbon nanotubes. Chem. Phys. Lett. **380**(1), 201–205 (2003)
327. H.X. Wu, X.Q. Qiu, R.F. Cai, S.X. Qian, Poly(N-vinyl carbazole)-grafted multiwalled carbon nanotubes: Synthesis via direct free radical reaction and optical limiting properties. Appl. Surf. Sci. **253**(11), 5122–5128 (2007)
328. G. Wang, Y. Ding, F. Wang, X. Li, C. Li, Poly(aniline-2-sulfonic acid) modified multiwalled carbon nanotubes with good aqueous dispersibility. J. Colloid and Interface Sci. **317**(1), 199–205 (2008)
329. R. Sainz, A.M. Benito, M.T. Martínez, J.F. Galindo, J. Sotres, A.M. Baró, B. Corraze, W.K. Maser, A soluble and highly functional polyaniline–carbon nanotube composite. Nanotechnology **16**(5), S150–S154 (2005)
330. R. Wang, P. Cherukuri, J.G. Duque, T.K. Leeuw, M.K. Lackey, C.H. Moran, V.C. Moore, H.K. Schmidt, SWCNT PEG-eggs: Single-walled carbon nanotubes in biocompatible shell-crosslinked micelles. Carbon **45**(12), 2388–2393 (2007)
331. J.U. Lee, J. Huh, K.H. Kim, C. Park, W.H. Jo, Aqueous suspension of carbon nanotubes via non-covalent functionalization with oligothiophene-terminated poly(ethylene glycol). Carbon **45**(5), 1051–1057 (2007)

332. A. Liu, T. Watanabe, I. Honma, J. Wang, H. Zhou, Effect of solution pH and ionic strength on the stability of poly(acrylic acid)-encapsulated multiwalled carbon nanotubes aqueous dispersion and its application for NADH sensor. Biosens. Bioelectron. **22**(5), 694–699 (2006)
333. Q.-j. Meng, X.-x. Zhang, S.-h. Bai, X.-c. Wang, Preparation and characterization of poly (methyl methacrylate)-functionalized carboxyl multi-walled carbon nanotubes. Chin. J. Chem. Phys. **20**(6), 660–664 (2007)
334. A. Liu, I. Honma, M. Ichihara, H. Zhou, Poly(acrylic acid)-wrapped multi-walled carbon nanotubes composite solubilization in water: Definitive spectroscopic properties. Nanotechnology **17**(12), 2845–2849 (2006)
335. Z. Liu, G. Bai, Y. Huang, F. Li, Y. Ma, T. Guo, X. He, X. Lin, H. Gao, Y. Chen, Microwave absorption of single-walled carbon nanotubes/soluble cross-linked polyurethane composites. J. Phys. Chem. C **111**, 13696–13700 (2007)
336. C.Y. Hong, Y.Z. You, C.Y. Pan, A new approach to functionalize multi-walled carbon nanotubes by the use of functional polymers. Polymer **47**(12), 4300–4309 (2006)
337. W. E. Ford, J. Wessels, Soluble carbon nanotubes. EP1428793, 2004
338. W. E. Ford, J. Wessels, A. Yasuda, Soluble carbon nanotubes. WO04052783, 2004
339. W. E. Ford, J. Wessels, A. Yasuda, Soluble carbon nanotubes. US20060014375, 2006
340. R. Yerushalmi-Rozen, O. Regev, Method for the preparation of stable suspensions and powders of single carbon nanotubes. WO02076888, 2002
341. H. Yu, T. Cao, L. Zhou, E. Gu, D. Yu, D. Jiang, Layer-by-layer assembly and humidity sensitive behavior of poly(ethyleneimine)/multiwall carbon nanotube composite films. Sensors & Actuators: B. Chemical **119**(2), 512–515 (2006)
342. A.J. Miller, R.A. Hatton, S.R.P. Silva, Water-soluble multiwall-carbon-nanotube-polythiophene composite for bilayer photovoltaics. Appl. Phys. Lett. **89**, 123115 (2006)
343. B.K. Kuila, S. Malik, S.K. Batabyal, A.K. Nandi, In-situ synthesis of soluble poly (3-hexylthiophene)/Multiwalled carbon nanotube composite: Morphology, structure, and conductivity. Macromolecules **40**, 278–287 (2007)
344. C.-Y. Hong, Y.-Z. You, C.-Y. Pan, Synthesis of water-soluble Multiwalled carbon nanotubes with grafted temperature-responsive shells by surface RAFT polymerization. Chem. Mater. **17**, 2247–2254 (2005)
345. F. Cheng, P. Imin, C. Maunders, G. Botton, A. Adronov, Soluble, discrete Supramolecular complexes of single-walled carbon nanotubes with Fluorene-based conjugated polymers. Macromolecules **41**(7), 2304–2308 (2008)
346. D. Banerjee, A. Jha, K.K. Chattopadhyay, Synthesis and characterization of water soluble functionalized amorphous carbon nanotube-poly(vinyl alcohol) composite. Macromol. Res. **20**(10), 1021–1028 (2012)
347. M.J. Kim, J. Lee, D. Jung, S.E. Shim, Surface modification of carbon nanotube by poly (ethylene glycol) for the preparation of poly(vinyl alcohol) nanocomposite. Journal of Macromolecular Science, Part A: Pure and Applied Chemistry **47**(6), 588–594 (2010)
348. S.W. Tam, W. Rashimi, S. Rashidi, L.C. Abdullah, M. Khalid, Stability and characterization of CNTnanofluids using polyvinyl alcohol dispersant. J. Eng. Sci. Technol. 1–15 (2016)
349. J. Wongon, S. Thumsorn, N. Srisawat, Poly(vinyl alcohol)/multiwalled carbon nanotubes composite Nanofiber. Energy Procedia **89**, 313–317 (2016)
350. I. Kalinina, K. Worsley, C. Lugo, S. Mandal, E. Bekyarova, R.C. Haddon, Synthesis, dispersion, and viscosity of poly(ethylene glycol)-functionalized water-soluble single-walled carbon nanotubes. Chem. Mater. **23**(5), 1246–1253 (2011)
351. X.-Q. Li, C.-Y. Sun, Q. Zhao, Z.-S. Wu, D.-K. Xu, W.-Y. Zhong, Oxidation, modification and dispersibility of single-walled carbon nanohorns. *Gaodeng Xuexiao Huaxue Xuebao*. Chem. J. Chin. Univ. **34**(4), 763–770 (2013)
352. Y. Ji, Y.Y. Huang, R. Rungsawang, E.M. Terentjev, Dispersion and alignment of carbon nanotubes in liquid crystalline polymers and elastomers. Adv. Mater. **22**(31), 3436–3440 (2010)

353. H. Ltzen, M. Wirts-Rtters, A. Hartwig, Structural studies of aromatic surfactants for dispergation of multiwall carbon nanotubes. Soft Materials **10**(4), 462–471 (2012)
354. H. Li, T. Zhang, G. Liang, Y. Zhang, X. Wang, *vivo* evaluation of acute toxicity of water-soluble carbon nanotubes. Toxicol. Environ. Chem. **93**(3), 603–615 (2011)
355. A. Di Crescenzo, M. Aschi, A. Fontana, Toward a better understanding of steric stabilization when using block copolymers as stabilizers of single-walled carbon nanotubes (SWCNTs) aqueous dispersions. Macromolecules **45**(19), 8043–8050 (2012)
356. J. Määttä, S. Vierros, P.R. Van Tassel, M. Sammalkorpi, Size-selective, Noncovalent dispersion of carbon nanotubes by PEGylated lipids: A coarse-grained molecular dynamics study. J. Chem. Eng. Data **59**, 3080–3089 (2014)
357. C. Semaan, M. Schappacher, A. Soum, Dispersion of carbon nanotubes through amphiphilic block copolymers: Rheological and dielectrical characterizations of poly(ethylene oxide) composites. Polym. Compos. **33**(1), 1–9 (2012)
358. M. Nazhipkyzy, S.D. Mohan, F.J. Davis, G.R. Mitchell, Carbon nanotubes in electrospun polyethylene oxide nanofibres: A potential route to conducting nanofibers. J. Phys. Conf. Ser. **646**, conference 1, 012007 (2015)
359. N.M. Uddin, F.M. Capaldi, B. Farouk, Molecular dynamics simulations of the interactions and dispersion of carbon nanotubes in polyethylene oxide/water systems. Polymer **52**, 288–296 (2011)
360. Z. Guo, Y. Feng, D. Zhu, S. He, H. Liu, X. Shi, J. Sun, M. Qu, Light-switchable single-walled carbon nanotubes based on host-guest chemistry. Adv. Funct. Mater. **23**(40), 5010–5018 (2013)
361. P.D. Petrov, G.L. Georgiev, A.H.E. Müller, Dispersion of multi-walled carbon nanotubes with pyrene-functionalized polymeric micelles in aqueous media. Polymer **53**(24), 5502–5506 (2012)
362. M. Matsuoka, M. Yamamoto, K. Adachi, Y. Tsukahara, T. Konno, Dispersion behavior of multi-walled carbon nanotubes with pyrene-containing linear and graft polymers as non-covalent surface modifiers. Designed Monomers and Polymers **13**(4), 387–397 (2010)
363. J. Li, F. Yang, G. Guo, D. Yang, J. Long, D. Fu, J. Lu, C. Wang, Preparation of biocompatible multi-walled carbon nanotubes as potential tracers for sentinel lymph nodes. Polym. Int. **59** (2), 169–174 (2010)
364. J. Zhang, A. Wang, pH- and thermo-responsive dispersion of single-walled carbon nanotubes modified with poly(N-isopropylacrylamide-co-acrylic acid). J. Colloid Interface Sci. **334**(2), 212–216 (2009)
365. A. Kubotera, R. Saito, Dispersion of carbon nanoparticle in water with poly(acrylic acid)-poly(amideimide) copolymers. Colloid Polym. Sci. **294**, 941–946 (2016)
366. K.C. Etika, M.A. Cox, J.C. Grunlan, Tailored dispersion of carbon nanotubes in water with pH-responsive polymers. Polymer **51**(8), 1761–1770 (2010)
367. Z. Guo, H. Yin, Y. Feng, S. He, Functionalization of single-walled carbon nanotubes with thermo-responsive poly(N-isopropylacrylamide): Effect of the polymer architecture. RSC Adv. **6**, 37953–37964 (2016)
368. D. Wang, L. Chen, Temperature and pH-responsive "smart" carbon nanotube dispersions. Methods Mol. Biol. **625**, 27–38 (2010)
369. G. Li, H. Wang, H. Zheng, R. Bai, A facile approach for the fabrication of highly stable superhydrophobic cotton fabric with multi-walled carbon nanotubes-azide polymer composites. Langmuir **26**(10), 7529–7534 (2010)
370. Y. Kwon, W. Shim, S.-Y. Jeon, J.-H. Youk, W.-R. Yu, Improving dispersion of multi-walled carbon nanotubes and graphene using a common non-covalent modifier. Carbon Letters **20**, 53–61 (2016)
371. P. Jiménez, P. Castell, R. Sainz, A. Ansón, M.T. Martínez, A.M. Benito, W.K. Maser, Carbon nanotube effect on polyaniline morphology in water dispersible composites. J. Phys. Chem. B **114**(4), 1579–1585 (2010)

372. X. Chen, J. Wang, J. Zou, X. Wu, X. Chen, F. Xue, Mechanical and thermal properties of functionalized multiwalled carbon nanotubes and multiwalled carbon nanotube-polyurethane composites. J. Appl. Polym. Sci. **114**(6), 3407–3413 (2009)
373. G.-J. Sohn, H.-J. Choi, I.-Y. Jeon, D.W. Chang, L. Dai, J.-B. Baek, Water-dispersible, sulfonated hyperbranched poly(ether-ketone) grafted multiwalled carbon nanotubes as oxygen reduction catalysts. ACS Nano **6**(7), 6345–6355 (2012)
374. J. Pang, G. Xu, Y. Tan, F. He, Water-dispersible carbon nanotubes from a mixture of an ethoxy-modified trisiloxane and pluronic block copolymer F127. Colloid. Polym. Sci. 1–11 (2010)
375. M.-F. Liu, Y.-L. Chen, B. Zhu, Y. Han, W.-G. Huang, C. Du, Z.-S. Bo, Synthesis of polyfluorenes bearing lateral pyreneterminated alkyl chains for dispersion of single-walled carbon nanotubes. Chinese Journal of Polymer Science (English Edition) **30**(3), 405–414 (2012)
376. F. Jakubka, S.P. Schießl, S. Martin, J.M. Englert, F. Hauke, A. Hirsch, J. Zaumseil, Effect of polymer molecular weight and solution parameters on selective dispersion of single-walled carbon nanotubes. ACS Macro Lett. **1**(7), 815–819 (2012)
377. W.J. Bodnaryk, N.A. Rice, A. Adronov, Influence of regiochemistry in the selective dispersion of metallic carbon nanotubes using electron poor conjugated polymers. Polym. Chem. **7**, 3832–3837 (2016)
378. D. Fong, W.J. Bodnaryk, N.A. Rice, S.R. Saem, J.M. Moran-Mirabal, A. Adronov, Influence of polymer electronics on selective dispersion of single-walled carbon nanotubes. Chem. Eur. J. **22**, 14560–14566 (2016)
379. T. Lei, X. Chen, G. Pitner, H.-S.P. Wong, Z. Bao, Removable and recyclable conjugated polymers for highly selective and high-yield dispersion and release of low-cost carbon nanotubes. J. Am. Chem. Soc. **138**, 802–805 (2016)
380. N. Berton, F. Lemasson, A. Poschlad, et al., Selective dispersion of large-diameter semiconducting single-walled carbon nanotubes with pyridine-containing copolymers. Small **10**(2), 360–367 (2014)
381. S.H. Min, H.-I. Kim, K.-s. Kim, et al., Selective dispersion of single-walled carbon nanotubes by binaphthyl based conjugated polymers: Integrated experimental and simulation approach. Polymer **96**, 63–69 (2016)
382. L. Hong, Y. Takagaki, H. Yoshikawa, N. Nakashima, Efficient dispersion of "super-growth" single-walled carbon nanotubes using a copolymer of Naphathalene Diimide and poly (dimethylsiloxane). Bull. Chem. Soc. Jpn. **89**(2), 183–191 (2016)
383. M. Su, Y.N. Wei, L.Y. Qi, Y.C. Pang, Y.Z. Guo, P. Li, J. Li, H.P. Li, Remarkable crystallization morphologies of poly(4-vinylpyridine) on single-walled carbon nanotubes in CO_2-expanded liquids. Express Polym Lett **5**(12), 1102–1112 (2011)
384. X.L. Wu, P. Liu, Poly(vinyl chloride)-grafted multi-walled carbon nanotubes via Friedel-crafts alkylation. Express Polym Lett **4**(11), 723–728 (2010)
385. F. Bourdiol, F. Mouchet, A. Perrault, I. Fourquaux, L. Datas, C. Gancet, J.-C. Boutonnet, E. Pinelli, L. Gauthier, E. Flahaut, Biocompatible polymer-assisted dispersion of multi walled carbon nanotubes in water, application to the investigation of their ecotoxicity using Xenopus laevis amphibian larvae. Carbon **54**, 175–191 (2013)
386. Z. Iatridi, C. Tsitsilianis, pH responsive MWCNT-star terpolymer nanohybrids. Soft Matter **9** (1), 185–193 (2013)
387. T. Liu, G. Xu, J. Zhang, H. Zhang, J. Pang, Dispersion of carbon nanotubes by the branched block copolymer Tetronic 1107 in an alcohol-water solution. Colloid Polym. Sci. **291**(3), 691–698 (2013)
388. L. Moradi, A. Mohajeri, H. Naeimi, A.M. Rashidi, Amidation of multiwalled carbon nanotubes in mild and efficient conditions. J. Nanosci. Nanotechnol. **13**(3), 1923–1926 (2013)
389. E.C. Botelho, E.R. Edwards, B. Bittmann, T. Burkhart, Dispersing carbon nanotubes in phenolic resin using an aqueous solution. J. Braz. Chem. Soc. **22**(11), 2040–2047 (2011)

390. C.J. Collison, S. Spencer, A. Preske, C. Palumbo, A. Helenic, R. Bailey, S. Pellizzeri, A new model for quantifying the extent of interaction between soluble polyphenylene-vinylenes and single-walled carbon nanotubes in solvent dispersions. J. Phys. Chem. B **114**(34), 11002–11009 (2010)
391. T. Umeyama, K. Kawabata, N. Tezuka, Y. Matano, Y. Miyato, K. Matsushige, M. Tsujimoto, S. Isoda, M. Takano, H. Imahori, Dispersion of carbon nanotubes by photo- and thermal-responsive polymers containing azobenzene unit in the backbone. Chem. Commun. **46**(32), 5969–5971 (2010)
392. M. Imit, P. Imin, A. Adronov, π-conjugated polymers with pendant coumarins: Design, synthesis, characterization, and interactions with carbon nanotubes. Can. J. Chem. **94**, 759–768 (2016)
393. T. Lee, J. Park, K. Kim, et al., Facile tuning of a polymeric dispersant for single-walled carbon nanotube dispersion. RSC Adv. **5**, 69410–69417 (2015)
394. Z. Song, J. Dai, S. Zhao, Y. Zhou, et al., Aqueous dispersion of pristine single-walled carbon nanotubes prepared by using a vinylimidazole-based polymer dispersant. RSC Adv. **4**, 2327–2338 (2014)
395. G. Clave, G. Delport, C. Roquelet, et al., Functionalization of carbon nanotubes through polymerization in micelles: A bridge between the covalent and Noncovalent methods. Chem. Mater. **25**, 2700−2707 (2013)
396. L. Xu, Z. Ye, G. Siemann, Z. Noncovalent solubilization of multi-walled carbon nanotubes in common low-polarity organic solvents with branched Pd-diimine polyethylenes: Effects of polymer chain topology, molecular weight and terminal pyrene group. Polymer **55**, 3120–3129 (2014)
397. J. Mao, Q. Liu, X. Lv, Z. Liu, Y. Huang, Y. Ma, Y. Chen, S. Yin, A water-soluble hybrid material of single-walled carbon nanotubes with an amphiphilic poly(phenyleneethynylene): Preparation, characterization, and photovoltaic properties. J. Nanosci. Nanotechn. **7**(8), 2709–2718 (2007)
398. J. Chen, H. Liu, Polymer and method for using the polymer for solubilizing nanotubes, in *US7244407*, (2007)
399. G.J. Wang, S.Z. Huang, Y. Wang, L. Liu, J. Qiu, Y. Li, Synthesis of water-soluble single-walled carbon nanotubes by RAFT polymerization. Polymer **48**(3), 728–733 (2007)
400. Y.-Z. You, C.-Y. Hong, C.-Y. Pan, Directly growing ionic polymers on multi-walled carbon nanotubes via surface RAFT polymerization. Nanotechnology **17**(9), 2350–2354 (2006)
401. G. Xu, W.-T. Wu, Y. Wang, W. Pang, P. Wang, Q. Zhu, F. Lu, Synthesis and characterization of water-soluble multiwalled carbon nanotubes grafted by a thermoresponsive polymer. Nanotechnology **17**(10), 2458–2465 (2006)
402. Y. Ni, H. Hu, E.B. Malarkey, B. Zhao, V. Montana, R.C. Haddon, V. Parpura, Chemically functionalized water soluble single-walled carbon nanotubes modulate neurite outgrowth. J. Nanosci. Nanotechn. **5**(10), 1707–1712 (2005)
403. B. Zhao, H. Hu, A. Yu, D. Perea, R.C. Haddon, Synthesis and characterization of water soluble single-walled carbon nanotube graft copolymers. J. Amer. Chem. Soc. **127**(22), 8197–8203 (2005)
404. H. Zhang, H.X. Li, H.M. Cheng, Water-soluble multiwalled carbon nanotubes functionalized with sulfonated polyaniline. J. Phys. Chem. B **110**(18), 9095–9099 (2006)
405. Y. Ma, S.R. Ali, L. Wang, P.L. Chiu, R. Mendelsohn, H. He, In situ fabrication of a water-soluble, self-doped polyaniline nanocomposite: The unique role of DNA functionalized single-walled carbon nanotubes. J. Amer. Chem. Soc. **128**(37), 12064–12065 (2006)
406. Z. Li, Y. Dong, M. Häussler, J.W.Y. Lam, Y. Dong, L. Wu, K.S. Wong, B.Z. Tang, Synthesis of, light emission from, and optical power limiting in soluble single-walled carbon nanotubes functionalized by disubstituted polyacetylenes. J. Phys. Chem. B **110**(5), 2302–2309 (2006)
407. B.-Z. Tang, H. Xu, *Preparation, Alignment, and Optical Properties of Soluble Polyphenylacetylene-Wrapped Carbon Nanotubes.*, Preprints of Second East Asian Polymer

Conference, Hong Kong, 12-16 Jan. 1999 (Hong Kong University of Science and Technology, Hong Kong, 1999), pp. 141–148

408. B.Z. Tang, H. Xu, Preparation, alignment, and optical properties of soluble poly (phenylacetylene)-wrapped carbon nanotubes. Macromolecules **32**, 2569–2576 (1999)
409. J. Chen, H. Liu, W. Weimer, M. Halls, D. Waldeck, G. Walker, Noncovalent engineering of carbon nanotube surfaces by rigid, functional conjugated polymers. J. Amer. Chem. Soc. **124** (31), 9034–9035 (2002)
410. Z. Wang, Q. Liu, H. Zhu, H. Liu, Y. Chen, M. Yang, Dispersing multi-walled carbon nanotubes with water-soluble block copolymers and their use as supports for metal nanoparticles. Carbon **45**(2), 285–292 (2007)
411. K.A.S. Fernando, Y. Lin, B. Zhou, M. Grah, R. Joseph, L.F. Allard, Y.-P. Sun, Poly(ethylene-co-vinyl alcohol) functionalized single-walled carbon nanotubes and related nanocomposites. J. Nanosci. Nanotechn. **5**(7), 1050–1054 (2005)
412. M.C. Paiva, B. Zhou, K.A.S. Fernando, Y. Lin, J.M. Kennedy, Y.-P. Sun, Mechanical and morphological characterization of polymer-carbon nanocomposites from functionalized carbon nanotubes. Carbon **42**(14), 2849–2854 (2004)
413. N. Zhang, J. Xie, V.K. Varadan, Soluble functionalized carbon nanotube/poly(vinyl alcohol) nanocomposite as the electrode for glucose sensing. Smart Mater. Struct. **15**(1), 123–128 (2006)
414. B. Zhou, Y. Lin, L.M. Veca, K.A.S. Fernando, B.A. Harruff, Y.-P. Sun, Luminescence polarization spectroscopy study of functionalized carbon nanotubes in a polymeric matrix. J. Phys. Chem. B **110**(7), 3001–3006 (2006)
415. J.Y. Choi, S.W. Han, W.S. Huh, L.S. Tan, J.B. Baek, In situ grafting of carboxylic acid-terminated hyperbranched poly(ether-ketone) to the surface of carbon nanotubes. Polymer **48** (14), 4034–4040 (2007)
416. L. Qu, L.M. Veca, Y. Lin, A. Kitaygorodskiy, B. Chen, A.M. McCall, J.W. Connell, Y.-P. Sun, Soluble nylon-functionalized carbon nanotubes from anionic ring-opening polymerization from nanotube surface. Macromolecules **38**, 10328–10331 (2005)
417. C. Liu, B. Liu, M.B. Chan-Park, Synthesis of polycaprolactone-polyimide-polycaprolactone triblock copolymers via a 2-step sequential copolymerization and their application as carbon nanotube dispersants. Polym. Chem. **8**, 674–681 (2017)
418. M. Loginov, N. Lebovka, E. Vorobiev, Laponite assisted dispersion of carbon nanotubes in water. J. Colloid Interface Sci. **365**(1), 127–136 (2012)
419. O. Yaroshchuk, S. Tomylko, O. Kovalchuk, N. Lebovka, Liquid crystal suspensions of carbon nanotubes assisted by organically modified Laponite nanoplatelets. Carbon **68**, 389–398 (2014)
420. A. Ivanoska-Dacikj, G. Bogoeva-Gaceva, A. Buzarovska, Clay improved dispersion of carbon nanotubes in different solvents. *Contributions, Sec. Nat. Math. Biotech. Sci.* MASA **36**(1), 5–10 (2015)
421. K.Y. Leong, N. Mohd Hanafi, R. Risby Mohd Sohaimi, N.H. Amer, The effect of surfactant on stability and thermal conductivity of carbon nanotube based nanofluids. Thermal Sci. **20** (2), 429–436 (2016)
422. A. Amiri, M. Shanbedi, H. Eshghi, et al., Highly dispersed Multiwalled carbon nanotubes decorated with ag nanoparticles in water and experimental investigation of the Thermophysical properties. J. Phys. Chem. C **116**(5), 3369–3375 (2012)
423. W. Rashmi, A.F. Ismail, I. Sopyan, et al., Stability and thermal conductivity enhancement of carbon nanotube nanofluid using gum Arabic. J. Exp. Nanosci. **6**(6), 567–579 (2011)
424. B. Wang, Y. Han, K. Song, T. Zhang, The use of anionic gum arabic as a dispersant for multi-walled carbon nanotubes in an aqueous solution. J. Nanosci. Nanotechnol. **12**(6), 4664–4669 (2012)
425. A. Indhuja, K.S. Suganthi, S. Manikandan, K.S. Rajan, Viscosity and thermal conductivity of dispersions of gum arabic capped MWCNT in water: Influence of MWCNT concentration and temperature. J. Taiwan Inst. Chem. Eng. **44**(3), 474–479 (2013)

426. M.T. Kim, H.S. Park, D. Hui, K.Y. Rhee, Carbon nanotube modification using gum arabic and its effect on the dispersion and tensile properties of carbon nanotubes/epoxy nanocomposites. J. Nanosci. Nanotechnol. **11**(8), 7369–7373 (2011)
427. L. Lu, W. Chen, Large-scale aligned carbon nanotubes from their purified, highly concentrated suspension. ACS Nano **4**(2), 1042–1048 (2010)
428. M.G. Adsul, D.A. Rey, D.V. Gokhale, Combined strategy for the dispersion/dissolution of single walled carbon nanotubes and cellulose in water. J. Mater. Chem. **21**(7), 2054–2056 (2011)
429. O. Rochez, G. Zorzini, J. Amadou, M. Claes, A. Richel, Dispersion of multiwalled carbon nanotubes in water by lignin. J. Mater. Sci. **48**(14), 4962–4964 (2013)
430. N.-Y. Teng, I. Dallmeyer, J.F. Kadla, Effect of softwood Kraft lignin fractionation on the dispersion of Multiwalled carbon nanotubes. Ind. Eng. Chem. Res. **52**(19), 6311–6317 (2013)
431. S.M. Goodman, N. Ferguson, A.B. Dichiara, Lignin-assisted double acoustic irradiation for concentrated aqueous dispersions of carbon nanotubes. RSC Adv. **7**, 5488–5496 (2017)
432. B. Shi, X. Zhuang, X. Yan, J. Lu, H. Tang, Adsorption of atrazine by natural organic matter and surfactant dispersed carbon nanotubes. J. Environ. Sci. **22**(8), 1195–1202 (2010)
433. X. Zhou, L. Shu, H. Zhao, et al., Suspending multi-walled carbon nanotubes by humic acids from a peat soil. Environ. Sci. Technol. **46**(7), 3891–3897 (2012)
434. K. Kristiansen, H. Zeng, P. Wang, J.N. Israelachvili, Microtribology of aqueous carbon nanotube dispersions. Adv. Funct. Mater. **21**(23), 4555–4564 (2011)
435. O.V. Kharissova, H.V. Rasika Dias, B.I. Kharisov, B. Olvera Pérez, V.M. Jiménez Pérez, The greener synthesis of nanoparticles. Trends Biotechnol. **31**(4), 240–248 (2013)
436. A.J. Uddin, A. Watanabe, Y. Gotoh, T. Saito, M. Yumura, From “strong” to “much stronger”: Utilization of green tea extract dispersant for SWCNT-reinforced polymer composites. Macromol. Mater. Eng. **297**(11), 1114–1123 (2012)
437. A.J. Uddin, A. Watanabe, Y. Gotoh, T. Saito, M. Yumura, Green tea-aided dispersion of singlewalled carbon nanotubes in non-water media: Application for extraordinary reinforcement of nanocomposite fibers. Text. Res. J. **82**(9), 911–919 (2012)
438. J. Cao, W. Dun, H. Qu, Evaluation of the addition of various surfactant-suspended carbon nanotubes in MEEKC with an in situ-synthesized surfactant system. Electrophoresis **32**(3–4), 408–413 (2011)
439. S. Yuan, L. Zeng, Y. Zhuang, Q. Hou, M. Song, Functionalized single-walled carbon nanotubes for the improved solubilization and delivery of curcumin. Fullerenes, Nanotubes and Carbon Nanostructures. **24**(1), 13–19 (2016)

Chapter 4
Biological/Biochemical Methods

4.1 Sugars and Biomolecules

Soluble CNTs and especially water-soluble ones, functionalized with biomolecules, could get many applications in medicinal chemistry; that is why a host of efforts has been dedicated to CNT treatment with biologically active species, *sugars*, and their derivatives. Thus, as a first step before incorporation into poly(amide–imide) matrix, MWCNTs were modified with glucose and fructose (reaction scheme 4.1) carbohydrates as biomolecules to obtain Gl-MWCNTs and Fr-MWCNTs, leading to water-soluble and biocompatible products [1].

O.V. Kharissova, B.I. Kharisov, *Solubilization and Dispersion of Carbon Nanotubes*, DOI 10.1007/978-3-319-62950-6_4

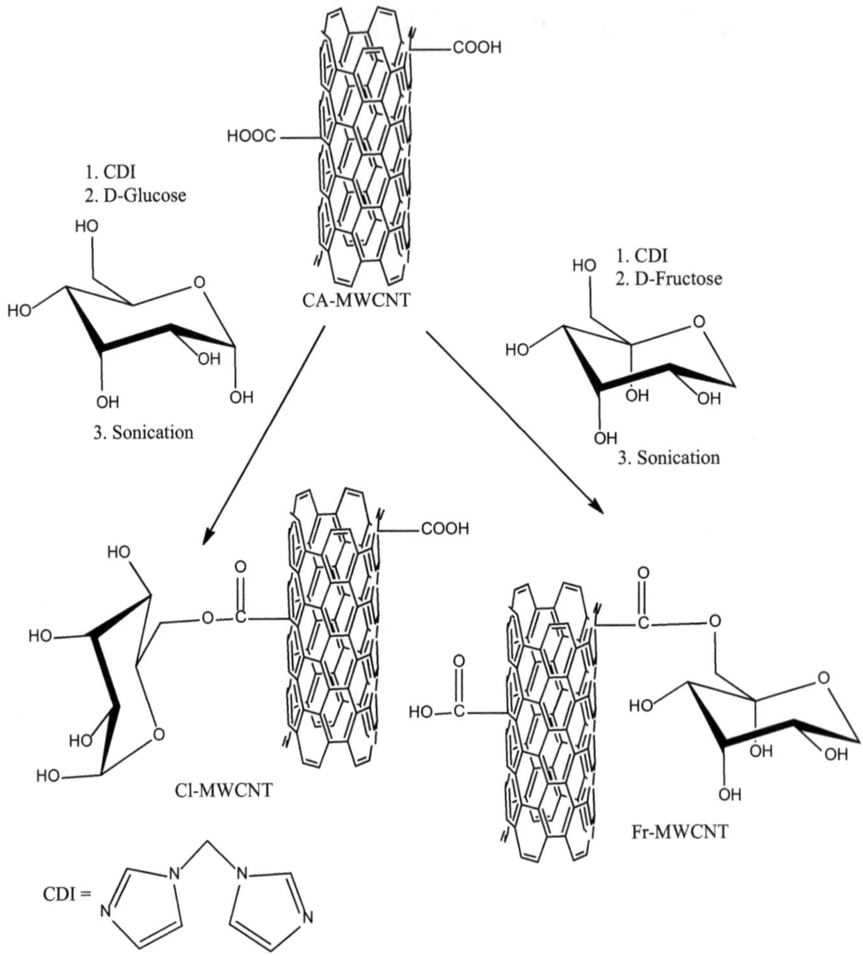

A covalent microwave-assisted functionalization of pristine SWCNTs directly with three *sugar azides* (reaction scheme 4.2), 2,3,4,6-tetra-O-acetyl-β-D-glucopyranosyl, 2,3,4,6-tetra-O-acetyl-β-D-galactopyranosyl, or 2,3,4,6-tetra-O-acetyl-β-D-mannopyranosyl azide [2], was carried out for SWCNTs prepared with the HiPCO method (high-pressure carbon monoxide process). Deacetylation of the functionalized tubes by sodium methoxide yielded nitrogen-linked, sugar-functionalized CNTs that formed stable dispersions in water. The water solubility was found to be highest for galactopyranosyl and lowest for gluco- and mannopyranosyl derivatives varying between 0.6 and 1.3 mg/ml. Based on the water solubility, it was calculated that ca. 16–35% of functionalized SWCNTs dispersed in water depending on the reactant sugar azide from which galactopyranosyl azide was most reactive. These functionalized SWCNTs were not soluble in ethanol, methanol, or ethyl acetate and thereby showed similar solubility properties as the corresponding untreated sugars.

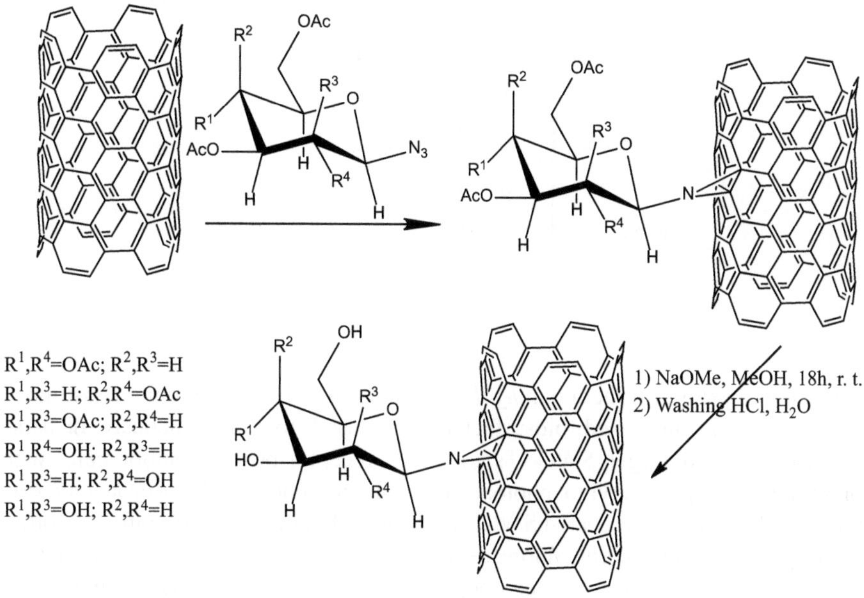

Functionalization of HiPCO SWCNTs by sugar azides under microwave conditions yielding nitrogen-linked functionalized CNTs. Deacetylation of the product with sodium methoxide produces water-dispersible nitrogen-linked b-D-pyranosyl-functionalized carbon nanotubes.

A variety of biomolecules have been applied for SWCNT functionalization and dispersion, especially DNA (see below); these investigations are obviously directed mainly to biomedical purposes. Among these compounds, in order to get stable high-concentration SWCNT suspensions in various alcohols, *cholic acid* **4.1** (one of the most important human bile acids) was applied as a useful additive [3, 4] as well as its sodium salt [5]. SWCNT solubility in alcohols was found to depend on its concentration. The best solvent for all alkanol–cholic acid mixtures was found to be EtOH-surfactant mixture (0.018 mol/kg), decreasing in the row ethanol–isopropanol–*t*-butanol–butanol–propanol.

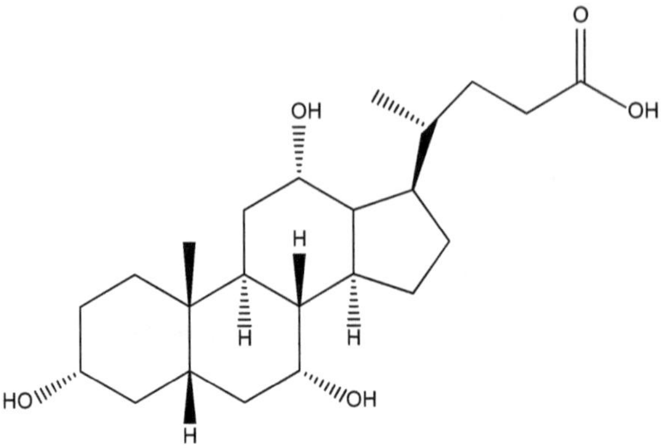

4.1

A single, yet multifunctional, *hyaluronic acid* (**4.2**, HA, an anionic, nonsulfated glycosaminoglycan)-based biosurfactant was used to simultaneously disperse nanocarbons and target SWCNTs to CD44 receptor-positive tumor cells with prompt uptake [6]. In vivo photoacoustic, fluorescence, and positron emission tomography imaging of coated SWCNTs displayed high tumor-targeting capability while providing long-term, fluorescence molecular imaging of targeted enzyme events. A HA derivative, pyrene-conjugated hyaluronan (Py-HA), facilitated the exfoliation of low-dimensional nanomaterials including graphite, hexagonal boron nitride (h-BN), and molybdenum disulfide (MoS_2) and the dispersion of CNTs [7] and carbon nano-onions (CNOs) in water with the assistance of sonication.

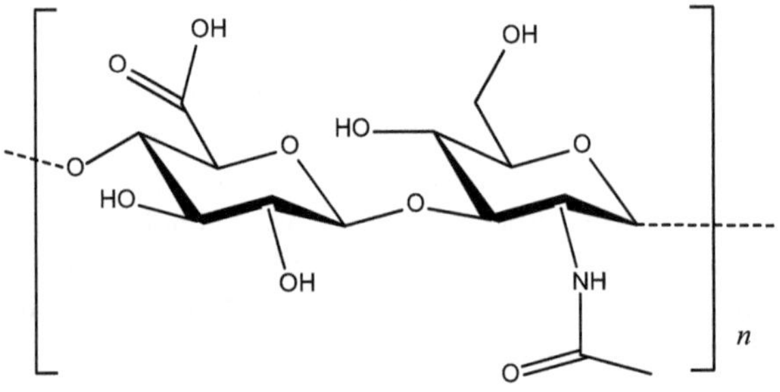

4.2
HA, Hyaluronan (hyaluronic acid or hyaluronate)

A dispersion of SWCNTs in an aqueous medium was prepared [8] by functionalizing the SWCNT with *D-glucosamine* **4.3**. The grafting resulted in a good dispersion of the SWCNTs in water of an amount less than 1 mg·ml^{-1}. β-1,3-Glucan polysaccharides have potential for producing gene carriers and bio-nanomaterials. Carboxylic curdlan (CurCOOH) bearing the β-1,3-polyglucuronic acid structure was prepared [9] from one of them, β-1,3-glucan polysaccharide curdlan (Cur), by one-step oxidation using a 4-acetamido-TEMPO/NaClO/$NaClO_2$ system as the oxidant. Its further complexation with SWCNTs resulted in a water-soluble 1D architecture, which formed dispersion in aqueous solution, stable for several months and much more stable than SWCNT complexes of the similar negatively charged polyacrylic acid (PAA) and polymethacrylic acid (PMAA). It was shown that in the complex, SWCNTs are effectively wrapped by a small amount of CurCOOH, enabling them to avoid electrostatic repulsion.

Also, MWCNTs were well dispersed in an aqueous solution of the cyanobacterial polysaccharide, *sacran*, with an ultrahigh molecular weight > 10 million g/mol [10]. MWCNT powder was put into aqueous solutions of various polysaccharides including sacran and was dispersed under sonication. Cryogenic transmission electron microscopic (Cryo-TEM) studies directly demonstrated the existence of MWCNTs in the supernatant, and high-resolution TEM observation revealed that MWCNTs covered by sacran chains made their efficient dispersion in water.

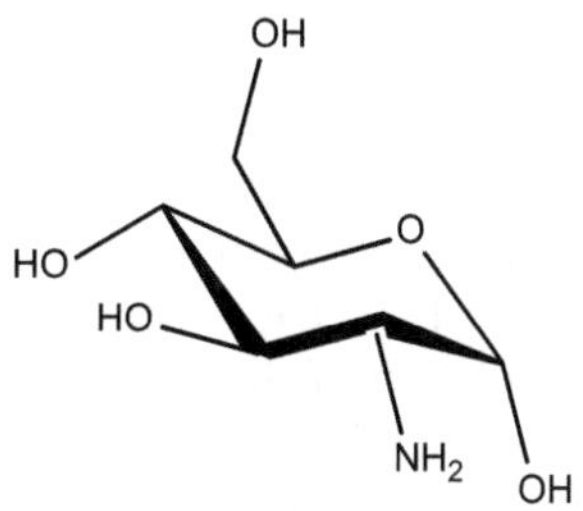

4.3
D-glucosamine

Carbohydrates [11] were also reported for CNT dispersion. Thus, a *starch* (**4.4**)/ carboxylated multiwall carbon nanotube (CCNT) composite (CCNT–starch) was prepared [12] by covalently grafting a natural polymer starch onto the surfaces of CCNT. It was revealed that the covalent bonds between -OH groups of soluble starch and CCNT were formed in CCNT–starch; CCNTs were covered with the grafted starch about 7.7 and 12.8 wt.% in CCNT–starch, respectively. The grafted starch facilitated the dispersion of CCNT–starch in water and chitosan films because of the hydrophilic polysaccharide structure of starch components. CCNT–starch containing 12.8 wt.% starch in water could form homogeneous solutions without obvious aggregation of CCNTs. In addition, phase transfer of MWCNTs from an aqueous phase into an organic phase was achieved using the hydrogen-bonding interactions between the disaccharide group of a sugar-based amphiphile and oxygenated functional groups on the surface of MWCNTs [13]. The dispersion of MWCNTs in a wide range of organic solvents was enabled.

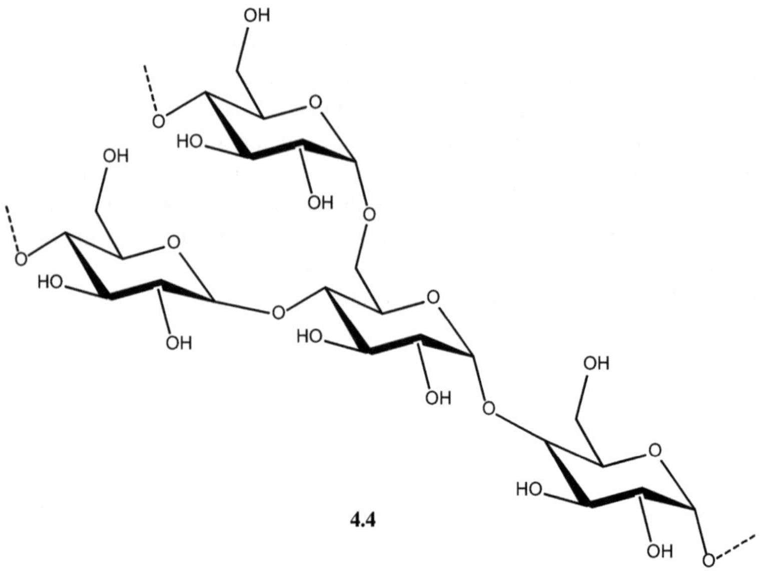

4.4

Starch

The *chitosan* **4.5**, a very popular object in the nanotechnology, has been used to improve CNT solubility [14–16]. Thus, the dispersion of CNTs in water of different pH and in chitosan aqueous solution of three acids, acetic acid, formic acid, and hydrochloric acid, was investigated [17]. Chitosan is soluble in water of pH ≤ 3 and could well disperse untreated CNT and acid-treated CNT, both of which had poor dispersion in water of pH ≤ 3. Among the three acids, hydrochloric acid gave the smallest particle size of the CNT dispersion. Without chitosan, the dispersibility of the acid-treated CNT in aqueous solutions of three acids was in the order of acetic acid > formic acid > hydrochloric acid. Debundling and selective dispersion of semiconducting SWCNTs were demonstrated [18] using a neutral pH water-soluble chitosan derivative, N-acetylated chitosan **4.6** (NACHI), which was synthesized by controlled N-acetylation of chitosan using acetic anhydride (reaction scheme 4.3). The SWCNT-NACHI supernatant solution demonstrated semiconductor-enriched property owing to the preferential adsorption of N-groups of the NACHI on semiconducting nanotubes with a fairly weak charge transfer.

Another chitosan derivative, neutral pH water-soluble chitosan-hydroxyphenyl acetamide (CHPA), prepared by functionalizing the amino groups of chitosan with 4-hydroxyphenyl acetic acid, was also found to be an efficient biocompatible dispersant to effectively debundle and individually disperse SWCNTs in a neutral aqueous solution [19]. A so-called wet-grinding-assisted ultrasonication (GU) method was reported [20], in which wet grinding of MWCNTs in chitosan solution was carried out before ultrasonication. Interestingly, the dispersion quality of chitosan/MWCNT suspension prepared by wet-grinding-assisted ultrasonication is much better than that by ultrasonication or wet grinding alone. The authors also found that wet grinding could improve the water wettability of MWCNTs and eliminate the barrier of air layer around MWCNTs to ultrasonic waves.

A new amperometric biosensor for hydrogen peroxide was developed [21] based on cross-linking horseradish peroxidase (HRP) by glutaraldehyde with multiwall carbon nanotubes/chitosan (MWCNTs/chitosan) composite film coated on a glassy carbon electrode. The biosensor had good repeatability and stability for the determination of H_2O_2. In general, the CNTs–CS composite provides a suitable biosensing matrix due to its good conductivity, high stability, and good biocompatibility [22]. Additionally to H_2O_2, these composites can detect different substrates (ABTS, catechol, and O_2), possessing high affinity and sensitivity, durable long-term stability, and facile preparation procedure. Chitosan–CNT composites were also obtained in the form of fibers with a wet-spinning method [23].

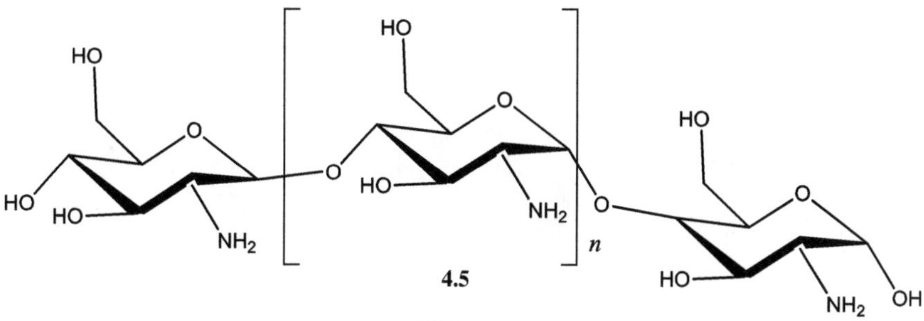

4.5

Chitosan

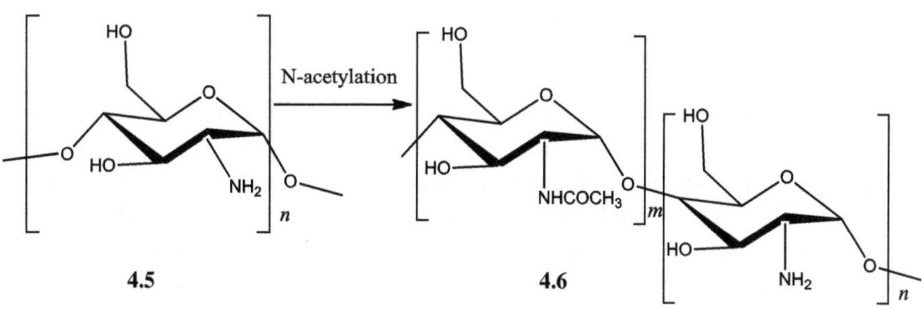

Controlled N-acetylation of chitosan using acetic anhydride.

An environmentally friendly group of dispersants, *cyclodextrins* **4.7–4.9** (CD), was revealed as a preferable alternative to functionalize CNTs without changing their pristine structure and properties. MWCNTs were functionalized using β-CD **4.8** [24]. Different concentrations of β-CD served as the dispersants, and different nonaqueous media (ethanol, acetic acid, and water) were studied to achieve required degree of functionalization for MWCNTs, showing that the degree of functionalization increases with increasing β-CD ratios. Its derivative, diaminofunctionalized β-CD, was covalently bound [25] with carboxylic acid-functionalized MWCNTs via amide linkages using a water-soluble condensation agent at r.t. The obtained product, in which ~70 wt.% β-cyclodextrin was attached on the surface of MWCNTs, denoted as β-cyclodextrin-modified MWCNTs was also highly dispersible in an aqueous medium.

Also, the functionalized SWCNTs, having substantial solubility in dimethyl sulfoxide and dimethylformamide, were obtained from oxidized SWCNTs by acyl halogenation with thionyl chloride and dimethylformamide and further interaction with methyl-β-cyclodextrin and octadecylamine (ODA) simultaneously [26]. Such systems offer considerable advantages over polymer-based composites due to their biocompatibility and noncovalent coupling which can potentially preserve the unique properties of the tubes [27]. The mechanism of interaction for such systems has been proposed to be dominated by hydrophobic and hydrophilic interactions along the surface of the tube. Pyrene-modified CD was found to contribute to CNT dispersion via π–π stacking [28].

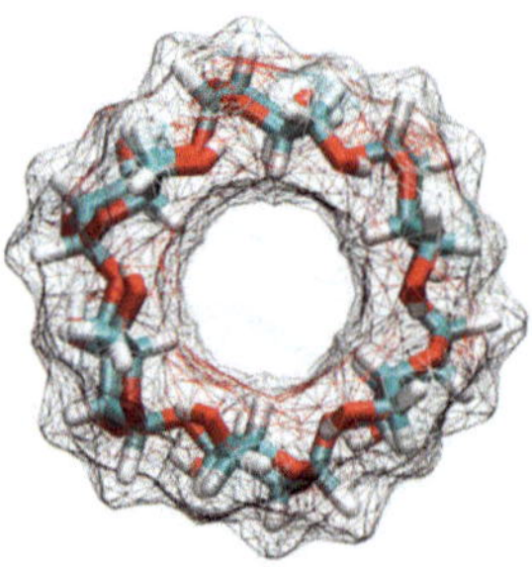
4.7 (α-CD)

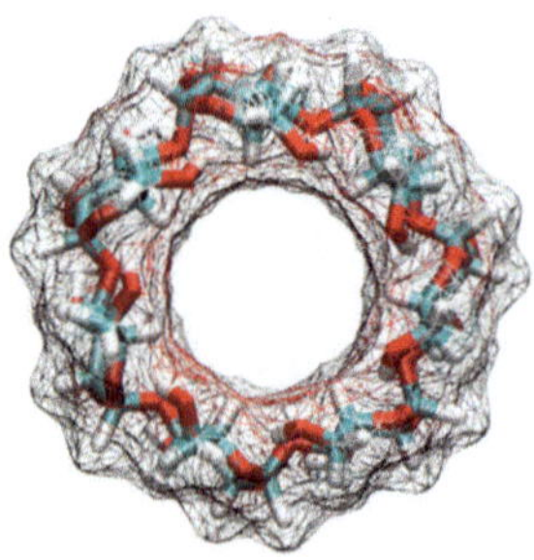
4.8 (β-CD)

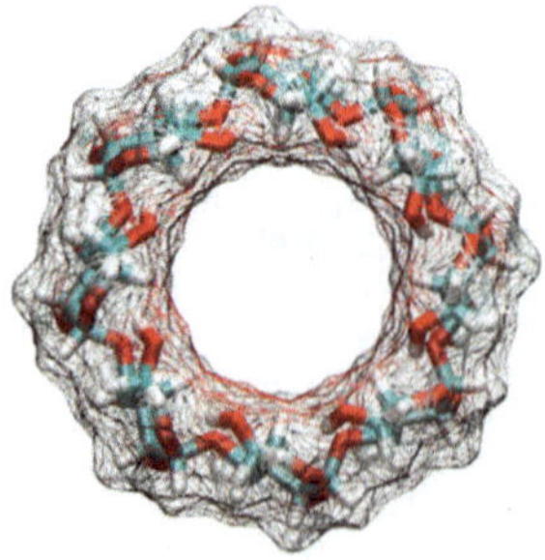
4.9 (γ-CD)

Lysine is sometimes used for CNT dispersion [29]. Thus, MWCNTs were first functionalized with lysine **4.10** under microwave irradiation [30], and then the water-dispersed CNTs were obtained by diazonium-assisted functionalization. A diazonium reaction occurred between lysine molecules and CNT, resulting in the attachment of lysine to the nanotube surface, so covalent functionalization procedure could significantly increase aqueous solubility. Lysine analogue, poly(-ε-benzyloxycarbonyl-L-lysine) (PZLL), prepared by ring-opening polymerization of ε-carbobenzoxy-L-lysine N-carboxyanhydride (Z-L-Lys NCA) [31], after the removal of the benzyloxycarbonyl group, led to water-soluble poly-L-lysine (**4.11**, PLL). When the mass ratio of MWCNTs to PLL was 1:50 and the pH value of the water solution of PLL is 5, MWCNTs could be well dispersed in water. Aqueous solution of high-molecular-weight poly-L-lysine is also capable to disperse CNTs [32].

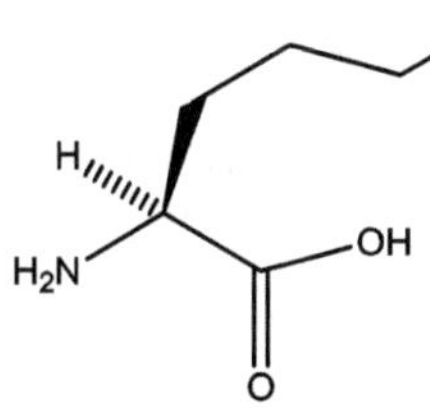

4.10
Lysine

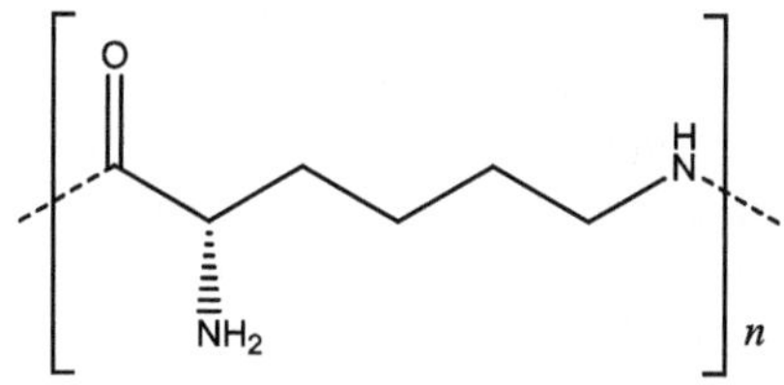

4.11
Poly-L-lysine

Biologically active *peptides* can be easily linked through a stable covalent bond to CNTs [33]. Thus, a CNT-bound peptide from the foot-and-mouth disease virus retained the structural integrity; this is immunogenic, eliciting antibody responses of the right specificity [34]. In order to explore the utilization of CNTs in solvent and the affinities of CNTs for different peptides, binding free energies of peptides to SWCNTs were calculated and analyzed [35]. Simulation of interactions between different peptides and SWCNTs was carried out using molecular dynamics methods, and estimation of the binding free energies of peptides onto the outer surface of the SWCNTs was based on thermodynamics theory. A good agreement between theoretical and experimental results was observed. A composite material was developed with SWCNTs and artificially designed peptides [36], which were designed to form a β-sheet structure that would be suitable for wrapping SWCNTs. The composite SWCNT-peptide showed good dispersibility in aqueous media and was considerably stable even in the absence of an excess amount of peptide in the media. The authors suggested the potential of the SWCNT-peptide composite as a molecular platform on which a desirable structure and/or function can be constructed for biomedical and industrial application.

In a related research [37], the noncovalent modification of MWCNTs immersed in aqueous solution using the ionic-complementary peptide EFK16-II orients hydrophilic functional groups toward the solution phase and enables them to form highly stable dispersions in water. This stability can be attributed to the electrostatic repulsion between self-assembled peptides on the MWCNTs. Tissue culture plates previously contacted with EFK16-II-modified MWCNTs were shown to have enough biocompatibility for growth and attachment of cells. The biocompatibility and enhanced electrical conductivity that should result from the modification with these EFK16-II-MWCNT suspensions open up their use in a number of potential biomedical applications such as the design of bioelectrode interfaces and fabrication of biosensors with high sensitivity.

Cholesterol-based peptide carboxylates were also found to be efficient dispersing agents for SWCNTs as well as graphene in water [38]. The dipeptide carboxylate comprising of two alanine residues showed 80% SWCNTs dispersion which is ~2 fold higher than that obtained by using the common surfactant, SDBS. The dipeptide amphiphiles also efficiently dispersed the 2D allotrope of carbon, graphene, in water. Acidification of the nanohybrids with HCl converted the carboxylates to the water-insoluble carboxylic acids leading to the precipitation of carbon nanomaterials. Most importantly, addition of an equivalent amount of NaOH resulted in the restoration of stable aqueous dispersion of SWCNT/graphene.

Comparing [39] dispersion capacity of peptides, DNA, low-molecular-weight surfactants, and a water-soluble polymer, the peptide aptamer, A2 (IFRLSWGTYFS, Fig. 4.1), exhibited the highest dispersion capability below the critical micelle concentration at a concentration of 0.02 w/v%. Aromatic groups of this peptide aptamer and SWCNT walls provide binding capacity due to $\pi-\pi$ interaction between them. In addition, cholesterol-based dipeptide carboxylates **4.12–4.16** were applied for the pH-responsive *reversible dispersion and precipitation* of SWCNTs (Fig. 4.2) in water specifically at tumorigenic environmental pH (6.0–6.5) showing an excellent pH-responsive drug release in this range [40]. These nanocomposites possess higher ability to fight with cancer cells in comparison with native drug, and their action is selective (normal cells are less affected). Other lipids are reported in [41, 42].

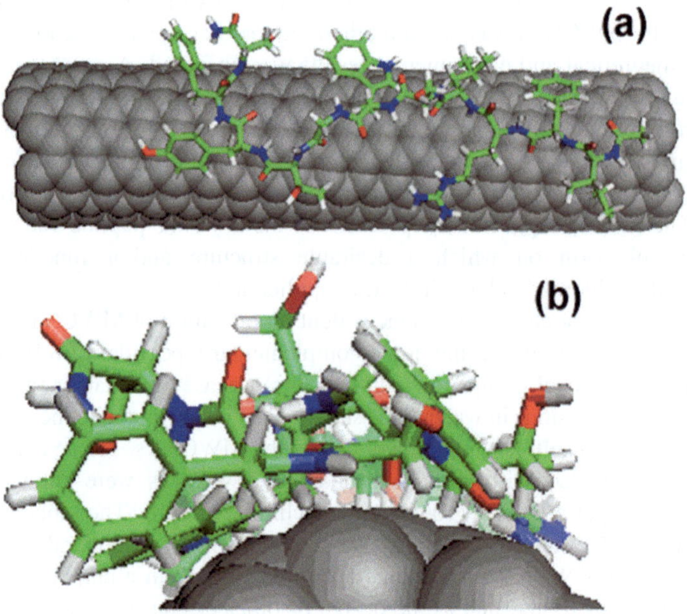

Fig. 4.1 Representative conformation of A2 on a SWCNT. (**a**) Side view and (**b**) front view (Reproduced with permission of *American Chemical Society*)

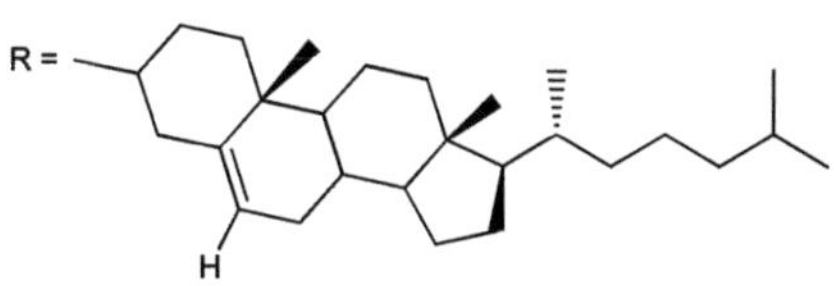

% dispersion

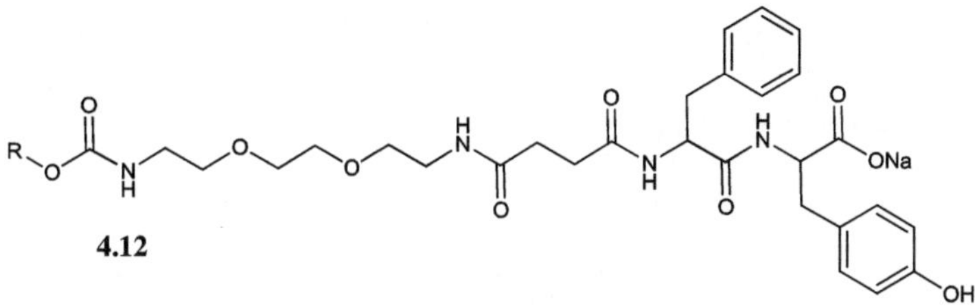

4.12 67%

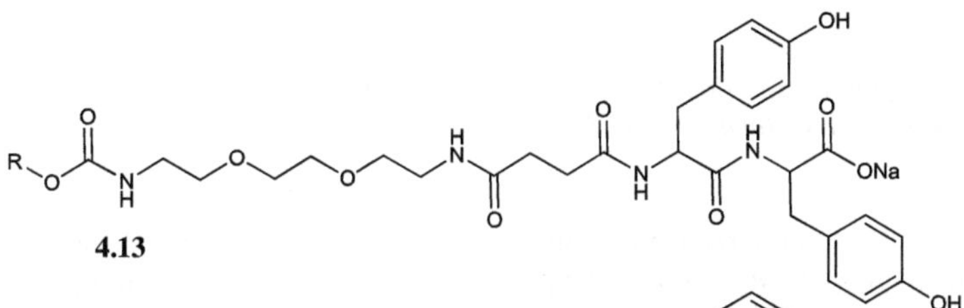

4.13 72%

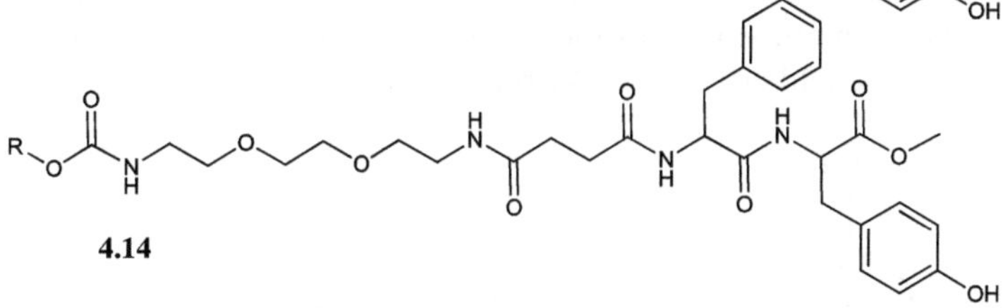

4.14 insoluble in water

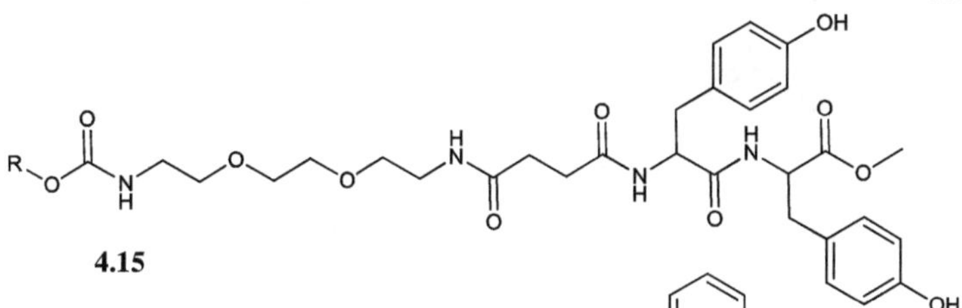

4.15 insoluble in water

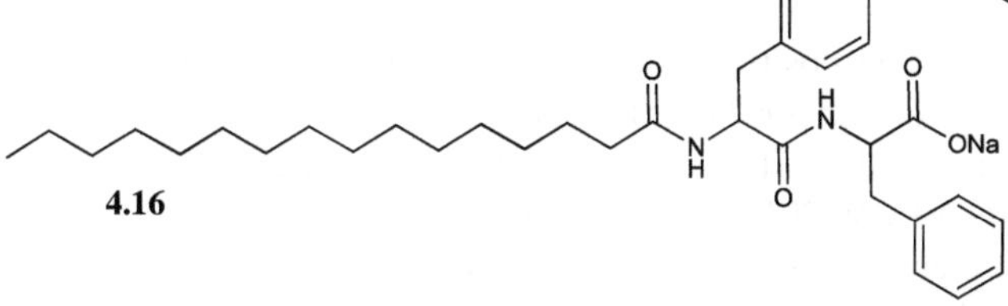

4.16 38%

Structures of synthesized amphiphiles used for dispersing SWCNT and the corresponding % dispersion by the amphiphiles

An intriguing strategy to manipulate reversibly the aggregation and dispersion of CNTs in solution via formation of intermolecular i-motif (four-stranded C-quadruplex) structures in a pH-dependent manner was offered [43]. Firstly, single-stranded (ss) DNAs containing two stretches of *cytosine* **4.17** (C-)rich

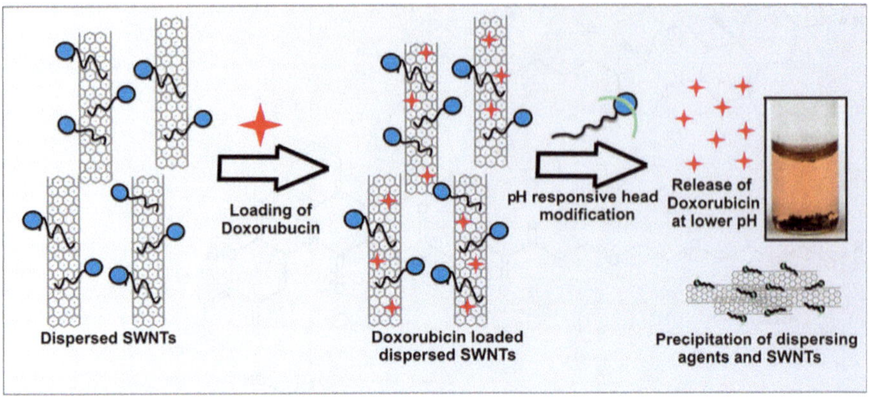

Fig. 4.2 Schematic representation of pH-responsive SWCNT precipitation and drug release. (Reproduced with permission of *John Wiley & Sons*)

domains are covalently linked to carbon nanotubes. At pH 8.0, DNAs are at random coil state, which enhance the dispersion of MWCNTs in water; after changing pH to 5.0, the intermolecular i-motif structures formed by the C-rich ssDNAs on neighboring carbon nanotube could drive the MWCNT aggregate.

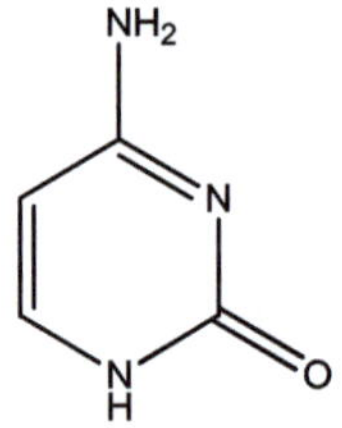

4.17
Cytosine

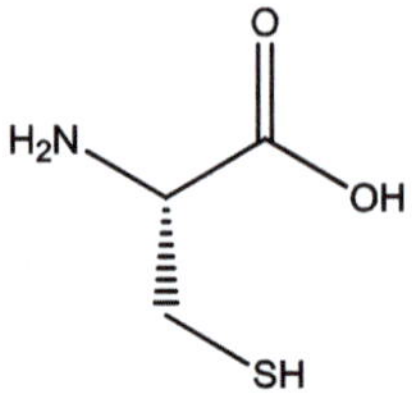

4.18
Cysteine

The dispersion stability, viscosity, shear stress, and thermal conductivity of MWCNTs in the presence of *gum arabic* (MWCNT-GA, see also previous section on natural products) as well as functionalized MWCNT with *cysteine* **4.18** (MWCNT-Cys) and silver (MWCNT-Ag) were investigated [44]. To realize good dispersibility and high thermal enhancement, MWCNTs were first functionalized by cysteine and then silver ions covalently attached through the thiol group of cysteine. For the enhancement of thermal conductivity, the results suggest that the covalent functionalization by Ag is more effective than noncovalent functionalization. *Polyhistidine* (Polyhis) [45] was also used to efficiently disperse MWCNTs. The optimum dispersion MWCNT-Polyhis was obtained by sonicating for 30 min 1.0 mg·mL^{-1} MWCNTs in 0.25 mg·mL^{-1} Polyhis solution prepared in 75:25 (v/v) ethanol/0.200M acetate buffer solution pH 5.00. The glassy carbon electrode modified with MWCNT-Polyhis dispersion was used to quantify dopamine or uric acid at nanomolar levels, even in the presence of large excess of ascorbic acid.

DNA functionalization of CNTs holds interesting prospects [46] in various fields including solubilization in aqueous media, nucleic acid sensing, gene therapy, and controlled deposition on conducting or semiconducting substrates [47] or for determination of polymerase chain reaction (PCR) efficiency [48]. Among a variety of other investigations, double-stranded calf thymus DNA (deoxyribonucleic acid from calf thymus, dsDNA) was found to disperse bamboo-like MWCNTs (bCNT) [49]. The drastic treatment for dispersing the bCNT (45-min sonication in a 50% (v/v) ethanol–water solution) produced a partial denaturation and a decrease in the length of dsDNA that facilitates the dispersion of CNT and made possible an efficient electron transfer of guanine residues to the electrode. The electron transfer of redox probes and guanine residues was more efficient at GCE modified with bCNT dispersed in dsDNA than at GCE modified with hollow CNT (hCNT) dispersed in dsDNA, demonstrating the importance of the presence of bCNT. Stable dispersions of single-walled and nonassociated carbon nanotubes in aqueous lysozyme solution were investigated [50] by analyzing the stabilizing effect of both protein concentration and pH. Lysozyme is presumably adsorbed onto the SWCNT surface through hydrophobic and π–π interactions. It was observed that electrophoretic mobility and dielectric increment increase with the amount of added lysozyme, up to saturation. This occurs at a definite LYS/SWCNT mass ratio. At higher lysozyme content, depletion phenomena, opposing to the stabilization process, occurred.

Also, a solid-state mechanochemical reaction was used for obtaining DNA-wrapped nanotubes of both MWCNTs and SWCNTs, resulting a high aqueous solubility of the products with a stability of >6 months [51]. It was established that the nanotubes were cut into shorter lengths and were fully covered with DNA, in which the product is intact. DNA–SWCNTs were also prepared by another method, the *layer-by-layer technique* [52], where poly(diallyldimethylammonium) (PDDA), a positively charged polyelectrolyte, and DNA as a negatively charged counterpart macromolecule are alternatively deposited on the water-soluble oxidized SWCNTs (reaction scheme 4.4). (We note that the same technique was applied also to fabricate a thin film from RNA-dissolved SWCNT solutions [53].) As a representative application of the obtained DNA composite, an electrode modified by DNA/PDDA/SWCNT particles exhibited larger electrocatalytic oxidation current in an aqueous solution of Ru-$(bpy)_3^{2+}$; moreover, DNA–CNTs can be used as sensors for NO_2 detection. Other recent achievements on DNA immobilization on the CNTs are reported in [54–58].

SWCNTs $\xrightarrow{H_2SO_4/HNO_3,\ )))}$ Oxidized SWCNTs $\xrightarrow{PDDA}$ PDDA/SWCNTs $\xrightarrow{DNA}$ DNA/PDDA/SWCNTs

Scheme for fabrication of DNA-modified SWCNTs.

Also, DNA–SWCNT hybrids were also investigated from the reverse point of view: *aggregation* of dispersed SWCNTs from aqueous dispersions. It was shown [59] the kinetics of SWCNTs' aggregation in aqueous media strongly depends on the overall surface charge. SWCNTs having a greater number of surface charges showed faster aggregation. At the beginning, the microfilaments appear; they then grow forming larger aggregates, which are difficult to redisperse into water. The electrostatic interactions dominate instead of van der Waals interactions the interactions among dispersed SWCNTs in aqueous media.

It was shown that high solubility of CNTs in water can be reached by functionalization with other *amino acids* [60, 61]. Thus, in case of reaction of $NH_2(CH_2)_nCO_2H$ with fluoronanotubes, the solubility in water is controlled by the length of the hydrocarbon side chain [62]. The authors showed that the 6-aminohexanoic acid CNT derivative is soluble in aqueous solution (0.5 mg mL^{-1}) between pH 4 and 11, whereas the glycine and 11-aminoundecanoic acid derivatives are insoluble across all pH values. Highly water-dispersed MWCNTs (stable concentration as high as 10 mg/ml in deionized water) were obtained by attaching the lysine $HO_2CCH(NH_2)(CH_2)_4NH_2$ onto MWCNTs by producing acyl chloride on the carboxylic groups associated with the nanotubes [63]. The functionalized MWCNTs can be dispersed in water under a wide range of pH values (5–14).

Functionalization of MWCNTs with amino acid molecule using microwave irradiation, followed by reaction with a carboxylic acid moiety, allowed direct attachment by an amide bond [64]. The process was resulted in a high degree of functionalization as well as dispersibility in organic solvents like N,N-′-dimethylacetamide. Complexes composed of poly-l-tyrosine **4.19** (pLT) and SWCNTs were produced [65]. It was found that the complexes remained stably dispersed in the water at least for 2 weeks. Stable aqueous SWCNT dispersion up to 92% was also achieved using amino acid-based amphiphiles through a structure–property investigation [66]. The nanohybrids showed remarkable serum stability and biocompatibility to mammalian cells. The ability of N-(fluorenyl-9-methoxycarbonyl) (Fmoc) amino acids to disperse CNTs was demonstrated [67].

Four Fmoc-protected aromatic amino acids, W, H, *tyrosine* (Y), and phenylalanine (F), were selected, with glycine (G) as a nonaromatic control. The interactions of these molecules with SWCNTs and MWCNTs were studied by examining their ability to disperse CNTs in phosphate buffer saline (PBS). In addition, all-atom molecular dynamics simulations were carried out for aqueous *flavin* mononucleotide **4.20** (FMN), which was found experimentally to efficiently separate and sort SWCNTs based on diameter and chirality [68, 69]. Several important conclusions had been made by authors: to stabilize carbon nanotubes in water, it is necessary to employ dispersing agents that (1) strongly adsorb on the nanotube surface; (2) present hydrophilic groups, better if rigid, that extend toward the aqueous phase; (3) are not very mobile on the nanotube surface; and (4) show aggregates with structure dependent on nanotube diameter and chirality. We note that these conditions can be obviously applied not only for the FMN case, but can be considered as universal requirements in the development of dispersion techniques.

HSVM technique was applied to obtain water-soluble SWCNT–nucleotide composites, whose solubility depends significantly on the number of phosphate groups and the kinds of bases employed [70]. The complex amylose–nanotube system was studied by molecular dynamics simulation for elucidation of the mode of interaction between the initially separated amylose and SWCNT fragments, which can be either wrapping or encapsulation [71]. The authors found that amylose molecules can be used to bind with nanotubes due to the dominance of the van der Waals force and, thus, favor noncovalent functionalization of carbon nanotubes.

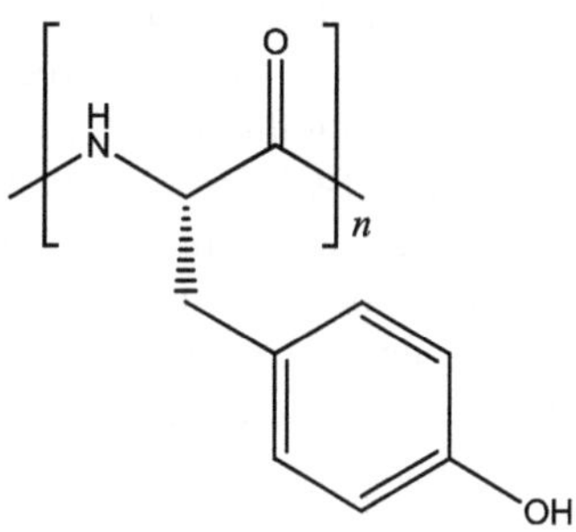

4.19
Poly-l-tyrosine

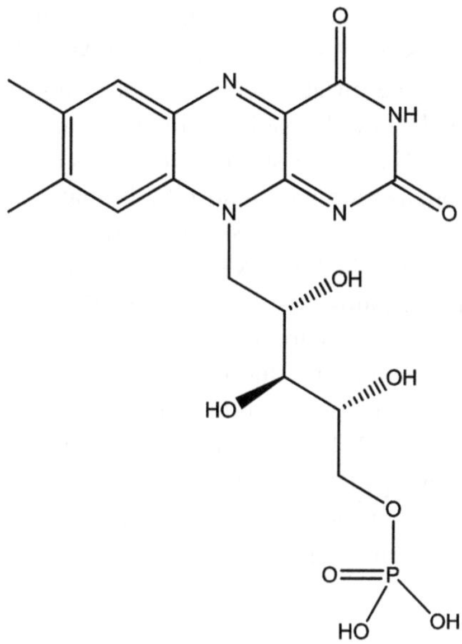

4.20
Flavin mononucleotide

The dynamics of *protein*-assisted CNT dispersion in water were studied [72]. It was found that in equilibrium, only a small fraction of the dispersants is indeed adsorbed to the nanotube surface, while there is a fast exchange process between the adsorbed and free protein molecules. A noncovalent approach was developed [73] for the functionalization of MWCNTs using the *hydrophobin*, HFBI. Owing to the amphipathic nature, HFBI can be adopted onto the surface of MWCNTs to form HFBI–MWCNT nanocomposite with good dispersion in water. Furthermore, a glucose biosensor was developed based on HFBI–MWCNTs by a one-step casting method. The resulting biosensor displayed high sensitivity, wider linear range, low detection limit, and fast response for glucose detection, which implicated that the HFBI–MWCNT nanocomposite film holds great promise in the design of electrochemical devices, such as sensors and biosensors.

Different other proteins such as bovine serum albumin, cytochrome *c*, horseradish peroxidase, plasmid DNA, ATP, mixture of deoxyribonucleoside triphosphates, compounds of vitreous humor extract, and sodium humate were used to solubilize CNTs in water aided by sonication [74–78]. Among other proteins used, the egg white lysozyme dispersed SWCNTs, whereas papain and pepsin could not [79]. The authors concluded that the main driving force to the hydrophobic interactions between the sidewall of the SWCNT and the inner hydrophobic domain was exposed to the solvent during the three-dimensional change of the protein induced by sonication. A gonadotrophin-releasing hormone (GnRH), which was overexpressed in the plasma membrane of several types of cancer cells, was covalently anchored onto the surface of the oxidized MWCNTs via an amide linkage [80]. Sidewall coverage of MWCNTs by the GnRH was about 0.7% of the available surface area. It was also shown that the GnRH–MWCNTs entered the cells and showed toxicity in the malignant cells.

The CNTs were successfully suspended in aqueous buffer solutions by their functionalization with a specific bifunctional molecule that is "sticky" to proteins 1-pyrene butanoic acid succidymidyl ester (1-pbase) [81]. Among other reported proteins, attached to CNTs making them soluble, we note Cy5-labeled goat anti-rabbit IgG (anti-IgG-Cy5), chemically bonded to CNTs via a two-step process of diimide-activated amidation and observed successfully using fluorescence microscopy, obtaining the fluorescent image of highly oriented f-CNTs at the first time [82], the foot-and-mouth disease virus (FMDV), leading to mono- and *bis*-derivatized CNTs **4.22** and **4.23** starting from the precursor **4.21** possessing free amino groups [83] and more [84–87]. Protein–CNT nanocomposites can be potentially applied for biosensor and biofuel cell applications. A comparison of protein-dispersing media for various engineered carbon nanoparticles is given in [88].

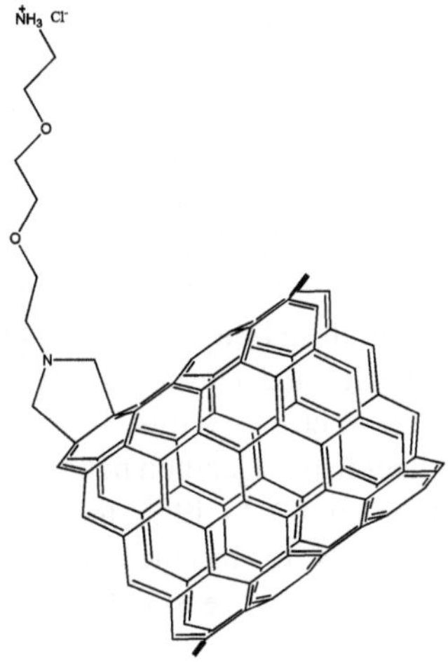

4.21

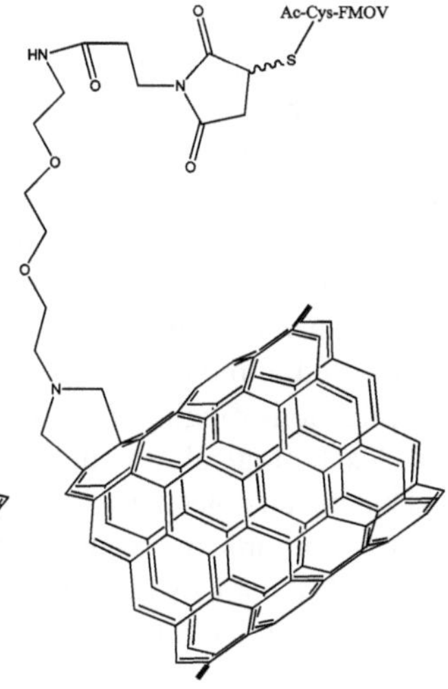

4.22

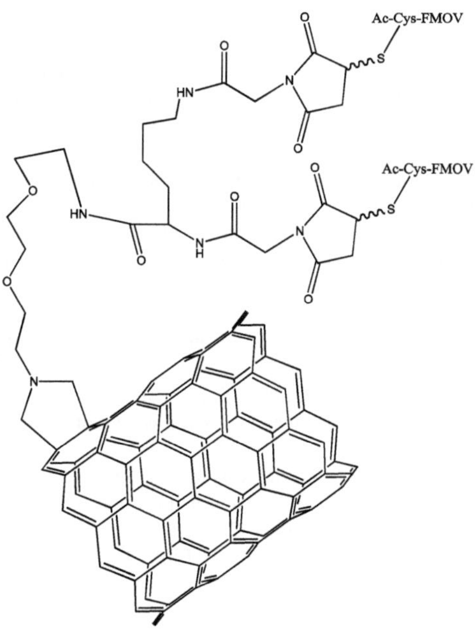

4.23

Enzymes, belonging to proteins, have also been extensively studied as solubilizing agents for CNTs. Thus, a variety of enzyme–MWCNT conjugates in aqueous solutions, their high activity and stability, and reusability were reported in [89–91]. The formed products are soluble in aqueous buffer, retained a high fraction of their native activity, and stable at higher temperatures relative to their solution phase counterparts. With use of noncovalent functionalization by the surfactant Triton X-100, the hydrophobic surfaces of the CNTs are changed to hydrophilic and interact with the hydrophilic surface of Biliverdin IX beta reductase enzyme, creating a water-soluble complex with real interaction between the enzyme and CNT–Triton conjugates [92]. We would like to note glucose oxidase [93] among other enzymes [94] forming soluble composites with CNTs. Some protein conjugates possess the attributes of both soluble enzymes (high activity and low diffusional resistance) and immobilized enzymes (high stability) making them attractive choices for applications ranging from diagnostics and sensing to drug delivery.

More complex biological systems (fungi, microbes, virus, bacteria) have not been reported having an objective to solubilize/disperse carbon nanotubes, although there is a row of reports on biodegradation of CNTs [95], their antimicrobial activity [96], or CNT composites with these cells, for instance, a CNT composite with a single-cell fungi, the *Saccharomyces cerevisiae* yeast extract [97].

References

1. S. Mallakpour, V. Behranvand, Improved solubilization of multiwalled carbon nanotubes (MWCNTs) in water by surface functionalization with D-glucose and D-fructose: Properties comparison of functionalized MWCNTs/alanine-based poly(amide–imide) nanocomposites. High Perform. Polym. **28**(8), 936–944 (2016)
2. H. Leinonen, M. Pettersson, M. Lajunen, Water-soluble carbon nanotubes through sugar azide functionalization. Carbon **49**(4), 1299–1304 (2011)
3. A.A. Dyshin, O.V. Eliseeva, G.V. Bondarenko, M.G. Kiselev, Dissolution of single-walled carbon nanotubes in alkanol–cholic acid mixtures. Russ. J. Phys. Chem. A **89**(9), 1628–1632 (2015)
4. A.A. Dyshin, O.V. Eliseeva, G.V. Bondarenko, M. Kolker, G. Kiselev, Dispersion of single-walled carbon nanotubes in dimethylacetamide and a dimethylacetamide–cholic acid mixture. Russ. J. Phys. Chem. A **90**(12), 2434–2439 (2016)
5. C. Yao, C. Carlisi, Y. Li, D. Chen, J. Ding, Y.-L. Feng, Interaction potency of single-walled carbon nanotubes with DNAs: A novel assay for assessment of hazard risk. PLoS One **11**(12), e0167796 (2016)
6. M. Swierczewska, K.Y. Choi, E.L. Mertz, X. Huang, F. Zhang, L. Zhu, H.Y. Yoon, J.H. Park, A. Bhirde, S. Lee, X. Chen, A facile, one-step nanocarbon functionalization for biomedical applications. Nano Lett. **12**(7), 3613–3620 (2012)
7. F. Zhang, X. Chen, R.A. Boulos, F. Md Yasin, H. Lu, C. Raston, H. Zhang, Pyrene-conjugated hyaluronan facilitated exfoliation and stabilisation of low dimensional nanomaterials in water. Chem. Commun. **49**(42), 4845–4847 (2013)
8. S. Nagarajan, M.K. Dhinakaran, V.G. Kumar, N. Rajaram, T.M. Das, K.A. Padmanabhan, On the use of glycosylated single-walled carbon nanotubes as a coolant additive. Nanosci. Nanotechnol. Lett. **3**(4), 477–482 (2011)
9. L.T.N. Lien, T. Shiraki, A. Dawn, Y. Tsuchiya, D. Tokunaga, S.-I. Tamaru, N. Enomoto, J. Hojo, S. Shinkai, A pH-responsive carboxylic β-1,3-glucan polysaccharide for complexation with polymeric guests. Org. Biomol. Chem. **9**(11), 4266–4275 (2011)
10. M.K. Okajima, A. Kumar, A. Fujiwara, T. Mitsumata, D. Kaneko, T. Ogawa, H. Kurata, S. Isoda, T. Kaneko, Anionic complexes of MWCNT with supergiant cyanobacterial polyanions. Biopolymers **99**(1), 1–9 (2013)

11. H. Dohi, S. Kikuchi, S. Kuwahara, T. Sugai, H. Shinohara, Synthesis and spectroscopic characterization of single-wall carbon nanotubes wrapped by glycoconjugate polymer with bioactive sugars. Chem. Phys. Lett. **428**(1), 98–101 (2006)
12. L. Yan, P.R. Chang, P. Zheng, Preparation and characterization of starch-grafted multiwall carbon nanotube composites. Carbohydr. Polym. **84**(4), 1378–1383 (2011)
13. L. Li, W. Feng, P. Ji, Dispersion of carbon nanotubes in organic solvents initiated by hydrogen bonding interactions. AICHE J. **58**(10), 2997–3002 (2012)
14. M. Rinaudo, Chitin and chitosan: Properties and applications. Prog. Polym. Sci. **31**(7), 603–632 (2006)
15. H. Yang, S.C. Wang, P. Mercier, D.L. Akins, Diameter-selective dispersion of single-walled carbon nanotubes using a water-soluble, biocompatible polymer. Chem. Commun. (13), 1425–1427 (2006)
16. D. Moura, J.F. Mano, M.C. Paiva, N.M. Alves, Chitosan nanocomposites based on distinct inorganic fillers for biomedical applications. Sci. Technol. Adv. Mater. **17**(1), 626–643 (2016)
17. Y.-T. Shieh, H.-M. Wu, Y.-K. Twu, Y.-C. Chung, An investigation on dispersion of carbon nanotubes in chitosan aqueous solutions. Colloid Polym. Sci. **288**(4), 377–385 (2010)
18. C.K. Najeeb, J. Chang, J.-H. Lee, M. Lee, J.-H. Kim, Preparation of semiconductor-enriched single-walled carbon nanotube dispersion using a neutral pH water soluble chitosan derivative. J. Colloid Interface Sci. **354**(2), 461–466 (2011)
19. C.K. Najeeb, J.-H. Lee, J.-H. Kim, D. Kim, Highly efficient individual dispersion of single-walled carbon nanotubes using biocompatible dispersant. Colloids Surf. B Biointerfaces **102**, 95–101 (2013)
20. C. Tang, T. Zhou, J. Yang, Q. Zhang, F. Chen, Q. Fu, L. Yang, Wet-grinding assisted ultrasonic dispersion of pristine multi-walled carbon nanotubes (MWCNTs) in chitosan solution. Colloids Surf. B Biointerfaces **86**(1), 189–197 (2011)
21. L. Qian, X. Yang, Composite film of carbon nanotubes and chitosan for preparation of amperometric hydrogen peroxide biosensor. Talanta **68**(3), 721–727 (2006)
22. Y. Liu, X. Qu, H. Guo, H. Chen, B. Liu, S. Dong, Facile preparation of amperometric laccase biosensor with multifunction based on the matrix of carbon nanotubes-chitosan composite. Biosens. Bioelectron. **21**(12), 2195–2201 (2006)
23. G.M. Spinks, S.R. Shin, G.G. Wallace, P.G. Whitten, S.I. Kim, S.J. Kim, Mechanical properties of chitosan/CNT microfibers obtained with improved dispersion. Sensors Actuators B Chem **115**(2), 678–684 (2006)
24. A.L. Ahmad, Z.A. Jawad, S.C. Low, S.H. Sharif Zein, The functionalization of beta-cyclodextrins on multi walled carbon nanotubes: Effects of the dispersant and non aqueous media. Curr. Nanosci. **9**(1), 93–102 (2013)
25. V.I. Bhoi, T. Imae, M. Ujihara, C.N. Murthy, Surface immobilization of carbon nanotubes by β-cyclodextrins and their inclusion ability. J. Nanosci. Nanotechnol. **13**(4), 2604–2612 (2013)
26. J.-G. Yu, K.-L. Huang, S.-Q. Liu, J.-C. Tang, Preparation and characterization of soluble methyl-β-cyclodextrin functionalized single-walled carbon nanotubes. Phys. E. **40**(3), 689–692 (2008)
27. A. Casey, G.F. Farrell, M. McNamara, H.J. Byrne, G. Chambers, Interaction of carbon nanotubes with sugar complexes. Synth. Met. **153**(1), 357–360 (2005)
28. A. Feng, L. Peng, B. Liu, S. Liu, S. Wang, J. Yuan, Electrochemical redox switchable dispersion of single-walled carbon nanotubes in water. ACS Appl. Mater. Interfaces **8**(17), 11024–11030 (2016)
29. H. Lu, L. Zou, Y. Wei, et al., The preparation of lysine modified multi-walled carbon nanotubes and the study of its dispersion properties. IOP Conf. Series Mater. Sci. Eng **87**, 012048 (2015)
30. A. Amiri, H.Z. Zardini, M. Shanbedi, M. Maghrebi, M. Baniadam, B. Tolueinia, Efficient method for functionalization of carbon nanotubes by lysine and improved antimicrobial activity and water-dispersion. Mater. Lett. **72**, 153–156 (2012)

31. T. Zhang, L. Zou, X. Ling, X. Li, Dispersion of multi-walled carbon nanotubes modified with poly-L-lysine in water. Appl. Mech. Mater. **275-277**, 1785–1788 (2013)
32. A. Ansun-Casaos, L. Grasa, D. Pereboom, et al., In-vitro toxicity of carbon nanotube/polylysine colloids to colon cancer cells. IET Nanobiotechnol. **10**(6), 374–381 (2016)
33. Z. Li, T. Kameda, T. Isoshima, et al., Solubilization of single-walled carbon nanotubes using a peptide aptamer in water below the critical micelle concentration. Langmuir **31**(11), 3482–3488 (2015)
34. D. Pantarotto, C.D. Partidos, R. Graff, J. Hoebeke, J.-P. Briand, M. Prato, A. Bianco, Synthesis, structural characterization, and immunological properties of carbon nanotubes functionalized with peptides. J. Am. Chem. Soc. **125**(20), 6160–6164 (2003)
35. Y. Cheng, G.R. Liu, Z.R. Li, C. Lu, Computational analysis of binding free energies between peptides and single-walled carbon nanotubes. Physica A Stat. Mech. Appl. **367**, 293–304 (2006)
36. Y. Hashida, T. Umeyama, J. Mihara, H. Imahori, M. Tsujimoto, S. Isoda, M. Takano, M. Hashida, Development of a novel composite material with carbon nanotubes assisted by self-assembled peptides designed in conjunction with β-sheet formation. J. Pharm. Sci. **101**(9), 3398–3412 (2012)
37. M. Sheikholeslam, M. Pritzker, P. Chen, Dispersion of multiwalled carbon nanotubes in water using ionic-complementary peptides. Langmuir **28**(34), 12550–12556 (2012)
38. S. Dutta, T. Kar, S. Brahmachari, P.K. Das, pH-responsive reversible dispersion of biocompatible SWCNT/graphene- amphiphile hybrids. J. Mater. Chem. **22**(14), 6623–6631 (2012)
39. Z. Li, T. Kameda, T. Isoshima, E. Kobatake, T. Tanaka, Y. Ito, M. Kawamoto, Solubilization of single-walled carbon nanotubes using a peptide aptamer in water below the critical micelle concentration. Langmuir **31**, 3482−3488 (2015)
40. M. Ghosh, S. Brahmachari, P. Kumar Das, pH-responsive single walled carbon nanotube dispersion for target specific release of doxorubicin to cancer cells. Macromol. Biosci. **14**, 1795–1806 (2014)
41. J. Posseckardt, J. Zhang, M. Mertig, Mobility of a supported lipid bilayer on dispersed single-walled carbon nanotubes. Phys. Status Solidi A **213**(6), 1427–1433 (2016)
42. J. Määttä, S. Vierros, P.R. Van Tassel, M. Sammalkorpi, Size-selective, noncovalent dispersion of carbon nanotubes by PEGylated lipids: A coarse-grained molecular dynamics study. J. Chem. Eng. Data **59**, 3080−3089 (2014)
43. E. Cheng, Y. Yang, D. Liu, pH-controlled carbon nanotube aggregation/dispersion based on intermolecular I-motif DNA formation. J. Nanosci. Nanotechnol. **10**(11), 7282–7286 (2010)
44. A. Amiri, M. Shanbedi, H. Eshghi, S.Z. Heris, M. Baniadam, Highly dispersed multiwalled carbon nanotubes decorated with ag nanoparticles in water and experimental investigation of the thermophysical properties. J. Phys. Chem. C **116**(5), 3369–3375 (2012)
45. P.R. Dalmasso, M.L. Pedano, G.A. Rivas, Dispersion of multi-wall carbon nanotubes in polyhistidine: Characterization and analytical applications. Anal. Chim. Acta **710**, 58–64 (2012)
46. K. Umemura, K. Izumi, S. Oura, Probe microscopic studies of DNA molecules on carbon nanotubes. Nano **6**, 180 (2016.) 14 pp
47. S. Daniel, T.P. Rao, K.S. Rao, S.U. Rani, G.R.K. Naidu, H.Y. Lee, T. Kawai, A review of DNA functionalized/grafted carbon nanotubes and their characterization. Sensors Actuators B Chem. **122**(2), 672–682 (2007)
48. R.M. Williams, S. Nayeem, B.D. Dolash, L.J. Sooter, The effect of DNA-dispersed single-walled carbon nanotubes on the polymerase chain reaction. PLoS One **9**(4), e94117 (2014)
49. E.N. Primo, P. Cañete-Rosales, S. Bollo, M.D. Rubianes, G.A. Rivas, Dispersion of bamboo type multi-wall carbon nanotubes in calf-thymus double stranded DNA. Colloids Surf. B Biointerfaces **108**, 329–336 (2013)
50. F. Bomboi, A. Bonincontro, C. La Mesa, F. Tardani, Interactions between single-walled carbon nanotubes and lysozyme. J. Colloid Interface Sci. **355**(2), 342–347 (2011)

51. D. Nepal, J.-I. Sohn, W.K. Aicher, S. Lee, K. Geckeler, Supramolecular conjugates of carbon nanotubes and DNA by a solid-state reaction. Biomacromolecules **6**(6), 2919–2922 (2005)
52. P. He, M. Bayachou, Layer-by-layer fabrication and characterization of DNA-wrapped single-walled carbon nanotube particles. Langmuir **21**(13), 6086–6092 (2005)
53. A. Ishibashi, Y. Yamaguchi, H. Murakami, N. Nakashima, Layer-by-layer assembly of RNA/single-walled carbon nanotube nanocomposites. Chem. Phys. Lett. **419**(4), 574–577 (2006)
54. Q.H. Yang, N. Gale, C.J. Oton, H. Li, I.S. Nandhakumar, Z.Y. Tang, T. Brown, W.H. Loh, Deuterated water as super solvent for short carbon nanotubes wrapped by DNA. Carbon **45** (13), 2701–2703 (2007)
55. A. Bianco, D. Pantarotto, K. Kostarelos, M. Prato. Non-covalent complexes comprising carbon nanotubes. WO05121799, 2005
56. N. Nakashima, S. Okuzono, H. Murakami, T. Nakai, K. Yoshikawa, DNA dissolves single-walled carbon nanotubes in water. Chem. Lett. **32**, 456–457 (2003)
57. M. Zheng, A. Jagota, D.S. Ellen, A.D. Bruce, S.M. Robert, R.L. Steve, E.R. Raymond, G.T. Nancy, DNA-assisted dispersion and separation of carbon nanotubes. Nat. Mater. **2**, 338–342 (2003)
58. Y. Noguchi, T. Fujigaya, Y. Niidome, N. Nakashima, Single-walled carbon nanotubes/DNA hybrids in water are highly stable. Chem. Phys. Lett. **455**, 249–251 (2008)
59. B. Koh, W. Cheng, The kinetics of single-walled carbon nanotube aggregation in aqueous media is sensitive to surface charge. C (J. Carbon Res) **2**(6), 11 (2016)
60. V. Georgakilas, N. Tagmatarchis, D. Pantarotto, A. Bianco, J.-P. Briand, M. Prato, Amino acid functionalisation of water soluble carbon nanotubes. Chem. Commun. (24), 3050–3051 (2002)
61. S. Mallakpour, A. Zadehnazari, Functionalization of multiwalled carbon nanotubes with S-valine amino acid and its reinforcement on amino acid-containing poly(amide-imide) bionanocomposites. High Perform. Polym **25**(8), 966–979 (2013)
62. L. Zeng, L. Zhang, A.R. Barron, Tailoring aqueous solubility of functionalized single-wall carbon nanotubes over a wide pH range through substituent chain length. Nano Lett. **5**(10), 2001–2004 (2005)
63. N. Hu, G. Dang, H. Zhou, J. Jing, C. Chen, Efficient direct water dispersion of multi-walled carbon nanotubes by functionalization with lysine. Mater. Lett. **61**(30), 5285–5287 (2007)
64. S. Mallakpour, A. Zadehnazari, Functionalization of multi-wall carbon nanotubes with amino acid and its influence on the properties of thiadiazol bearing poly(amide-thioester-imide) composites. Synth. Met. **169**, 1), 1–1),11 (2013)
65. M. Kojima, T. Chiba, J. Niishima, T. Higashi, T. Fukuda, Y. Nakajima, S. Kurosu, T. Hanajiri, K. Ishii, T. Maekawa, A. Inoue, Dispersion of single-walled carbon nanotubes modified with poly-l-tyrosine in water. Nanoscale Res. Lett. **6**(1), X1–X6 (2011)
66. S. Brahmachari, D. Das, P.K. Das, Superior SWCNT dispersion by amino acid based amphiphiles: Designing biocompatible cationic nanohybrids. Chem. Commun. **46**(44), 8386–8388 (2010)
67. B.G. Cousins, A.K. Das, R. Sharma, Y. Li, J.P. McNamara, I.H. Hillier, I.A. Kinloch, R.V. Ulijn, Enzyme-activated surfactants for dispersion of carbon nanotubes. Small **5**(5), 587–590 (2009)
68. N.R. Tummala, B.H. Morrow, D.E. Resasco, A. Striolo, Stabilization of aqueous carbon nanotube dispersions using surfactants: Insights from molecular dynamics simulations. ACS Nano **4**(12), 7193–7204 (2010)
69. M. Jang, S. Kim, H. Jeong, S.-Y. Ju, Affinity-mediated sorting order reversal of single-walled carbon nanotubes in density gradient ultracentrifugation. Nanotechnology **27**, 41LT01 (2016.) (6 pp)
70. A. Ikeda, T. Hamano, K. Hayashi, Jun-IchiKikuchi, Water-solubilization of nucleotides-coated single-walled carbon nanotubes using a high-speed vibration milling technique. Org. Lett. **8**(6), 1153–1156 (2006)
71. Y.H. Xie, A.K. Soh, Investigation of non-covalent association of single-walled carbon nanotube with amylose by molecular dynamics simulation. Mater. Lett. **59**(8), 971–975 (2005)

72. A.E. Frise, E. Edri, I. Furó, O. Regev, Protein dispersant binding on nanotubes studied by NMR self-diffusion and cryo-TEM techniques. J. Phys. Chem. Lett. **1**(9), 1414–1419 (2010)
73. X. Wang, H. Wang, Y. Huang, Z. Zhao, X. Qin, Y. Wang, Z. Miao, Q. Chen, M. Qiao, Noncovalently functionalized multi-wall carbon nanotubes in aqueous solution using the hydrophobin HFBI and their electroanalytical application. Biosens. Bioelectron. **26**(3), 1104–1108 (2010)
74. M.C. Kum, K.A. Joshi, W. Chen, N.V. Myung, A. Mulchandani, Biomolecules-carbon nanotubes doped conducting polymer nanocomposites and their sensor application. Talanta **74**(3), 370–375 (2007)
75. O.M. Burlaka, Y.V. Pirko, O.F. Kolomys, et al., Functionalization of carbon nanotubes by different biomolecules for stable dispersion in water. Biotechnol. Acta **8**(4), 71–81 (2015)
76. C. Kuroda, H. Haniu, K. Ajima, et al., The dispersion state of tangled multi-walled carbon nanotubes affects their cytotoxicity. Nano **6**(219), 10 (2016)
77. M. Eguílaz, C.J. Venegas, A. Gutiérrez, Carbon nanotubes non-covalently functionalized with cytochrome c: A new bioanalytical platform for building bienzymatic biosensors. Microchem. J. **128**, 161–165 (2016)
78. G. Risuleo, C. La Mesa, Dispersability of carbon nanotubes in biopolymer-based fluids and their potential biotechnological applications. Trends Nanotechnol. Mater. Sci **1**(2), 7 (2016)
79. G. Ke, W. Guan, C. Tang, W. Guan, D. Zeng, F. Deng, Covalent functionalization of multiwalled carbon nanotubes with a low molecular weight chitosan. Biomacromolecules **8** (2), 322–326 (2007)
80. B.Z. Yu, J.S. Yang, W.X. Li, In vitro capability of multi-walled carbon nanotubes modified with gonadotrophin releasing hormone on killing cancer cells. Carbon **45**(10), 1921–1927 (2007)
81. R. Prakash, R. Superfine, S. Washburn, M.R. Falvo, Functionalization of carbon nanotubes with proteins and quantum dots in aqueous buffer solutions. Appl. Phys. Lett. **88**, 063102 (2006)
82. Z. Xu, P. Hu, S. Wang, X. Wang, Biological functionalization and fluorescent imaging of carbon nanotubes. Appl. Surf. Sci. **254**(7), 1915–1918 (2008)
83. D. Pantarotto, C.D. Partidos, J. Hoebeke, F. Brown, E. Kramer, J.-P. Briand, S. Muller, A. Bianco, Immunization with peptide-functionalized carbon nanotubes enhances virus-specific neutralizing antibody responses. Chem. Biol. **10**(10), 961–966 (2003)
84. A. Bianco, D. Pantarotto, M. Prato Functionalized carbon nanotubes, a process for preparing the same and their use in medicinal chemistry. WO04089819, (2004)
85. P.-C. Ke, Y. Wu, A.M. Rao. Lysophospholipids solubilized single-walled carbon nanotubes. WO07136404, 2007
86. H. Dai, R.J. Chen. Noncovalent sidewall functionalization of carbon nanotubes. US20050100960, 2005
87. P.G. Holder, M.B. Francis, Integration of a self-assembling protein scaffold with water-soluble single-walled carbon nanotubes. Angew. Chem. Int. Ed. Engl. **46**(23), 4370–4373 (2007)
88. M.C. Buford, R.F. Hamilton, A. Holan, A comparison of dispersing media for various engineered carbon nanoparticles. Part. Fibre Toxicol. **4**(1), 6 (2007)
89. P. Asuri, S.S. Karajanagi, E. Sellitto, D.-Y. Kim, R.S. Kane, J.S. Dordick, Water-soluble carbon nanotube-enzyme conjugates as functional biocatalytic formulations. Biotechnol. Bioeng. **95**(5), 804–811 (2006)
90. P. Asuri, S.S. Bale, R.C. Pangule, D.A. Shah, R.S. Kane, J.S. Dordick, Structure, function, and stability of enzymes covalently attached to single-walled carbon nanotubes. Langmuir **23**(24), 12318–12321 (2007)
91. B.C. Kim, I. Lee, S.-J. Kwon, et al., Fabrication of enzyme-based coatings on intact multi-walled carbon nanotubes as highly effective electrodes in biofuel cells. Sci Rep **7**, Article number: 40202 (2017)

92. M.i.h. Panhuis, C. Salvador-Morales, E. Franklin, G. Chambers, A. Fonseca, J.B. Nagy, W.J. Blau, A. Minett, Characterization of an interaction between functionalized carbon nanotubes and an enzyme. J. Nanosci. Nanotechnol. **3**(3), 209–213 (2003)
93. J. Li, Y.-B. Wang, J.-D. Qiu, D.-C. Sun, X.-H. Xia, Biocomposites of covalently linked glucose oxidase on carbon nanotubes for glucose biosensor. Anal. Bioanal. Chem. **383**(6), 918–922 (2005)
94. Y. Wang, Z. Iqbal, S.V. Malhotra, Functionalization of carbon nanotubes with amines and enzymes. Chem. Phys. Lett. **402**(1), 96–101 (2005)
95. M. Chen, X. Qin, G. Zeng, Biodegradation of carbon nanotubes, graphene, and their derivatives. Trends Biotechnol. (2017.) in press
96. M. Khazaee, A. Majumder, L. Baraban, G. Cuniberti, J. Opitz, D. Ye, Non-covalent modified multi-walled carbon nanotubes: Dispersion capabilities and interactions with bacteria. Biomed. Phys. Eng. Express **2**(5), 8 (2016.) Art. 055008
97. L. Valentini, S. Bittolo Bon, S. Signetti, M. Tripathi, E. Lacob, N.M. Pugnob, Fermentation based carbon nanotube multifunctional bionic composites. Sci Rep **6**, 27031 (2016)

Chapter 5
Special Studies and Characterization of CNT Dispersions

During more than 20-year period of investigations in the field of carbon nanotubes and their dispersions in water and organic solvents, a series of important observations, effects, and conclusions have been made, in particular after the CNT dispersions have been studied within solution. As it will be seen below, influence of solvent and external influence (i.e., UV light or the presence of ions), free radicals, and a row of other factors can improve or affect the CNT dispersibility, which could differ depending on CNT type and their purity.

5.1 Dispersion, Precipitation, and Purification Studies

5.1.1 Dispersion of CNTs in Solvents

Influence of solvent nature A series of studies have been carried out for pristine and, in some cases, OH–MWCNTs in distinct protic and aprotic solvents and their mixtures, generally without surfactants. As a representative example, the dispersion of hydroxylated MWCNTs was modified in non-protic acetonitrile solvent using a treatment by ethanol [1]. Either the MWCNT–OH powder was used as it is or previously treated by EtOH. These two types of powders were diluted by acetonitrile, and the solvation was followed by photoluminescence and Rayleigh scattering methods. Despite the well-known very low solubility of nanotubes, present results showed the presence of nanotube dimers in the solution with considerable concentration. Applying a qualitative model, $\Delta H = -46.6 \pm 12$ kJ/mol and $\Delta S = -29.9 \pm 7$ J/K mol enthalpy and entropy changes were obtained during formation of nanotube dimers. This highly negative entropy term is of great importance for the deposition of carbon nanotubes by liquid-phase epitaxy to enlarge the surface coverage.

O.V. Kharissova, B.I. Kharisov, *Solubilization and Dispersion of Carbon Nanotubes*, DOI 10.1007/978-3-319-62950-6_5

The analysis of the reported data [2–4] clearly shows that N-methyl-2-pyrrolidone (NMP), N-dodecyl-pyrrolidone (N12P), acetone, tetrahydrofuran (THF), N,N-dimethylformamide (DMF), N,N-dimethylacetamide (DMA), cyclohexyl-pyrrolidone (CHP), and dichloromethane are good solvents to debundle and disperse the MWCNTs. In contrast, much precipitation can be obviously observed for systems of the CNTs in water, ethanol, and toluene. Tyndall effect tests suggested that the upper dispersions of the carbon nanotubes in acetone, tetrahydrofuran, and dichloromethane with a concentration of 0.1 mg/mL and the dispersion of carbon nanotubes in N-methyl-2-pyrrolidone with a concentration of 0.05 mg/mL are colloidal systems rather than solutions.

If similar amounts of carboxylic acid and phenolic groups were introduced [5] to mostly already existing defects, the presence of acid oxygenated groups increased SWCNT dispersibility in NMP, DMF, and DMA but decreased in N12P and CHP. The absorption coefficients, however, decreased for all solvents above after oxidation, reflecting the weakening of the effective transition dipole of the π–π transition with even limited extension functionalization and solvent interaction. In case of mixtures of the organic solvents such as DMF and NMP [6], it was shown that mixing DMF and NMP can yield dispersion stabilities that are 60–115% greater than the pure solvents with a 50/50 mixture (v/v).

Influence of zeta potential for CNT dispersion in solvents is a permanent object for discussions. Thus, the correlation between the electrokinetic potential, dispersibility in solvents, surface energy, and oxygen content of CNTs affected by functionalization was studied [7]. Colloidal systems consisting of CNTs with varying degrees of dispersion were prepared and characterized to evaluate CNT dispersibility and suspension stability in solvents with different polarities. It was revealed that an absolute value of zeta potential at about 25 mV is closely related to the micro- and macroscopic dispersion of CNTs, whereas a high absolute value of 40 mV is regarded as an indication of high-quality CNT dispersion with much enhanced suspension stability in solvents. The absolute zeta potential value increases consistently with increasing degree of CNT functionality, the increase being most pronounced in a hydrophilic liquid such as water. The CNT dispersibility in a liquid is determined not only by their physical states, but also by the hydrophilicity and surface functionality of CNTs, all of which are reflected by zeta potential.

Dispersion and solubilization of CNTs in solvents is obviously related with the terms of *Hildebrand and Hansen solubility parameters*. As an example of such study, dispersions of produced HiPco SWCNTs in a series of organic solvents were prepared [8] by dilution with the aid of tip sonication. Mild centrifugation (~945 g) was carried out to remove large bundles. AFM studies revealed that the bundle size decreased as the dispersion was diluted. SWCNTs are easily dispersed in solvents with Hildebrand solubility parameter range from ~22–24 $MPa^{1/2}$ and Hansen polarity component (P) ~ 12–14 $MPa^{1/2}$. It was found that the degree of dispersion depends critically on sample preparation conditions and in particular sonication time; however, increased sonication also induces discernible changes to the

SWCNTs themselves and in itself influences their solubility, under which conditions no clear solubility parameters can be determined.

In a related research [9], the dispersion behavior of graphene oxide (GO) and oxidized carbon nanotubes (o-CNT) in a polar solvent, as well as the differences in the behavior related to the Hansen solubility parameter windows. In polar aprotic solvents, GO and o-CNT showed similar dispersion behavior. On the other hand, in polar protic solvents, such as ethanol and isopropanol, GO did not show dispersion stability whereas the o-CNTs did. This difference in the dispersion behavior between GO and o-CNTs resulted from the stronger hydrogen bonding between the GO interlayer induced by a large amount of oxygen functional groups and flexible 2D morphology with a large surface area. Also, in-depth studies of CNT solubility in a variety of solvents were carried out by author of [10]. Dispersions of SWCNTs in various solvents and aqueous surfactant emulsions were investigated to correlate the degree of dispersion state with *Hansen solubility parameters* $\delta^2_t = \delta^2_d + \delta^2_p + \delta^2_h$ [11], and it was found that the nanotubes were dispersed or suspended very well in the solvents with certain dispersive component (δ_d) values. They were precipitated in the solvents with high polar component (δ_p) values or hydrogen-bonding component (δ_h) values. The surfactants with a lipophilic group equal to and longer than decyl, containing 9 methylene groups and 1 methyl group, contributed to the dispersion of nanotubes in water.

5.1.2 Precipitation of CNTs from Dispersions

Sedimentation and precipitation of CNTs from dispersions is not considered as favorable process: on the contrary, main efforts are dedicated to CNT solubilization, and the value of a surfactant depended on how many months dispersion remains stable. However, these processes also need to be studied in order to avoid premature precipitations, the more especially as CNT precipitation could lead to unusual results. Thus, a general phenomenon of *length-dependent precipitation* (Fig. 5.1) of surfactant-dispersed CNTs by salts, polymers, and their combinations was reported [12]. For the cholate- and deoxycholate-dispersed SWCNTs, such polyelectrolytes as polystyrene sulfonate (PSS) and polymethacrylate (PMAA) were found to be most effective. If PMAA or PSS are added to CNT dispersion stepwise, the CNTs sediment in a length-dependent order: first CNTs with average length of 650 nm and then successively of 450 nm, 350 nm, and 250 nm. This effect could be useful for macroelectronics applications.

The dispersibility of SWCNTs was determined in a wide range of good solvents by measuring their concentration *remaining after centrifugation* [13], showing the best result for cyclohexyl-pyrrolidone (3.5 mg/mL). The authors of this fundamental research suggested (1) the surface energy to be the fundamental parameter determining the interaction of SWCNTs with solvents and (2) Hansen parameters are probably the more practical set of solubility parameters for SWCNTs in the conditions of the absence of published tables of surface energy solubility

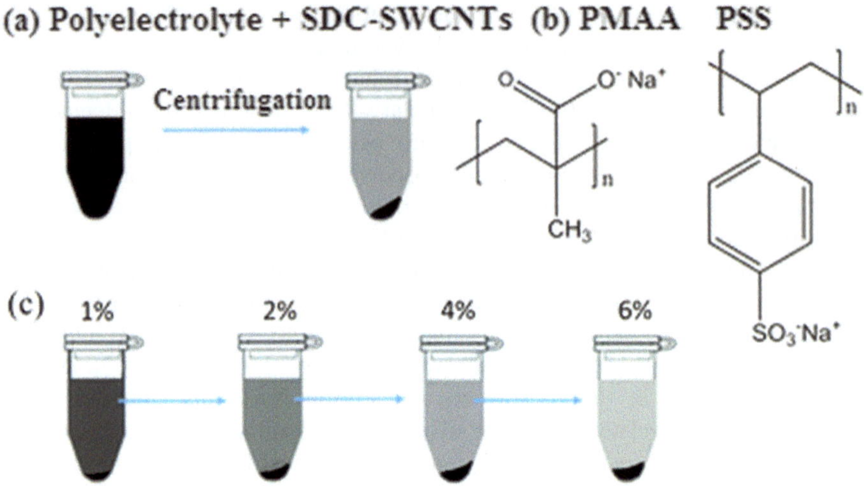

Fig. 5.1 (**a**) Schematic showing that adding polyelectrolyte (PMAA or PSS) to SDC-dispersed SWCNTs leads to nanotube precipitation. (**b**) Molecular structures of PMAA and PSS. (**c**) Length sorting by forward sequential polymer addition and SWCNT precipitation (Reproduced with permission from the *Royal Society of Chemistry*)

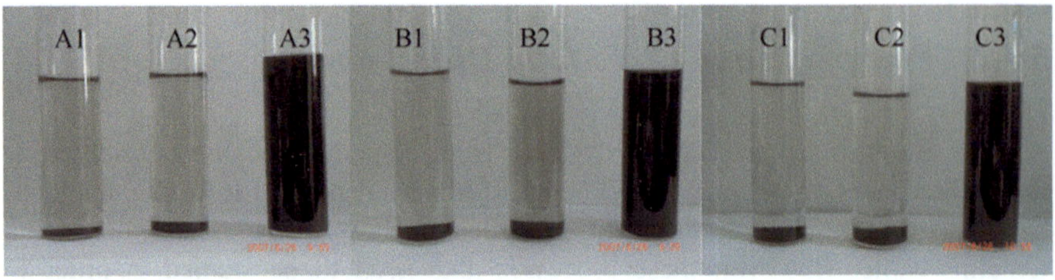

Fig. 5.2 Visual examination of oxidized CNTs after quiescent settling for 2 h with different salt concentrations: (*A1*) 400 mmol/L NaCl, (*A2*) 300 mmol/L NaCl, (*A3*) 150 mmol/L NaCl, (*B1*) 0.80 mmol/L $CaCl_2$, (*B2*) 0.60 mmol/L CaCl2, (*B3*) 0.40 mmol/L $CaCl_2$, (*C1*) 0.10 mmol/L $AlCl_3$, (*C2*) 0.05 mmol/L $AlCl_3$, (*C3*) 0.025 mmol/L $AlCl_3$ (Reproduced with permission from the *Elsevier Science*)

parameters. Also, the stability of CNTs, oxidized using nitric acid, and their *precipitation in water by salts* (Fig. 5.2) were investigated [14]. During an aging time of 30 days, oxidized CNT dispersion showed characteristic UV-vis absorbance peak of individual CNTs at 252 nm, and CNT concentration in the dispersion after 30 days of aging is 85% of the initial concentration. The precipitation value follows the order of NaCl > KCl > $1/2(MgCl_2)$ > $1/2(CaCl_2)$ > $1/3(FeCl_3)$ > $1/3(AlCl_3)$, being little dependent on anion nature. In case of cations, the trivalent cations were found to have the highest precipitation efficiency. In whole, the precipitation depends on such factors, discussed in detail in this report, as cation radius and charge, effect of hydrolyzation (e.g., for Fe^{3+} and Al^{3+}), and properties of cation hydrated shells.

5.1.3 Comparison of Surfactant Abilities

Comparative studies of surfactants belonging to distinct groups of compounds have been carried out [15]. Thus, a systematic study was carried out to establish the surfactant-assisted dispersion mechanism of carbon nanoclusters (size ≈150 nm, zeta potential ≈ − 15 mV), obtained from soot, in water. An array of surfactants, SDS, CTAB, TX-100, sodium cholate (bile salt), and dipalmitoyl phosphatidylcholine (DPPC, a clinical pulmonary surfactant preparation called Survanta), were used [16] in a wide range of concentration (0.01CMC–2CMC) to probe the dispersion mechanism. It was revealed that the adsorption of surfactant molecules on the nanoparticle surface was interplay of ionic, hydrophobic, and π–π stacking forces. The CTAB molecules (*cationic*) formed a bilayer on the carbon nanoclusters providing robust dispersion stability, whereas SDS molecules (*anionic*) were poorly adsorbed through hydrophobic interactions. TX-100 molecules (*neutral*) stabilized the dispersion via hydrophobic and π–π stacking interactions. Sodium cholate was adsorbed on nanoclusters mostly through hydrophobic interaction and generated large asymmetric complexes. DPPC, a gemini surfactant, formed a rigid monolayer around the carbon nanocluster even at nanomolar concentration and provided excellent stability to the dispersion.

In a related research [17], to test the dispersion of CNTs in biocompatible solutions, five known biocompatible dispersants were selected that are widely used for nanomaterial toxicity evaluation studies. SWCNTs and MWCNTs were both dispersed in these dispersants and their macrodispersion evaluated using a light absorbance method. Distilled water and dimethylsulfoxide (DMSO) both showed a poor macrodispersion of only 1–13% for the various CNT concentrations. In 1,2-dipalmitoyl-sn-glycero-3-phosphocholine (DPPC), the 0.02 and 0.1% MWCNTs showed a macrodispersion of 11 and 74%, respectively, while the 0.02 and 0.1% SWCNTs showed a macrodispersion of 15 and 16%, respectively. In 0.5% bovine serum albumin (BSA), the 0.02, 0.1, and 0.5% MWCNTs showed a very good macrodispersion of 32, 53, and 70%, respectively, yet the 0.02% SWCNTs only showed a macrodispersion of 17%. In 1% Tween 80, the 0.02–0.5% SWNCTs exhibited a good macrodispersion of 27–81%, whereas the 0.02–05% MWCNTs only showed a macrodispersion of 13–23%. Also, SWCNTs were dispersed [18] in water using a range of natural (gum arabic, amylose, Suwannee River natural organic matter) and synthetic (polyvinylpyrrolidone, Triton X-100) dispersing agents that attach to the CNT surface noncovalently via different physisorption mechanisms. The charge and the average effective hydrodynamic diameter of suspended SWCNTs as well as the concentration of exfoliated SWCNTs in the dispersion were found to remain relatively stable over a period of 4 weeks. A strong dependence of the toxicity of SWCNT suspensions on the toxicity of the dispersant and point to the potential of noncovalent functionalization with nontoxic dispersants was suggested as a method for the preparation of stable aqueous suspensions of biocompatible CNTs.

5.1.4 Combinations of Surfactants

An intriguing approach (Fig. 5.3) for the stabilization of double-walled carbon nanotubes (DWNT) in aqueous media was developed [19]. A *low-molecular-weight* surfactant was used in the first stage for the debundling of the nanotubes followed by substitution with a *higher-molecular-weight* surfactant or nonionic surfactants. A combination of different types of surfactants {SDS, Tween 20 (**5.1**), Surfhope 1216 {sucrose ester of lauric acid (C_{12} fatty chain)}, Montanov 82 (emulsifying alkylpolyglucoside), cholates} and a polymer surfactant (gum arabic, high-molecular-weight (HMW) surfactant) was used for the aqueous dispersion of DWNTs. The surfactant SDS was first used for exfoliation and surface coating of the carbon nanotubes. It can penetrate easily inside the aggregates of carbon nanotubes and is useful for improving the dispersibility; however, it does not protect individual nanotubes from agglomeration during film drying. In order to achieve better stabilization, wrapping of CNT with more or less amphiphilic polymer chains was used. The hydrophobic part of the polymer is strongly anchored to the hydrophobic nanotube surface with a polymer layer of sufficient thickness, while the hydrophilic parts, when fully ionized, impart sufficient ionic charge to the CNT surfaces. The resulting dispersions showed better stability after centrifugation compared to regular dispersions with each surfactant alone. Low-molecular-weight surfactant molecules were shown to be successfully removed during the substitution process and were not found in dispersions. Mixtures of CTAB and SDS surfactants with different mixing ratios (especially 90:10 cationic rich and anionic rich of CTAB and SDS) were considered to find the best composition for dispersing of MWCNTs [20].

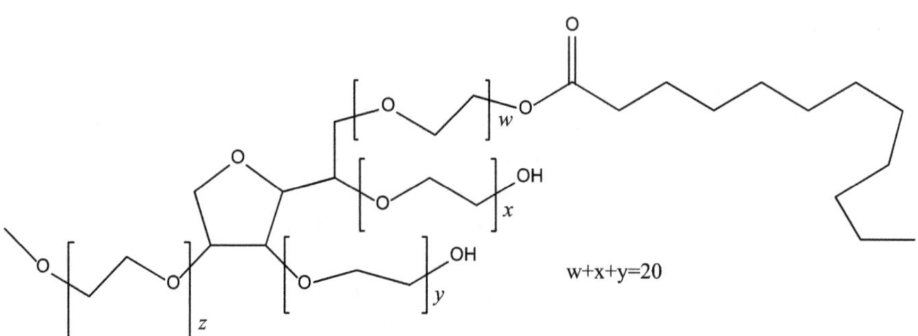

5.1

Polysorbate 20 (Tween 20)

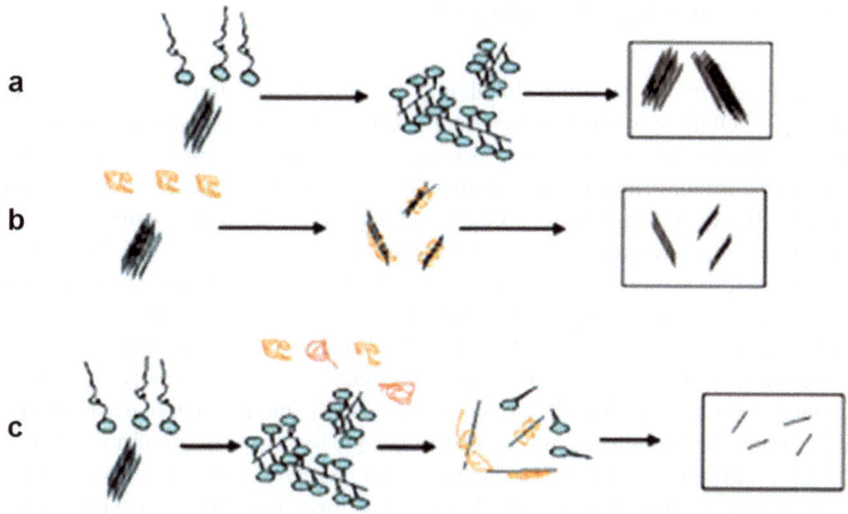

Fig. 5.3 Schematic illustration of the mechanisms of stabilization with different surfactants: (**a**) LMW surfactant; (**b**) HMW surfactant; (**c**) substitution of LMW surfactant by HMW surfactant (Reproduced with permission of the *Royal Society of Chemistry*; Datsyuk et al. [19])

Suspensions of MWCNTs (0–1 wt.%) in *binary liquid mixtures* (water content from 0 to 100%) water +1-cyclohexyl-2-pyrrolidone (CHP, **5.2**) were studied in the temperature range of 253–318 K, in the heating and cooling cycles [21]. It was established that dispersing quality of MWCNTs in a mixture of *good* (CHP) and *bad* (water) solvents may be finely regulated by adjustment of composition of these mixtures. Dispersion of MWCNTs was found to be almost ideal in the good solvent (pure CHP), meanwhile the introduction of bad solvent (water) into suspension caused formation of MWCNT aggregates. This aggregation was relatively unessential at $0 \leq X \leq 0.3$ ($X = [\text{water}]/([\text{CHP}] + [\text{water}])$). This fact can be explained by high viscosity and relatively high dispersing ability of CHP + water mixtures in the presence of CHP hydrates.

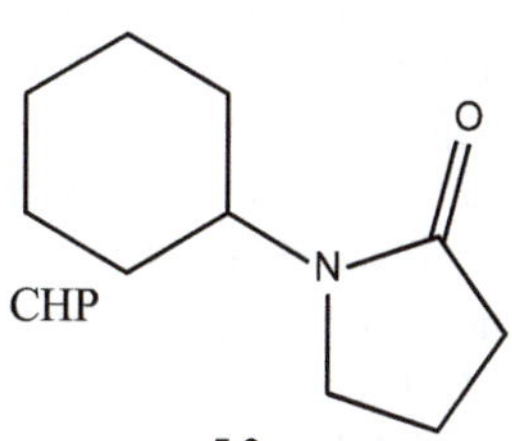

5.2

5.1.5 UV and Visible Light Influence

The photochemical reactions involving colloidal dispersions of carboxylated SWCNT–COOH in sunlight were examined [22]. Production of reactive oxygen species (ROS) during irradiation occurs and is evidence for potential further phototransformation and may be significant in assessing their overall environmental impacts. In aerated samples exposed to sunlight or to lamps that emit light only within the solar spectrum, the probe compounds, furfuryl alcohol (FFA), tetrazolium salts (NBT^{2+} and XTT), and *p*-chlorobenzoic acid (pCBA), were used to indicate production of $^1O_2$, $O_2\bullet^-$, and •OH, respectively. All three ROS were produced in the presence of SWCNT–COOH and molecular oxygen ($^3O_2$). $^1O_2$ production was confirmed by observing enhanced FFA decay in deuterium oxide, attenuated decay of FFA in the presence of azide ion, and the lack of decay of FFA in deoxygenated solutions. In a related report [23], it was also confirmed that, in certain conditions, CNTs could generate singlet oxygen ($^1O_2$) and hydroxyl radical (•OH) under irradiation of simulated sunlight, while superoxide radical ($O^{2\bullet-}$) was not detected. Also, SWCNTs were functionalized by the covalent attachment of 2-propanol-2-yl radicals, generated by photolysis of 2-hydroxy-2-methyl-1-phenyl-1-propanone under UV light, to their surface in THF solution [24]. A loss of Van Hove singularities and decrease of the intensity ratio of the G band and D band (IG/ID) were observed. The solubility in common organic solvents was improved, and the original electronic structure of the SWCNTs was retained without severe modification that damages the nanotubes.

5.1.6 Reactivity of Soluble CNTs Toward Active Oxygen Sources

Soluble SWCNTs were oxidized with singlet *oxygen* ($^1\Delta_g$), and the reaction progress was monitored utilizing FTIR and UV-vis-NIR spectroscopy [25]. The results indicated reversible covalent addition of oxygen to the walls of the nanotubes, most likely producing either the [2 + 2] or [4 + 2] cycloaddition product **5.3**. Dilute aqueous *ozone* solution with or without ultrasound was used to functionalize SWCNTs [26] (see also Chap. 3). Both O_3 and O_3/ultrasound treatments greatly increased the stability of SWCNTs in water. The oxidation pathway was proposed as follows: at the onset of the oxidation reaction, the C = C double bond was first converted to -C-OH which was then oxidized to -C = O and O = C-OH concurrently. Ozonating CNTs in fluorinated solvents (perfluorinated polyethers) to functionalize the sidewalls of the carbon nanotubes yielded functionalized CNTs with oxygen-containing functional moieties was reported in [27, 28]. Short (about 15 min) and long (about 3 h) exposures to ozone as well as cold (−78 °C.), r.t., and hot (50 °C) temperatures were tested.

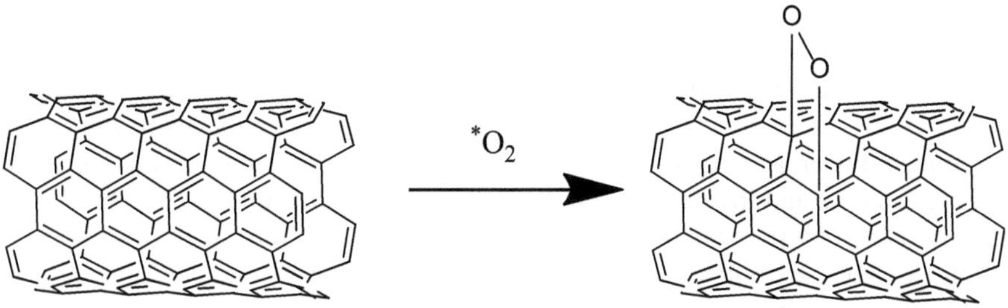

5.3

Products of SWCNTs oxidation with oxygen

The reaction of H_2O_2 with an aqueous suspension of water-soluble HiPco SWCNTs encased in the surfactant sodium dodecyl sulfate was studied [29]. Preliminary studies on the mechanisms suggested that H_2O_2 withdraws electrons from the SWCNT valence band by charge transfer, which suppresses the nanotube spectral intensity. CNTs were modified by oxidation with *peroxygen compounds* (inorganic peroxoacids, peroxycarboxylic acids of the formula $Q(C(O)OOH)_n$, hydroperoxides of the formula $Q(OOH)_n$, salts thereof, and combinations of any of the above, where Q is an alkyl, cycloalkyl, aryl, or heterocyclic group of C_1 to C_{12} and n is one or two) [30]. The oxidized CNTs included carbon- and oxygen-containing moieties, such as carbonyl, carboxyl, aldehyde, ketone, hydroxy, phenolic, esters, lactones, quinones, and derivatives thereof. Oxidation of the nanotubes increases the degree of dispersion of aggregates of nanotubes and aids in the disassembling of such aggregates. The dispersed nanotubes are used to prepare rigid structures and can be used in electrodes and capacitors.

5.1.7 Selected Studies on Carboxylated and Amine-Functionalized CNTs

The size-dependent colloidal behavior of aqueous dispersions of carboxylated multiwall carbon nanotubes (c-MWCNTs) was discussed [31]. The presence of carboxylic groups provided electrostatic stabilization in water, where the size affected agglomeration. While aspect ratio did not show any definite correlation, the hydrophobicity indices (HI), zeta potential, and aggregation kinetics showed dependence on the length of the c-MWCNTs where the shorter c-MWCNTs showed significantly lower HI values, smaller particle aggregates, higher zeta potential values, and higher critical coagulation concentrations (ccc) in the presence of electrolytes.

Two cell culture systems, human A549 pneumocytes and D384 astrocytoma cells, were used [32] to assess cytotoxicity of MWCNTs with varying degrees of functionalization and to understand their potential harmful effects. Laboratory-made highly functionalized hf-MW–NH_2 and less functionalized CNTs (MW–COOH and MW–NH_2) were tested in comparison with pristine MWCNTs, carbon black, and silica (SiO_2) by MTT assay and calcein/propidium iodide (PI) staining. hf-MW–NH_2 were found to be water soluble and easily dispersible in medium; they presented lower aggregate size range as well as considerably lower length to diameter ratios and low tendency to form aggregates compared to the other CNTs tested. The properties obtained by chemical functionalization, such as water solubility, high dispersibility, and low agglomeration tendency, were relevant factors in modulating cytotoxicity.

5.1.8 Purification of Soluble CNTs

Purification of soluble CNTs from impurities (carbon nanoparticles, graphite fragments, amorphous carbon, etc.) can be carried out made by several methods, in particular by *chromatography* [33, 34], flow field-flow *fractionation* [35], and *centrifugation* [36]. Thus, through a systemic study of a series of centrifuged solutions, the authors of [37] confirmed by Raman spectroscopy that heavily functionalized amorphous carbon was fractionated into the early centrifuged solutions, whereas lightly functionalized graphite fragments as well as polyhedral carbon and metal catalyst particles were fractionated into the late centrifuged solutions and centrifuged residue, and then highly pure and well-dispersed SWCNTs were collected from the middle centrifuged solutions. It is proposed that the purity, dispersibility, and aggregation state of SWCNTs can be qualitatively estimated by the relative intensity of their absorption features, the fine structure, and the slope of their absorption curves.

5.1.9 Other Particular Studies

A simple model describing the solubility of rods in solvents, expressing the final result explicitly in terms of the surface entropy and the enthalpy of mixing, was used to describe experimental results for the concentration of dispersed nanotubes in various solvents [38]. Qualitative agreement with these predictions is observed experimentally. However, the authors suggested that the fact that quantitative agreement is not found may be explained by solvent ordering at the nanotube surface. Also, the potential of asymmetrical flow field-flow fractionation (A4F) coupled to ultraviolet spectrometry (UV) and multi-angle light scattering (MALS) for the study of SWCNT dispersion in aqueous solutions containing a surfactant was demonstrated [39]. This technique is a powerful analytical tool that is able to evaluate SWCNT dispersion states in aqueous media and determine the presence or absence of aggregates, the numbers and sizes of different SWCNT populations, and the SWCNT size distribution.

Despite certain possibility of CNT destruction, aggressive acidic media continue to be an object of recent investigations. Thus, the *oxidation of MWCNTs in nitric acid* was monitored using sample weight, Raman spectrum, solubility, morphology, and alignment [40]. It was noted that high solubility (20–40 mg/ml) is obtained only after prolonged exposure (24–48 h) in concentrated acid (60%) with a considerable loss of the product (60–90%); the MWCNTs are strongly fragmented and covered by amorphous carbon after 48 h of oxidation. Moreover, it was found that the solubility correlates well with the area ratio of the G and D bands from the Raman spectrum. In a similar investigation, aqueous dispersions of SWCNTs, prepared using different dispersing agents, were also analyzed by Raman spectroscopy [41]. The influences of different dispersing agents and excitation wavelengths were discussed in comparison with UV-visible spectroscopic analysis data. The authors offered to use the most effective dispersing agent found in this study, sodium dodecylbenzene sulfonate, as a benchmark for future dispersion experiments. Also, direct imaging of SWCNTs, MWCNTs, and graphene, dissolved in *chlorosulfonic acid* and *oleum*, and the use of novel cryo-TEM and cryo-SEM specimen preparation are reported [42].

EPR studies on pristine, purified, shortened, and soluble SWCNTs in various solution phases showed that the soluble SWCNTs carry about 1 unpaired electron per 10,000 carbon atoms and give a free electron g-value [43]. *Density functional theory* calculations were carried out to study the effects of covalently binding isoniazid, an antitubercular compound to functionalized carbon nanotubes [44]. Binding energies, energies of solvation, and quantum chemical molecular descriptors were calculated. Significant differences were observed between SWCNTs and MWCNTs by investigating them in depth by conventional *electrochemical techniques* in solution [45, 46] Despite functionalization strongly modified the electronic properties of carbon nanotubes, the enrichment of the density of states of MWCNTs with respect to SWCNTs, due to larger tube diameters, is still appreciated. The bulk electronic properties of pyrrolidine-functionalized nanotubes (**5.4**) were obtained from cyclic voltammetry measurements and discussed in the light of quantum chemical calculations [47].

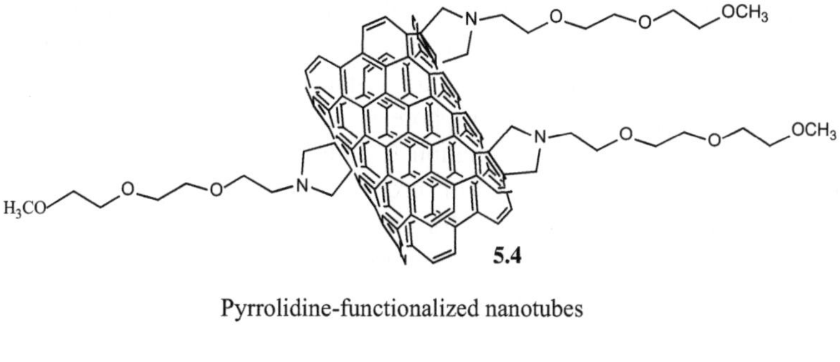

5.4

Pyrrolidine-functionalized nanotubes

Molecular dynamics simulations (MD) of liquid water near the external surface of a carbon nanotube bundle were presented in [48]. Hydrogen bonding, diffusive behavior, and rotational and vibrational motions are analyzed in the low- and high-density regimes. MD simulations of dispersion of CNTs in surfactant/polymer systems are also discussed [49]. *Supercritical water* in the presence and absence of oxygen was used for the first time for study of the opening and thinning of MWCNTs [50]. Among other investigations on CNTs, we note studies of optically sensing *sonication effects* [51], *biological effects* (for instance, the impact of dispersed SWCNTs on mitochondria [52]), and *toxicity*, depending on the MWCNT dispersion state [53], relations between *colloidal stability and percolation* phenomena in aqueous suspensions of MWCNTs [54], *destabilization* of surfactant-dispersed carbon nanotubes by anions (in particular, the effect was stronger using NaSCN than NaCl; see also the CNT precipitation above) [55], and *toxic effects* [56]. The *generation of* sp^3*-hybridized carbon atoms* in the sp^2-bonded lattice could also lead to dispersibility changes [57]. Conditions for the presence of surfactants inside the CNTs are established [58]. *Spectral methods* for dispersion characterization of carbon nanotubes in aqueous suspensions are described in [59]. Additionally, a method, a system, and an apparatus to determine a *concentration* of carbon nanotubes in a solution were proposed [60]. It was also established [61] on SDS example that dominant mechanisms of interaction of the surfactant molecules with the CNT surface in aqueous solutions depend on the *nature of counterions* (Li, Na, Cs). The authors suggested that a modification of CNT surfaces is related to the processes of a dynamic nature: adsorption of some of the counterions and restoring the integrity of a certain number of dissociated surfactant molecules.

5.2 Special Investigations on SWCNT Dispersion

Multiwalled carbon nanotubes (MWCNTs) and single-walled carbon nanotubes (SWCNTs) are frequently objects of comparison of their certain properties [62–66], in particular their dispersion in solvents, polymers [67, 68], or chlorosulfonic acid [69, 70]. In case of SWCNTs, in general the methods for their dispersion are similar to those used for MWCNTs and based, for instance, on physical (density gradient ultracentrifugation [71], gel electrophoresis [72], dielectrophoresis [73, 74], plasma, chromatography, and irradiation techniques) and chemical (ozonolysis; diazonium salts; functionalization with bromine, porphyrins, pyrene, DNA, peptides, and other biomolecules (needed for surface engineering of CNTs for targeting purposes [75]); various amines; polymers; use of dendrimers [76]; etc.) methods. Many of these methods have problems with effectiveness, dissolution sensitivity, and scalability. Quantitative evaluation of SWCNT solubilization is generally carried out by UV-visible absorbance of SWCNT suspensions [77, 78] as well as applying Design-Expert® software (Version 7.0.0, Stat-Ease, Inc. Minneapolis, USA) for additional calculations [79].

Several investigations, both theoretic and experimental, have been carried out to understand better the global dispersion processes of SWCNTs in aqueous and organic media and their particular aspects (principal methods for SWCNT "dissolution" are shown in Fig. 5.4). Thus, to *predict the dispersibility* of SWCNTs in a variety of organic solvents (**5.5–5.33**), a quantitative structure–activity/structure–property relationship (QSAR/QSPR) approach was used [80]. It was established that (1) heavier solvents (and small in size) most probably are better solvents for SWCNTs and (2) higher polarizability of the solvent molecule increases the dispersibility. The role of *ultracentrifugation* in dispersion processes was studied for arc discharge-produced SWCNTs and then functionalized by several routes, dispersed in different aqueous media containing dispersants (Table 5.1) and purified by ultracentrifugation [81]. Among other important results, the purity of all dispersions considerably improved; however, the centrifugation yield, the degree of purity, and the spectrum profile were influenced by the surface functional groups, the SWCNT type, and the dispersion medium. After SWCNT dispersion, the ultracentrifugation process always leads to substantially purified supernatant dispersions. Also, a good dispersion, orientation, and concomitant-polarized photoluminescence of SWCNTs in a *nematic chromonic liquid crystal* is also known [82], causing an *alignment* of individual SWCNTs (align parallel to the liquid crystal director) over stable macroscopically large domains.

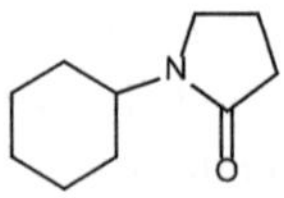
5.5

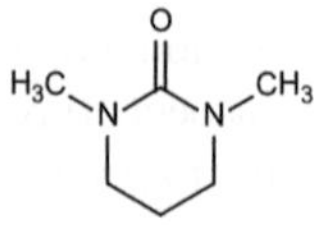
5.6

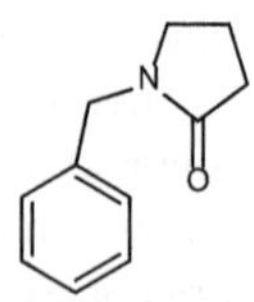
5.7

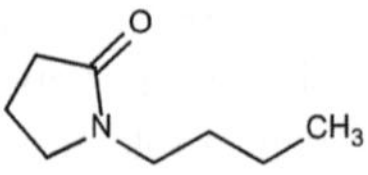
5.8

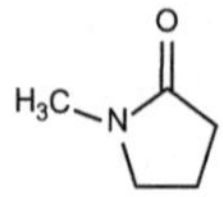
5.9

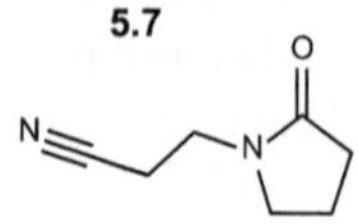
5.10

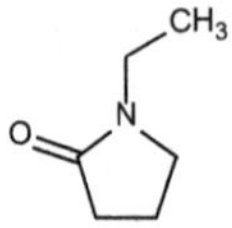
5.11

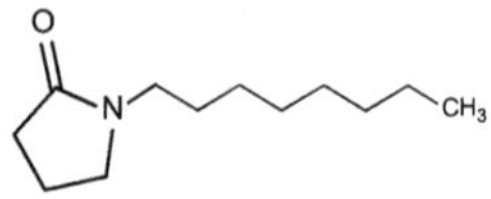
5.12

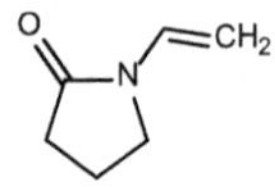
5.13

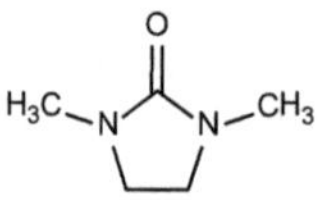
5.14

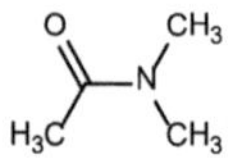
5.15

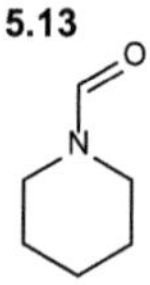
5.16

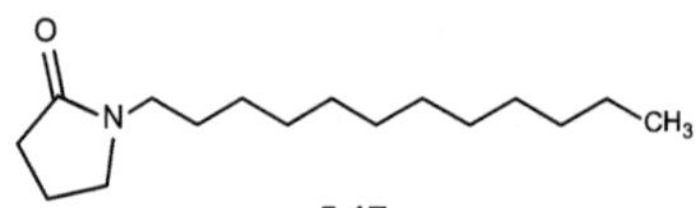
5.17

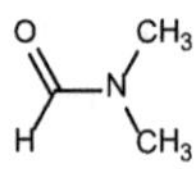
5.18

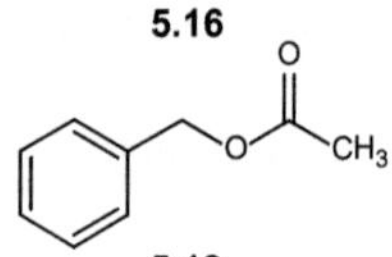
5.19

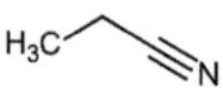
5.20

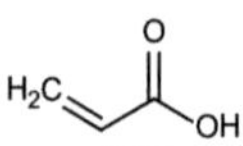
5.21

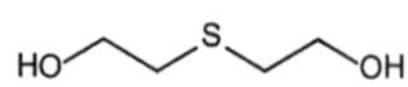
5.22

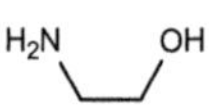
5.23

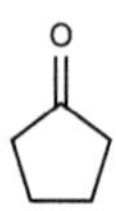
5.24

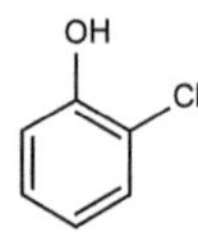
5.25

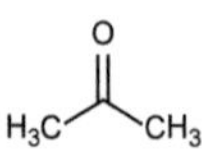
5.26

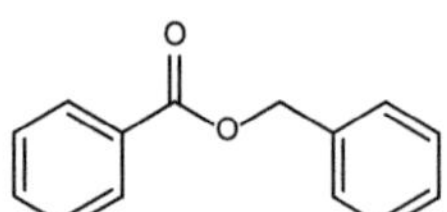
5.27

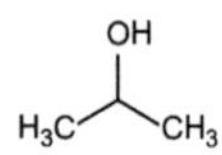
5.28

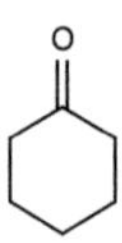
5.29

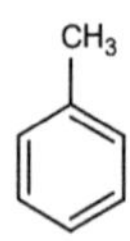
5.30

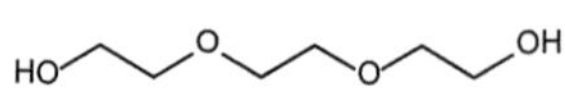
5.31

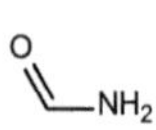
5.32

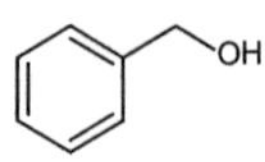
5.33

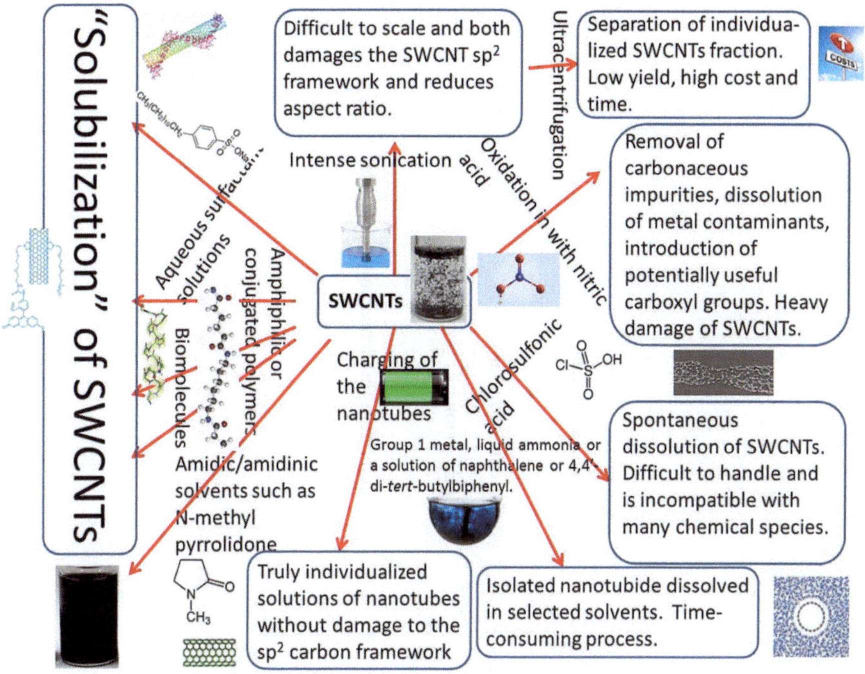

Fig. 5.4 Selected methods for SWCNT dispersion

SWCNT functionalization and, consequently, dispersion ability depend on the presence of impurities on nanotube wall surface, in particular other carbon allotropes. It was established that, applying standard nitric acid treatment, the functionalization can be increased by first *removal of the amorphous carbon* on SWCNT surface [83]. The amorphous carbon is the most reactive carbon form on SWCNT surface, and its presence leads to the formation of oxidation debris and undesirable prevention of SWCNT surface against functionalization. In addition, an intriguing *double-covalent* functionalization of SWCNTs (stepwise functionalization of the tube surface with two different organic moieties), involving both solubilizing ionic liquids and electroactive moieties, was reported [84]. Oxidized SWCNTs were first amidated with ionic liquid precursors and further treated with *n*-butyl bromide to afford SWCNTs functionalized with 1-butylimidazolium bromide. This allowed thus modified SWCNTs to get a relative solubility in water. Electron-acceptor units control the electronic properties of SWCNTs and imidazolium units modulate the solubility.

The *stability* of formed SWCNT, as well as MWCNT, dispersions is also important in terms of how to slow down or prevent CNTs reaggregation [85]. *Thermodynamic approach* (choosing a compatible solvent or surfactant) or *kinetic approach* (using a highly viscous solution/melt) was discussed. The following aspects were taken into account: (1) strong van der Waals binding energies for

Table 5.1 Dispersants used for modeling the dispersibility of SWCNTs

Abbreviation	Name	Structure	Category	C [Wt./vol. %]
SDBS	Sodium dodecylbenzenesulfonate	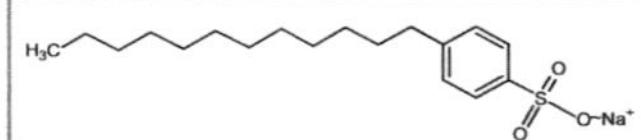	Anionic surfactant	
CTAB	Hexadecyltrimethylammonium bromide	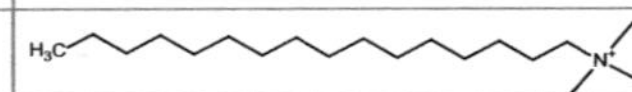	Cationic surfactant	0.5
Plu	Pluronic® F-68	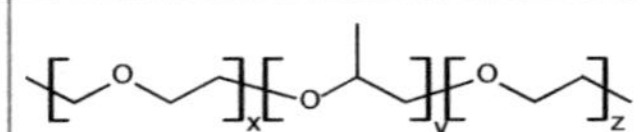	Block copolymer	1
GA	Gum arabic from acacia tree	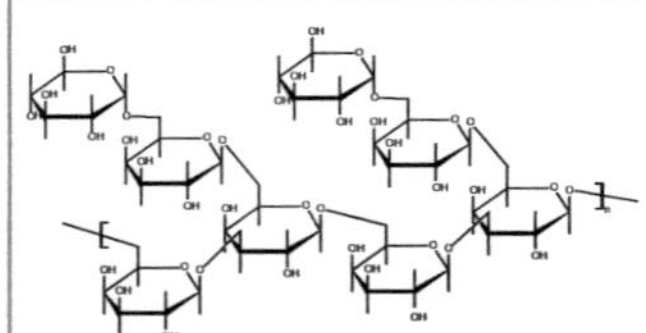	Natural gum (branched polysaccharide)	0.5

Reproduced with permission of the *Elsevier Science*

Table 5.2 Summary of dispersion mechanisms for various surfactants

Binding / Stabilization	Hydrophobic	Stacking/amine e-pair donation	Stacking
Electrostatic repulsion	*A*	*B*	*C*
	Sodium dodecyl sulfate and related salts Phospholipid	Single-stranded DNA, water-soluble proteins	Pyrene acid-derived salts
Steric hindrance	*D*	*E*	*F*
	Triton X/Pluronic range; Tween/Polysorbate range	Polyvinylpyrrolidone (aqueous, NMP)	Conjugated block-copolymers Pyrene siloxane (non-polar organic solvent) Conjugated polymer (polar organic solvent)

The surfactants are divided into six generic groups, labeled A–F. Those E–F are for dispersion in organic solvents, whereas the rest are for dispersion in aqueous solution (Reproduced with permission of the *MDPI*)

the CNT aggregates; (2) ratio between gravitational and Brownian forces; (3) apparently enhanced dispersion through "sonication cutting"; (4) the limiting tube length achievable by ultrasonication; (5) use of appropriate surfactants, for example, octadecylamine; (6) noncovalent dispersion techniques, using appropriate molecules of surfactant or solvent having favorable interaction with the curved graphene wall of the CNT surface; and (7) zeta potential of the surfactant/CNT complex, among others. The mechanisms of binding and stabilization are shown in Table 5.2.

An intriguing strategy to mitigate effects of unintentional filling by purposefully prefilling the SWCNTs is described [86], paying attention to effective and controllable *filling* of SWCNTs with a wide range of organic compounds through liquid-phase exposure prior to nanotube dispersion. Stable encapsulation was found to be achieved for all cases in which the compound is both small enough to fit into the SWCNT and sufficiently immiscible with the post-dispersion bulk liquid environment. In case of alkane and their derivatives, the encapsulation was further shown to enable tunable modulation of optical and other physical properties of the combined SWCNT/guest molecule complex such as density. Simple immersion in many liquids leads to controlled and specified filling, without energy-intensive processing or detrimental effects to SWCNT yield.

In addition, the effects of functionalization of SWCNTs were *DFT*-investigated [87] using a model, in which a finite, (5,0) zigzag nanotube segment containing 60 C atoms with hydrogen atoms added to the dangling bonds of the perimeter carbons. Three water-dispersible SWCNTs were functionalized with (a) formic acid, as a model of carboxylic acid; (b) isophthalic acid, as a model aromatic dicarboxylic acid; and (c) benzenesulfonic acid, as a model aromatic sulfonic acid. The functionalization with these organic molecules, either on the nanotube sidewall or tip sites where a hydrogen atom has been removed, was shown to be thermodynamically favorable. The enhanced dipole moment of the nanotubes due to functionalization was established in the improvement of their solubility in water. Also, *^{13}C NMR study* of highly soluble ^{13}C-enriched SWCNTs, functionalized with diamine-terminated oligomeric poly(ethylene glycol) (PEG(1500N), allowed detection of CNTs in solution [88].

5.3 Dispersion of Different Types of CNTs (SWCNTs, DWCNTs, and MWCNTs)

There are a few critical investigations, describing a comparative dispersion of CNTs depending on their type and shape. In general, but not always, all methods described above are just for all types of carbon nanotubes, SWCNTs, DWCNTs, and MWCNTs. Grade of dispersion could considerably differ depending on tube lengths, diameters, bundle sizes, and applied methods. Sometimes, nanotube type is important: thus, as it was shown above (see section "Natural Products"), MWCNTs can be dispersed using humic acid, but *not* SWCNTs. In a representative and very important report [89] in this area (described above in detail in the section "Mechanical and Related Methods"), 11 different dispersion techniques (Nanomizer, high-pressure jet mill, probe sonicator, ball mill, bead mill, paint shaker, ball collision mill, cone mill, rotor mill, high-shear batch disperser, and thin-film spin mixer) were used as a comparative performance for dispersion of long SWCNTs, short MWCNTs, and short SWCNTs to elucidate the most appropriate dispersion methods for the different types of CNTs.

A multidimensional stability analysis of five different types of long and short SWCNTs, DWNTs, and MWCNTs was carried out as a function of both buffer pH and salt concentration [90]. It was found that, in the same conditions, some CNT samples remain nearly unaltered, whereas others are significantly broken up, leading to altered CNT dimensions and surface properties (Fig. 5.5). In a related report [91], five kinds of CNTs (two SWCNTs and three MWCNTs) were dispersed in water by ultrasonic methods, and their geometrical properties were measured by image analysis from their TEM and SEM images. Flexibility of CNTs defined from the ratio of CNT fiber length and distance between the fiber ends was proposed. The flexibility of CNTs (both SWCNTs and MWCNTs) was found not to be influenced by the fiber length. On the other hand, the flexibility of MWCNTs depends on the tube diameter of CNTs. Similar investigations were carried out for CNT dispersion in polymers [92].

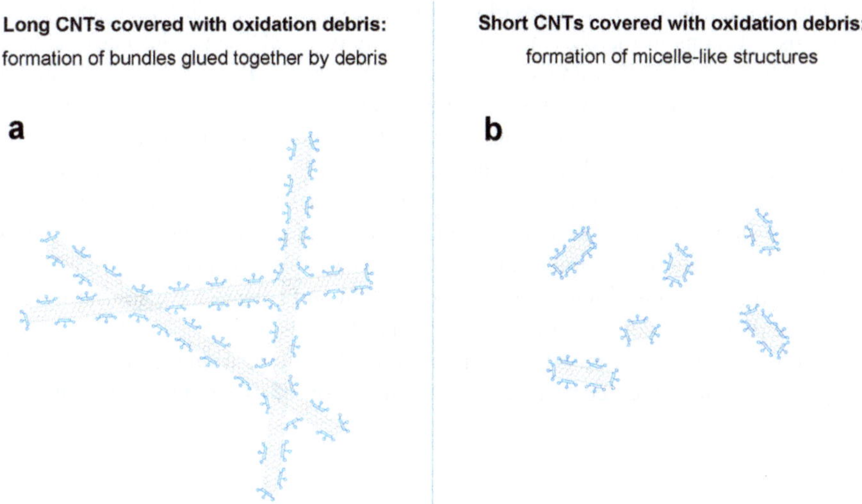

Fig. 5.5 Possible effect of oxidation debris on the bundling behavior of (**a**) long CNTs, suggesting the formation of bundles, and (**b**) short CNTs, suggesting the formation of micelle-like structures (Reproduced with permission of the *American Chemical Society*)

5.4 Characterization of CNT Dispersions Within Solution

Carbon nanotubes and other carbon allotropes are normally studied and characterized by a series of modern techniques, such as, first of all, electron microscopy methods SEM, TEM, and AFM, as well as XRD; UV-visible, Raman, and IR spectroscopy; and small-angle X-ray and neutron scattering, among rarer others (dynamic light scattering, photoluminescence, rheological and electrical percolation). Being processed in liquids (polymer melts, molten urea, water, organic solvents, solutions of surfactants), a lack of uniform characterization methods and standards for MWCNT and SWCNT dispersions ("macrodispersion" of CNT bundles or "nanodispersion" of individual CNTs, since a typical CNT dispersion contains both individual CNTs and small bundles) is clearly observed [93]. For SWCNTs, more expanded classification was offered, testifying that their following hierarchical structures with a broad range length scales can be found in a dispersion [94]: (1) individual tubes with different molecular structures; (2) SWCNT bundles, composed of multiple individual tubes organized into a 2D hexagonal lattice; (3) SWCNT aggregates formed by the topological entanglement or enmeshment of individual tubes and/or SWCNT bundles; and (4) SWCNT network that spans the overall dispersion sample.

To evaluate the degree of nanodispersion, a number of experimental methods above are applied. All they have certain limits, for instance, SEM and TEM can cause aggregation during sample preparation, SEM generally cannot distinguish between bundles and individually dispersed nanomaterials, UV-visible is only qualitative, and photoluminescence is limited to aqueous dispersions. A typical example of application of a series of methods is as follows: MWCNTs, dispersed in aqueous solutions in the presence of surfactants (e.g., SDS and Pluronic), natural organic matters (e.g., Suwannee River humic, Suwannee River fulvic acid), and a polysaccharide (alginate) via long-term magnetic stirring or ultrasonication, were characterized [95] with a variety of instrumental techniques – quartz crystal microbalance and AFM (to evaluate the deposition of MWCNTs with different dispersants on crystal sensor surfaces) and dynamic light scattering and asymmetrical flow field-flow fractionation (to determine the size and size distribution of prepared samples). Among other important results, it was established that MWCNTs suspended in the same solution by ultrasonication and stirring have similar hydrodynamic diameters but quite different size distributions.

In this section, we discuss the most common characterization methods for CNT dispersions, which are generally used in combination of two and more techniques.

5.4.1 UV-Visible Spectroscopy and Its Combinations

Spectral methods, in particular UV-visible spectroscopy, in relation with CNT dispersions are described in a recent comprehensive review [96]. In particular, it was emphasized that, by contrast to bundled CNTs, the completely dispersed CNTs are highly active in the 200 to 1200 nm wavelength region. The peak wavelengths are as follows: 253 nm for MWCNTs and 972 and 1710 nm for SWCNTs. The evaluation of the degree of dispersion in water under ultrasonication can be studied by recording the UV-vis spectra of the dispersions since individualized CNTs show characteristic bands in the UV region. At the beginning of sonication, maximum absorbance shows an increase and reaching a plateau depending on the surfactant concentration. A maximum absorption can be affected by CNT purity, since CNTs with more impurities are lesser strongly bound among each other making exfoliation simpler. A longer ultrasonication treatment or higher ultrasonication energy may lead to CNT damage and fragmentation.

Concentration Studies for MWCNTs [97] and SWCNTs The studies of the SWCNT concentration in dispersions showed [98] that this is an important indicator to evaluate the dispersibility of the surfactant-dispersed SWCNT suspension. Both the total concentration of the dispersed SWCNTs and the concentration of individually dispersed SWCNTs can be evaluated by the data of the absorbance and the resonance ratios of UV-vis-NIR absorption spectra, respectively. It was concluded that the wavelength range between 300 nm and 600 nm should be the most suitable for evaluation of the total concentration of dispersed SWCNTs in the suspension; also, wavelengths below 800 nm should be most suitable for evaluation of the concentration of individually dispersed SWCNTs in the suspension. In a related work [99], unknown concentrations of surfactant-dispersed SWCNTs in the supernatant solution, simply by measuring its optical absorbance, in combination with cryogenic transmission electron microscopy (cryo-TEM) and thermogravimetric analysis techniques. It was also established that the dispersion of higher SWCNT concentrations requires longer sonication time [100]. The optimal concentration of surfactant is 1.5 wt.% which is needed for effective dispersion; the concentration of SWCNTs that can be homogeneously dispersed in aqueous solution is about 0.4 mg/ml. In general, the following data could be got applying UV-visible spectroscopy to SWCNT dispersions:

- Relationship between UV-vis-NIR spectra and concentrations of dispersed SWCNTs
- Optimal original concentrations of SWCNTs dispersed
- Optimization of surfactant concentration for SWCNT dispersion
- Effects of sonication time on SWCNT dispersion
- Effects of sonication temperature on SWCNT dispersion

Separation Studies UV-visible-NIR absorption, in combination with fluorescence spectroscopy, dynamic light scattering, Raman scattering, and atomic force microscopy, was also used [101] for studying processes of separation of SWCNTs by length via centrifugation in a high-density medium. Slower separation was found to produce narrower distributions. The influence of a series of separation parameters was also studied: SWCNT concentration, rotor speed, added salt concentration, liquid density, the processing temperature, and surfactant concentration. Scale processing was 10 mg/day and can be applied in all SWCNT production methods.

Role of Functional Groups Dispersibility of chemically functionalized f-MWCNTs in water solution, adding small amount of a cationic ionic liquid (IL), 1-tetradecyl-3-methylimidazolium chloride, was investigated by UV-visible and FTIR spectroscopy and compared with unfunctionalized MWCNTs [102]. It was proved that the presence of functional groups on the sidewalls of MWCNTs considerably affected their dispersibility in water with IL as dispersing agent.

Evolution of Particle Size Dynamic light scattering (DLS) and UV-vis spectroscopy were combined for a real-time and quantitative monitoring MWCNT dispersion (Fig. 5.6) [103]. Two types of MWCNTs of different diameters and lengths were studied. The evolution of mean particle size distribution and transparency of MWCNTs suspension was identified (Fig. 5.7), determining the effective dispersion parameters and dispersion quality. For the pair "longer CNT/smaller diameter," the dispersion process was achieved at a steady 100 nm of agglomerate size with 40 min of sonication; meanwhile for the pair "shorter CNT/large diameter/less entangled" (60-min sonication), an agglomerate size of 100 nm was observed.

5.4.2 Raman Spectroscopy and FTIR

Raman spectroscopy is a powerful technique for the study of carbon nanotubes, providing information about nanotube type, the defects [104], purity, and tube alignment, and it assists in the distinction of the presence of MWCNTs relative to

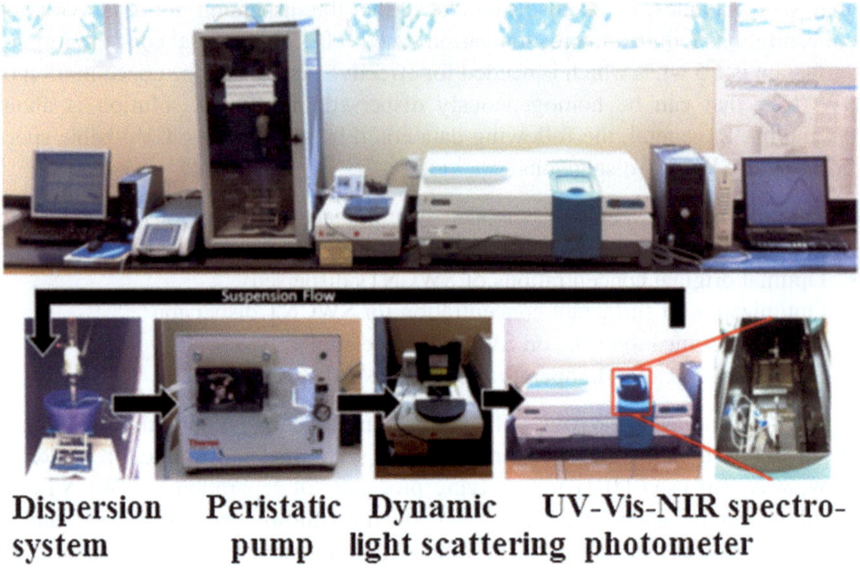

Fig. 5.6 Integrated setup of the in-line dispersion quality monitoring system (Reproduced with permission of *Springer*)

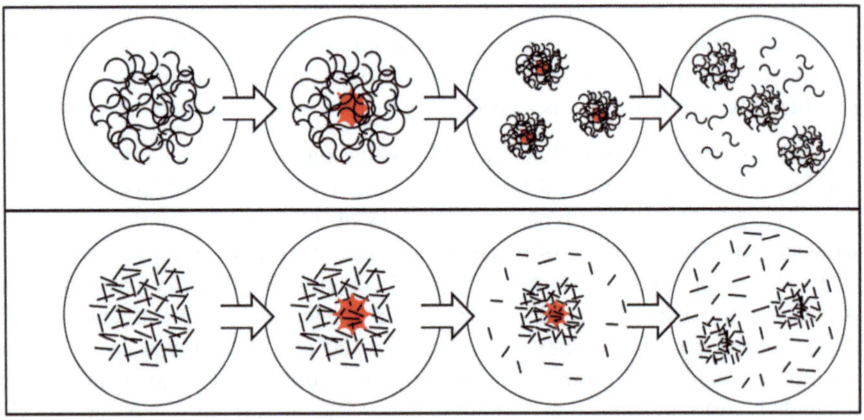

Fig. 5.7 Evolution of disaggregation of agglomerates of two different types of MWCNTs (up, outer diameter of 10 nm, 3 μm long; down, outer diameter of 13–16 nm, over 1 μm long) (Reproduced with permission of *Springer*)

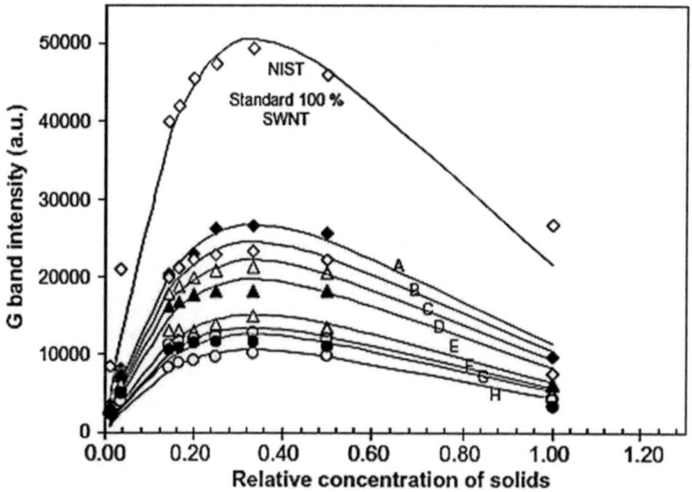

Fig. 5.8 G-band intensity of SWCNT suspensions of varying concentrations relative to those showing the same absorption intensity (Reproduced with permission of *Elsevier Science*)

other carbon allotropes [105, 106]. The effects of various parameters, such as surfactants, CNTs' geometry, sonication time, sonication method on CNTs' quality post-sonication, chemical functionalization, or doping, can be evaluated [107]. Indeed, we cannot imagine the nanocarbon chemistry without Raman spectroscopy [108], similarly like all nanotechnology cannot exist without electron microscopy studies and resulting images.

Purity Measurements and Doping The quantitative method to access purity of SWCNTs consists of measuring in Raman spectra the G-band intensity of liquid suspensions of SWCNT samples as a function of solid concentration (Fig. 5.8) [109]. The measurement of G/D ratio in liquid suspensions as a function of solid concentration provides a more practical and reliable method than others (electron microscopy, thermogravimetric analysis, and optical absorption). A quantitative analysis of the doping of individual CNTs in SDBS at the concentrations below ~0.07 g/l was conducted by varying the pH from 1 to 13 and monitoring the Raman spectrum [110]. The de-aggregation process can be monitored through changes in the absorbance as a function of concentration. Both metallic and semiconducting nanotubes can be doped by varying the pH of the dispersion.

Crystalline Properties DWCNTs in ethylene glycol (EG) and N, N-dimethylformamide (DMF) media without the addition of surfactant were investigated by Raman and FTIR spectroscopy by ultrasonication method to examine their crystalline perfection before and after dispersion (Figs. 5.9 and 5.10) [111]. Crystallinity and surface structure of DWCNTs were found to undergo minimal change. Upon sonication, DWCNT bundles were observed broken into smaller bundles and can be redispersed. In addition, homogenously dispersed and highly concentrated DWCNTs can carry out mass production of Pt/DWCNT catalysts with high electrochemical activity.

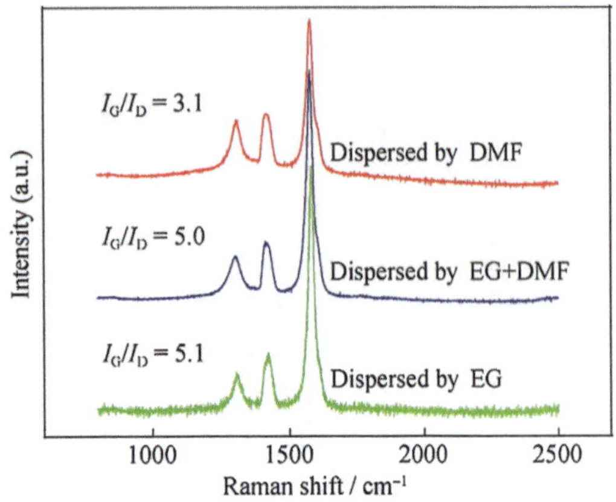

Fig. 5.9 Raman spectra of DWCNTs after dispersion in DMF, 5 vol.% DMF + 95 vol.% EG, and EG (Reproduced with permission of *Springer*)

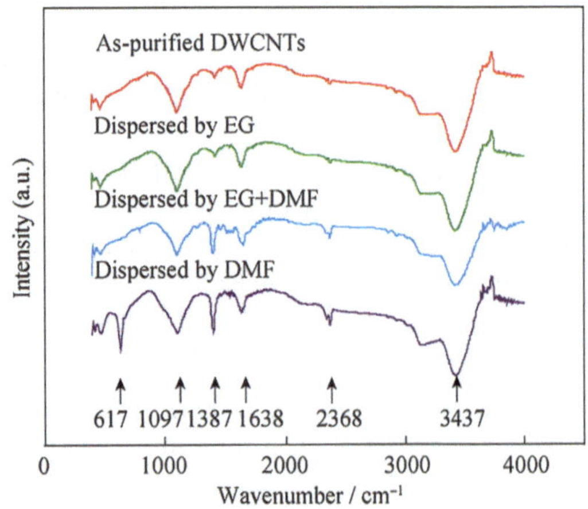

Fig. 5.10 FTIR spectra of DWCNTs before and after dispersion in DMF, 5 vol.% DMF + 95 vol.% EG, and EG (Reproduced with permission of *Springer*)

Structural, Electronic, and Other Properties of CNTs Studies on the structure and electronic properties of carbon nanotubes dispersed in a liquid phase showed that the radial breathing modes are upshifted in suspensions due to the molecular pressure of the solvent [112]. The frequency of these modes can be used to estimate the CNT diameters and charge transfer between CNTs and donors or acceptors. The influence of the dispersion on the vibrational properties of the carbon nanotube samples was studied by Raman spectroscopy [113]. Thus, MWCNTs were denatured in concentrated nitric acid and, after several intermediate treatments, were dispersed in 30 ml DMF solution. The denatured MWCNTs exhibited a small increase of frequency, but a clear decrease of the intensity of both the D and G bands, that can be explained by the lower MWCNT concentration in the solution. A low G/D intensity ratio was observed for both types of nanotubes (pristine and denaturate), corresponding to a low crystallinity of the MWCNT samples (lower G/D ratio due to the defects in the nanotube surfaces in denaturated samples).

Various functionalization effects have been studied by electron microscopy and Raman spectroscopy, for example, the noncovalent wrapping of lipid chains onto the graphitic surface of SWCNTs [114]. It was revealed that smaller diameter SWCNTs are preferentially dispersed by lipid molecules in the aqueous supernatant part of the prepared suspension. Among other reports, we note studies of DWCNTs, dispersed in an aqueous single-stranded DNA solution [115], and DWCNTs with different metallic and semiconducting configurations [116].

5.4.3 Small-Angle X-ray and Neutron Scattering

Small-angle scattering is a powerful technique for characterization of fractal objects, as well as clusters or aggregates of CNTs. Small-angle scattering (SAS) is the collective name given to the techniques of small-angle neutron (SANS), X-ray (SAXS), and light (SALS or just LS) scattering [117]. SAXS is sensitive to electron density variations in materials, so the internal structure of carbon nanotubes can be determined by calculating the electron density profile of the cross section [118]. To study CNT dispersions in water or molten phases, these methods have been mainly applied for SWCNTs. Thus, the concentrations of SWCNTs dispersed in molten poly(ethylene terephthalate) were studied using synchrotron radiation [119]. This method is useful to check the presence of branched rope-like structures of bundled nanotubes precluding the existence of isolated carbon nanotubes. X-ray scattering was also used to measure particle orientation in concentrated MWCNT suspensions in a Newtonian suspending fluid (uncured epoxy) under shear flow [120]. These dispersions exhibited shear thinning, approaching the matrix viscosity at high shear rates. The rheological and orientation data are discussed in terms of distortion, breakdown, and reformation of percolated MWCNT networks in these samples.

Noncovalent approaches to functionalize and solubilize CNTs were studied in detail using SAXS combined with TEM and UV-vis-NIR observations [121]. Different types of surfactants (cationic, nonionic, anionic, and biopolymer) were used to unzip CNT bundles. Specific wavelengths were elucidated in relation with the degree of CNT clustering. Also, noncovalently functionalized and individually isolated SWCNTs, well redispersible in water even after freeze drying, were prepared and studied by SANS [122]. These measurements showed that the SWCNTs do not form any network at the investigated range of concentrations.

As it is shown in other book chapters, surfactants are widely used to disperse CNTs, so the studies of their adsorption by SANS and SAXS provide important information for revealing more exact dispersion mechanisms and functions. Thus, the dispersion of SWCNTs in heavy water with the surfactant Triton X-100 was studied using SANS [123]. A competition, existing between maximization of surfactant adsorption onto SWCNT surfaces and a depletion interaction between SWCNT bundles mediated by surfactant micelles, was emphasized. In addition, a significant incoherent scattering from hydrogen was revealed in SWCNTs,

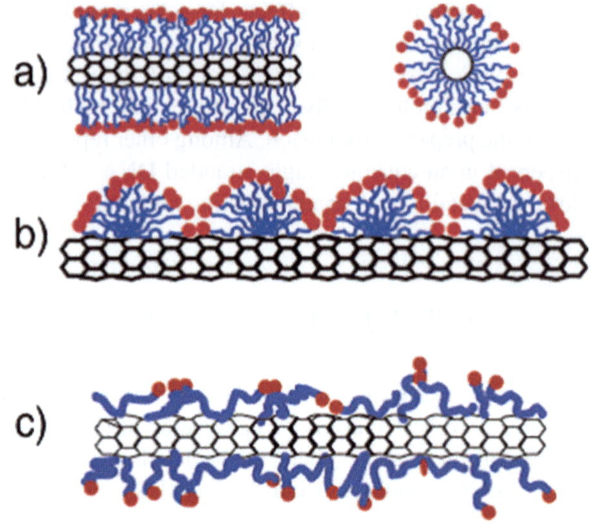

Fig. 5.11 Schematic representations of the mechanisms by which surfactants help disperse SWCNTs. (**a**) SWCNT encapsulated in a cylindrical surfactant micelle: *right*, cross section; *left*, side view. (**b**) Hemimicellar adsorption of surfactant molecules on a SWCNT. (**c**) Random adsorption of surfactant molecules on a SWCNT (Reproduced with permission of the *American Chemical Society*)

explained by acid and water residues left after the purification process. Adsorption of other well-known surfactant, SDS, on SWCNT surface was recently reported [124]. The SANS and SAXS studies showed the data can be fitted to a relatively simple core–shell cylinder model. This approach is consistent with a polydisperse nanotube core with 10 Å radius, surrounded by an adsorbed surfactant layer with 18 Å thickness and 0.5 volume fraction. The authors emphasized that these data are in contrast with earlier report [125], suggesting that SDS-dispersed nanotubes consist of single tubes surrounded by a thin and disordered SDS layer (Fig. 5.11). We also note (see below in AFM section) that SDS excess on CNT surface could affect AFM images and there is an approach allowing an improvement on their quality.

SANS was used with the "high-concentration" method to extract single-particle scattering from SWCNTs, labeled by covalently attaching -C_4H_9 or -C_4D_9 groups through free radical chemistry [126], which is able to attach a large mass of alkyl groups to the nanotubes, producing neutron contrast between different SWCNT batches. Such labeled SWCNTs can be used for measurements of the single-particle characteristics. The resulting single "particle" was not an individual SWCNT; indeed, this is a cluster of tubes despite vigorous sonication and applied dispersants. In a related study [127], the authors concluded that isolated SWCNTs can behave as rigid rods in suspension depending on the choice of surfactant. These rigid SWCNTs can also form a loose 3D network in suspension. This term appeared in other reports too. Thus, the microscopic network structure of SWCNTs, stabilized using NaDDBS as surfactant, in water was studied by SANS as a function of nanotube concentration in the semidilute (overlapping) regime [128]. Most of the samples exhibited rigid rod behavior, except at low SWCNT–surfactant ratios,

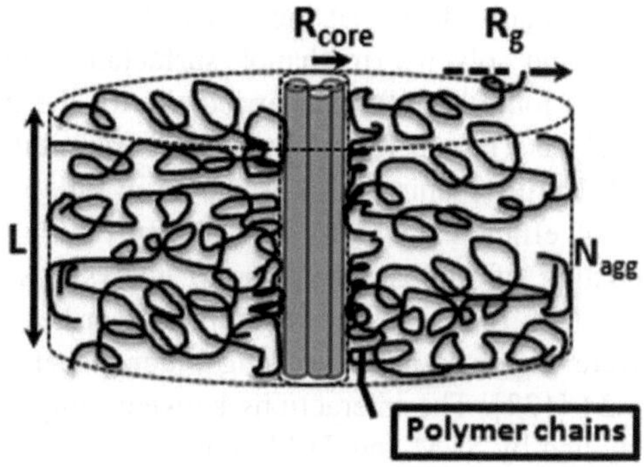

Fig. 5.12 Schematic drawing of the SWCNT/polymer core – chains model of length L (Reproduced with permission of the *American Chemical Society*)

when inadequately dispersed SWCNTs did not exhibit rigid rod behavior in the SANS profiles. At last, in a SANS study with the use of polymers, it was investigated [129] whether the coverage of Pluronic F127 polymers around the CNTs depends on the solution concentration in the range of 1–6% (w/w). The observations indicated that at these concentrations the SWCNT surface is fully saturated at about 14 chains per unit length of 100 Å (Fig. 5.12).

5.4.4 SEM Studies

Electron microscopy studies are absolutely obligatory in all nanotechnology works as a direct evidence of formed nanomaterials; they need to be additionally confirmed also by other techniques above. In case of its applications for CNT dispersions, their formation is frequently an intermediate step before further transformations resulting in nanocomposites, for example, (MWCNTs)/polyvinylpyrrolidone (PVP) composite nanofibers [130]. Since this step is very important and can influence on all further stages, its profound study by microscopy and other techniques is needed. Thus, scanning electron microscopy images were used for quantitative evaluation of carbon nanotube dispersion to improve the preparation of CNT-filled composites [131]. The CNTs, purified in a mixture of H_2SO_4/HNO_3 to remove impurities such as amorphous carbon, graphite particles, and metal catalysts, were blended homogenously with dispersant TX100 (polyoxyethylene nonylphenyl ether) and stirred in water to form a dispersion before adding epoxy resin. Two scalar values can be obtained from SEM images: the dispersion index (representing the dispersion state of CNTs) and distribution index (representing the distribution state of CNTs). The image analysis system can also be applied to various nanoparticle systems, such as nanofiber, nanobeads, and nanostructures.

Scanning electron microscopy was used to study carbon nanotubes dispersed in ionic liquids [132], whose advantages are as follows: (1) no damage of CNT surface and (2) the aggregation of the nanoparticles decreases, while the number of CNTs individually dispersed increases. A comparative study of the feasibility of three different solvents (methanol, surfactant Triton X-100 solution, and imidazolium-based ionic liquid) was carried out for the dispersion of MWCNTs prior to SEM analysis (Fig. 5.13). A typical representative set of images, corresponding to $[BMIm]PF_6$ ionic liquid, is shown in Fig. 5.14. Ionic liquids were found to be the most efficient in terms of stability of the dispersions; in addition, no metallic coating of the sample prior to analysis is needed for SEM image acquisition. Another representative example describes vertically aligned MWCNTs, which were dispersed in water using water-dispersible conjugated polymer nanoparticles **5.34** [133]. The interactions between polymer nanoparticles and MWCNTs were studied by SEM and TEM (Fig. 5.15) in combination with spectroscopy (UV-vis, fluorescence and Raman), confirming the attachment of polymer nanoparticles to the MWCNT sidewalls.

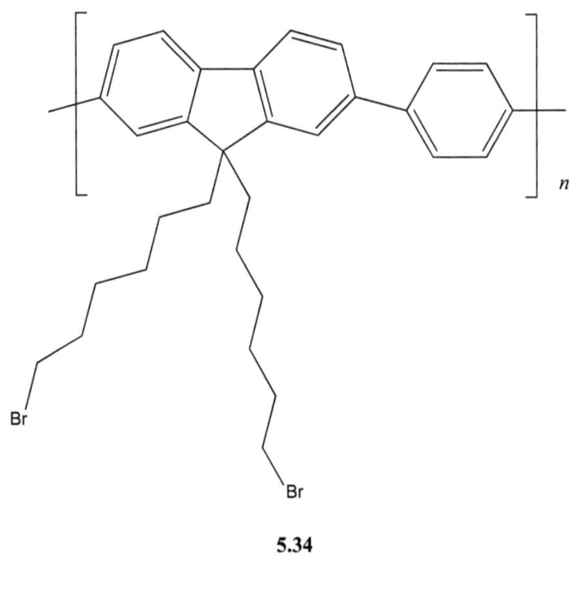

5.34

5.4.5 TEM Studies

Different steps, required for TEM analysis of dispersed nanomaterials, and methodologies to obtain homogeneous and stable dispersions of colloidal nanomaterials and powders are described in a recent book chapter [134]. In order to obtain complementary information on the morphology, size, electronic and crystallographic structure, and composition of nanomaterials, the application of TEM imaging methods, analytical TEM, and electron diffraction methods is reviewed. Various terms and approaches have been introduced. Thus, TEM-supported

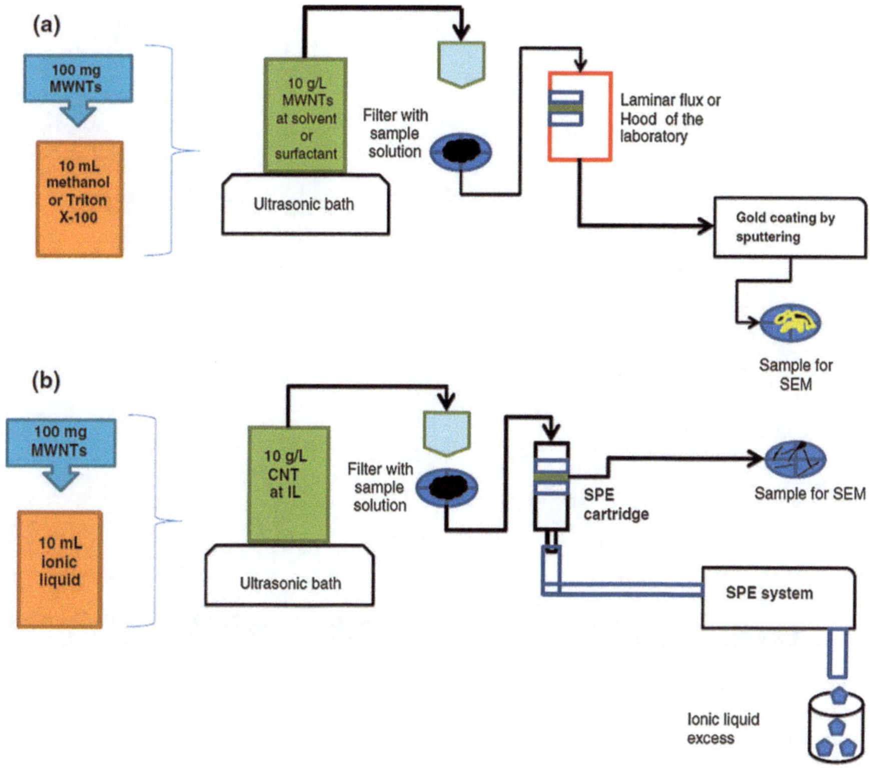

Fig. 5.13 Schemes of the procedures followed for the preparation of MWCNT samples prior to SEM analysis: (**a**) using methanol and a Triton X-100 solution as dispersing agents, (**b**) MWCNTs dispersed with the aid of ionic liquid BMIM PF_6 (Reproduced with permission of *Elsevier Science*)

experiments on a better understanding of the conditions for the surfactant-assisted dispersion of MWCNTs in DMF were carried out [135]. Authors defined an *ideal dispersion*, which uses the lowest amount of surfactant, supports the most MWCNTs, and exhibits acceptable stability for the end use. Two approaches (ultrasonication alone and redispersion of precipitate through a combination of *ultrasonication and centrifugation*) were proposed, where the second way was found to produce more stable dispersions.

Some other representative examples are as follows. Thus, dispersion of pristine SWCNTs in water by ultrasonication in the presence of a thiolated organosilane and subsequent ultracentrifugation was reported [136]. After dispersion, SWCNTs can be decorated with silica or gold nanoparticles (Figs. 5.16 and 5.17) due to the presence of thiol groups on SWCNT surface after functionalization. In addition, highly stable (4 months) MWCNT aqueous dispersions were observed using carminic acid (CA) as a dispersing agent (Fig. 5.18) [137]. The π–π stacking interactions between CA and MWCNTs provide good stability to the CA–CNT

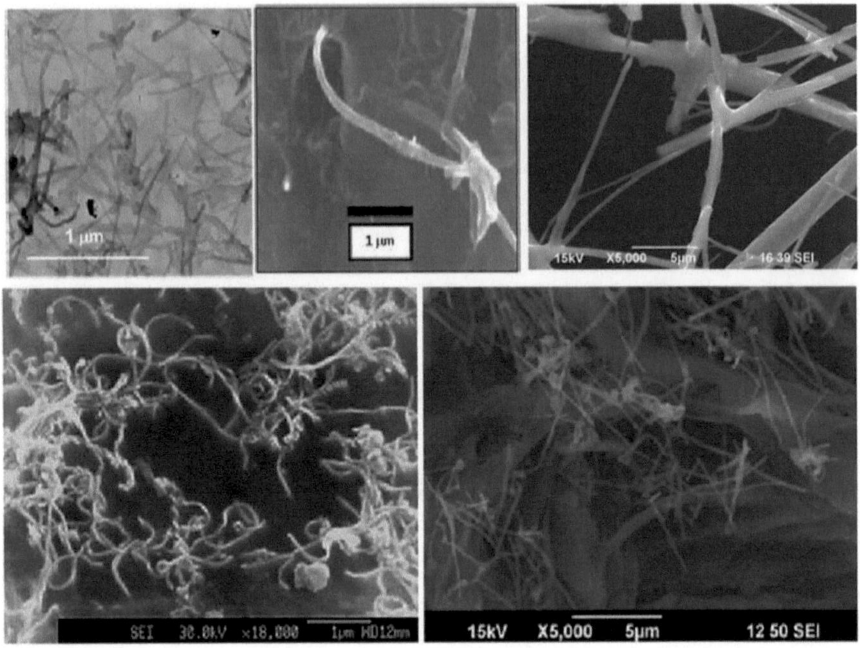

Fig. 5.14 SEM images of MWCNTs dispersed with the aid of [BMIm]PF_6 ionic liquid. A drop of the suspension was placed on a polycarbonate membrane and IL excess was removed (Reproduced with permission of *Elsevier Science*)

dispersion. Resulting CA–CNT could be used as an effective platform for immobilizing vital proteins and biomolecules.

5.4.6 AFM Studies

The atomic force microscopy studies are also useful for CNT dispersions for elucidation of some of their properties or reaction mechanisms, for instance, under dispersion of SWCNTs with supramolecular *Congo red* [138]. Sometimes, the presence of a surfactant could influence on image quality. Thus, the improvement measures to avoid unstable AFM images because of excess SDS molecules on SWCNT surface using sequential dialysis against pure water in order to remove these excess molecules from SDS–SWCNT hybrid dispersion were reported [139], showing that a 1:10^2 volume ratio of SDS–SWCNT dispersion to water in the dialysis is effective in realizing stable AFM observations. The hybrids were stable after dialysis even after 14 days of storage. On the other hand, CNTs could be affected by AFM tip. Thus, among other techniques used, an AFM nanomanipulation of ultrasonically dispersed and reflux-oxidized (Fig. 5.19) MWCNTs and SWCNTs by controlling the AFM tip with a NanoManipulator on

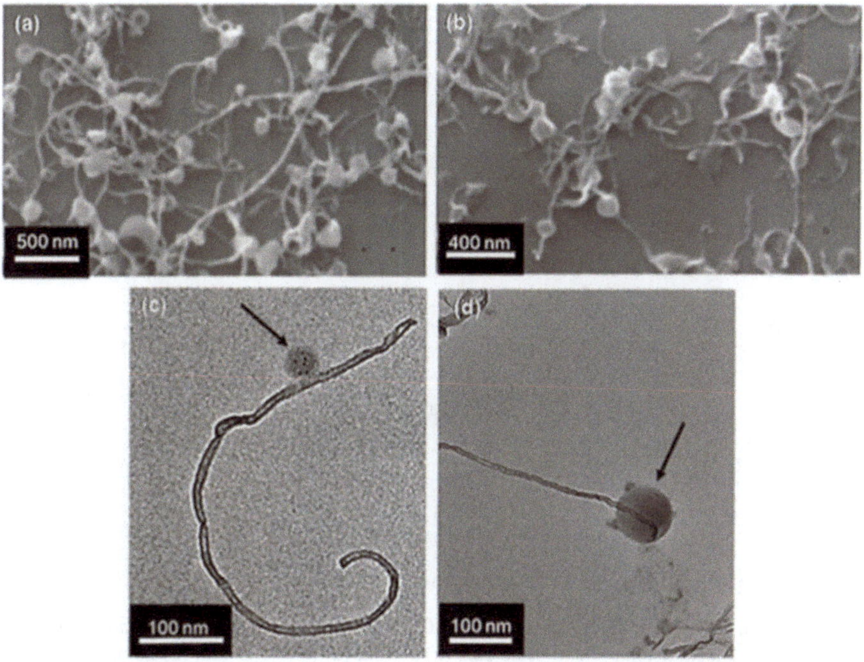

Fig. 5.15 SEM images of 40 nm (**a**) and 70 nm (**b**) polymer nanoparticles attached to CNTs; TEM images of 40 nm (**c**) and 70 nm (**d**) polymer nanoparticles (Reproduced with permission of *The Royal Society of Chemistry*; Baykal et al. [133])

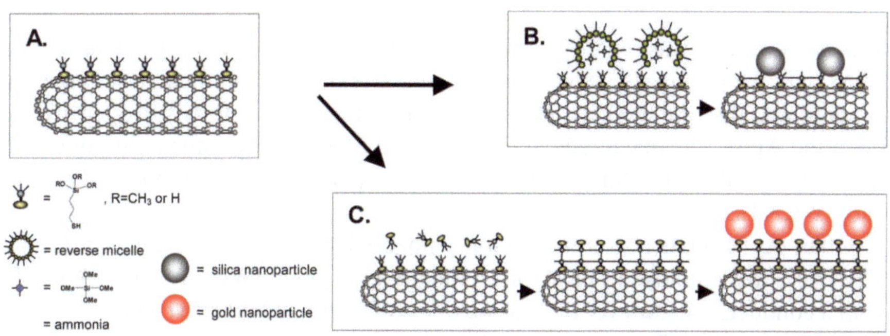

Fig. 5.16 Dispersion of pristine SWCNT by a thiolated organosilane and realization of supramolecular nanoassemblies. (**a**) SWCNT was dispersed in water by mercaptopropyl trimethoxysilane that adsorbed onto the sidewalls of the nanotubes through the thiol group leaving exposed the trimethoxy (or trihydroxy) silane group. (**b**) Silica nanoparticles were directly grown onto the sidewall of the functionalized SWCNT by a water-in-oil nanoemulsion system. (**c**) Gold nanoparticles were linked to the thiol groups extruding out of the SWCNT sidewalls after derivatization with a second layer of mptmos. Nanotubes, gold and silica nanoparticles, and the other molecules are not shown to scale (Reproduced with permission of the *American Chemical Society*)

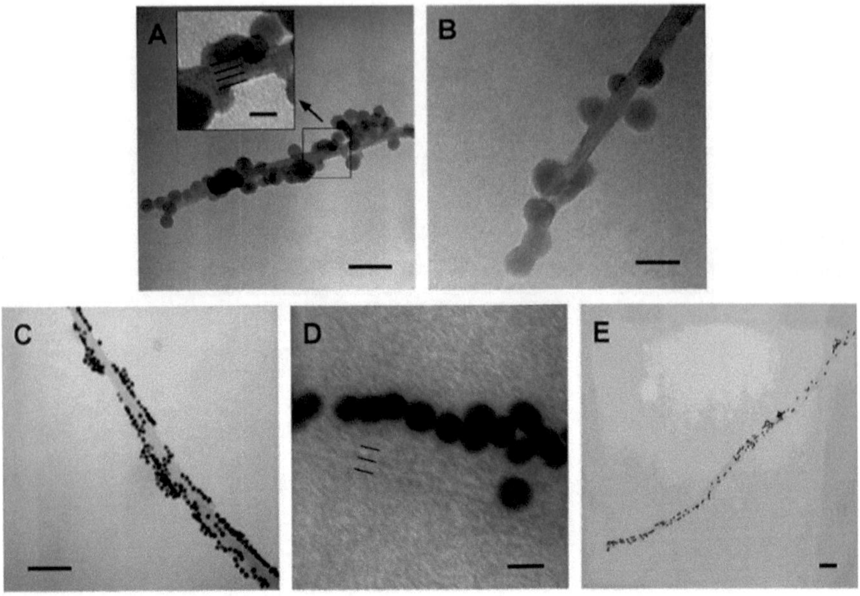

Fig. 5.17 (**a**, **b**) TEM images of dispersed SWCNT decorated with silica nanobeads by using a water-in-oil nanoemulsion system (scale bars (**a**) 50 nm, (**a**) (*inset*) 10 nm, (**b**) 20 nm). The inset shows that the fiber-like nanostructure exhibited equally spaced striations. (**c–e**) TEM images of dispersed SWCNT decorated with gold nanoparticles (scale bars (**a**) 100 nm, (**d**) 10 nm, (**e**) 100 nm). Approximately 3.5 nm-spaced striations are evidenced in (**d**) (Reproduced with permission of the *American Chemical Society*)

a silicon substrate was reported [140]. The diameter of the CNTs was confirmed by TEM and AFM images. It was shown that CNTs are deformed by combining bending and distortion when subjected to large mechanical forces exerted by the tip of the AFM. The magnitude of the force, required to deform the reflux-oxidized CNTs, is less than that for the ultrasonically dispersed CNTs, for both MWCNTs and SWCNTs.

Other AFM applications in relation with CNT dispersions are quite different. The AFM was employed to test the conductivity of both SWCNTs and MWCNTs with a conductive cantilever tip leading to the efficient and effective manufacturing of CNT-based ISFET (ion-sensitive field-effect transistor) for pH sensor applications [141]. CNT dispersion and their electrical property measurements are closely related with microchip fabrication. The quantitative characterization of the dispersion state (degree of dispersion) of SWCNTs using Raman spectroscopy and AFM was carried out [142]. Nanotubes were dispersed in SDS aqueous solution at seven different dispersion states by controlling ultrasonication and centrifugation parameters. This approach can be used to quantitatively characterize the dispersion state of SWCNTs, in a difference with only Raman studies, which provide qualitative measurements at 267 cm^{-1} and at the excitation wavelength of 785 nm.

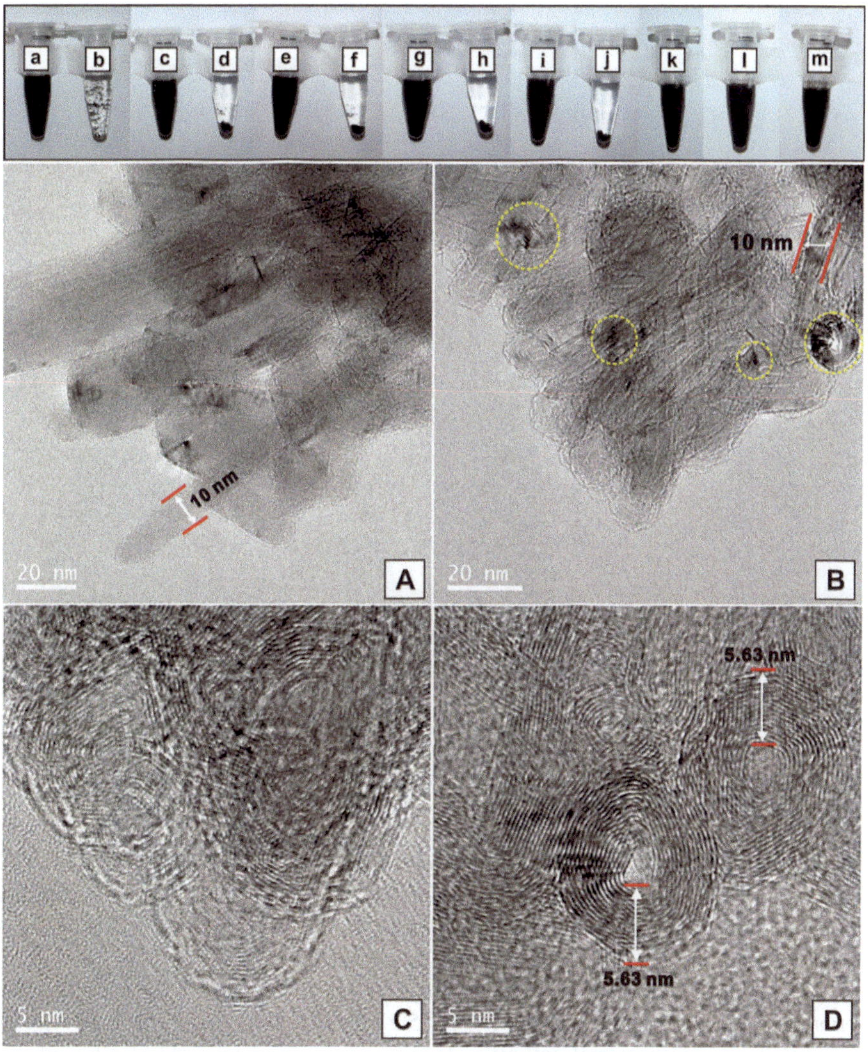

Fig. 5.18 Photographs of 1 mg/ml^{-1} of CA–CNT dispersion took after (*a*) 1 h; (*c*) 1, (*e*) 2, and (*g*) 3 weeks; and (*i*) 1, (*k*) 2, (*l*) 3, and (*m*) 4 months; CNT dispersion took after (*b*) 1 h; (*d*) 1, (*f*) 2, and (*h*) 3 weeks; and (*j*) 1 month. TEM images of (**a**) CNT and (**b–d**) CA–CNT at different magnifications. (**b**) The *dotted circles* in *yellow color* indicate the adsorbed CA molecules. The *arrows* in (**a**) and (**d**) indicate the OD and ID of CNT and CA–CNTs (Reproduced with permission of *Elsevier Science*)

Oligo(*m*-phenylene ethynylene)s can reversibly disperse (by functionalization via intermolecular π–π interactions) and release SWCNTs simply by changing the solvent (Fig. 5.20), consistent with a change from an unfolded state to a folded state [143]. Using AFM (Fig. 5.21) and a row of other methods, it was established that (1) the foldamer-dispersed SWCNTs are individually well dispersed and have a

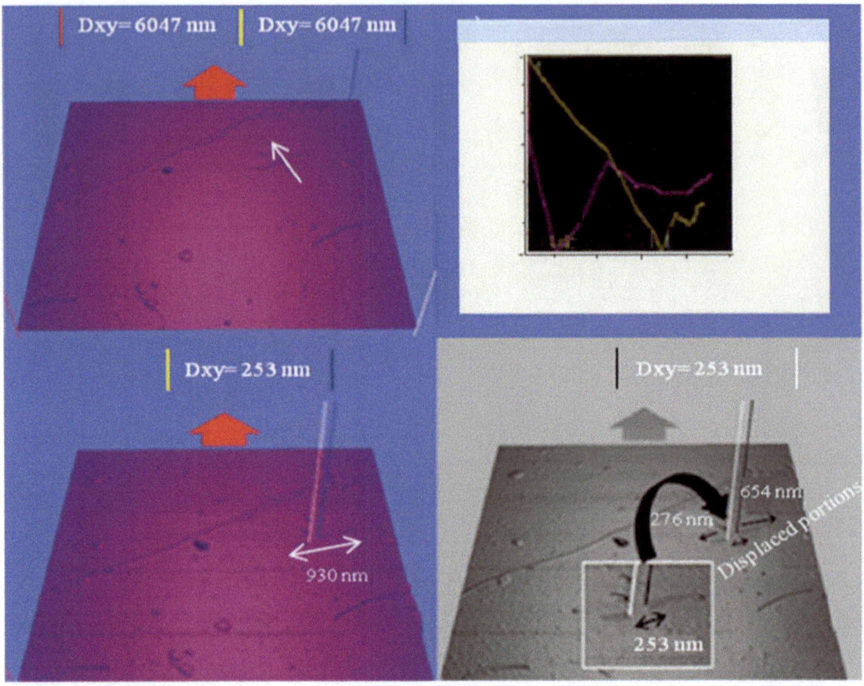

Fig. 5.19 Nanomanipulation of oxidized SWCNTs: (**a**) functionalized CNT arrow indicating the application of force, (**b**) force versus displacement plot, and (**c**, **d**) show final images after application of force showing the cut and lateral displacement (Reproduced with permission of *Springer*)

strong interfacial interaction with the foldamers and (2) the released SWCNTs appeared to be free of foldamers. Several possible optoelectronic applications were offered taking into account observed photoresponse of the transistors based on the foldamer-dispersed SWCNTs under illumination.

Antibacterial action of dispersed SWCNTs on *Escherichia coli* and *Bacillus subtilis* was investigated by AFM (Fig. 5.22) [144]. It is known that SWCNTs exhibit strong antibacterial activities, leading to cell damages. AFM allowed the monitoring dynamic changes in cell morphology and mechanical properties of two typical bacterial models above upon incubation with SWCNTs. It was observed that individually dispersed SWCNTs in solution develop nanotube networks on the cell surface, destroying the bacterial envelopes with leakage of the intracellular contents. It was concluded that the antibacterial activity of SWCNTs is the accumulation effect of large amount of nanotubes through interactions between SWCNT networks and bacterial cells.

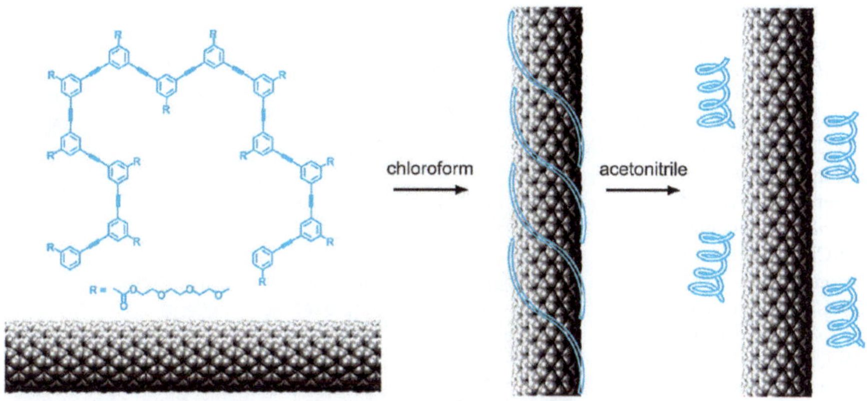

Fig. 5.20 Solution process for dispersion and release of SWCNTs by *m*PE-13mers (Reproduced with permission of the *American Chemical Society*)

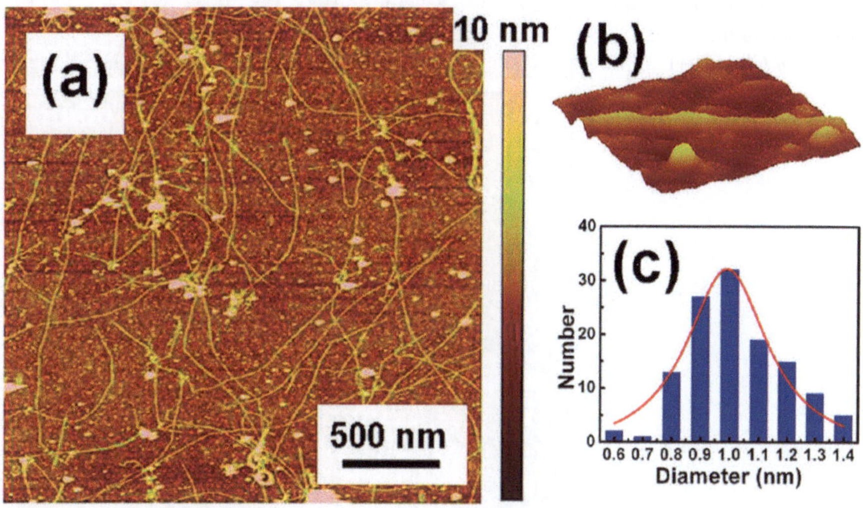

Fig. 5.21 (**a**) AFM image of individually separated SWCNT spin cast onto a SiO_2 substrate from a diluted SWCNT/mPE-13mer solution. (**b**) Typical 3D AFM image obtained over a single SWCNT (image size 120 nm × 120 nm). (**c**) Diameter distribution of the oligomer-wrapped SWCNTs (*blue*) and the Lorentzian fit (*red*) (Reproduced with permission of the *American Chemical Society*)

5.4.7 Electrical Conductivity Measurements

The dispersion dynamics of MWCNTs in distilled water was in situ monitored using instantaneous conductivity measurements (Fig. 5.23), showing the evolution with time and the existence of three distinct sonication phases [145]: (1) increase of conductivity fluctuations to a maximum (existence of large agglomerates of CNTs),

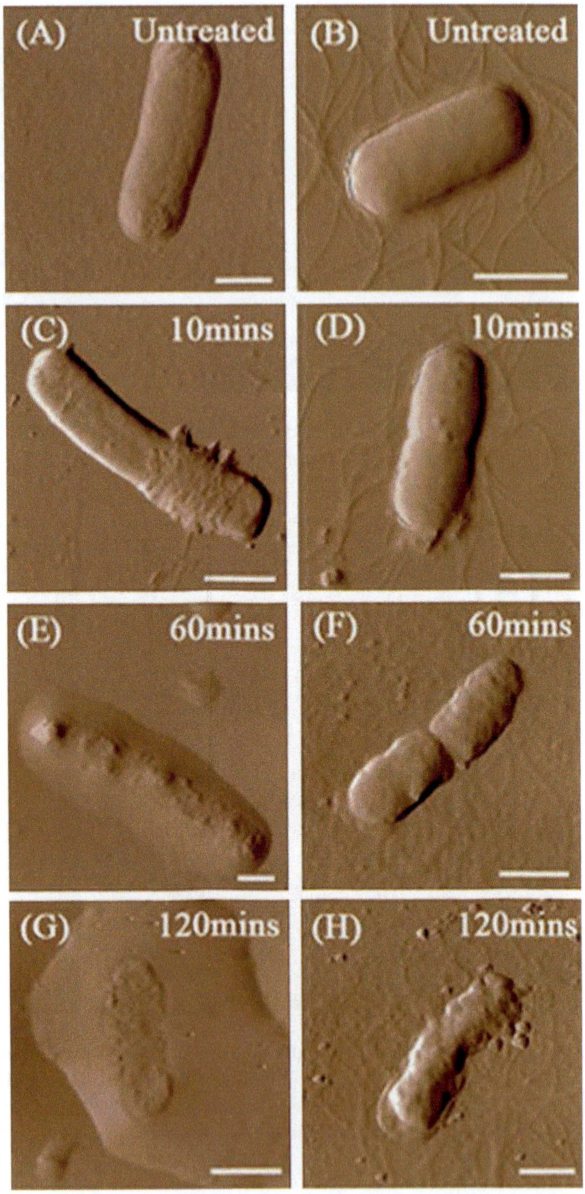

Fig. 5.22 AFM amplitude images of *E. coli* and *B. subtilis* (1 mL of bacterial suspensions, 106–107 cfu/mL) after incubation with SWCNT dispersions (10 mL, 5 mg/mL) over different time periods. (**a**) *E. coli* before contact with SWCNTs; (**b**) *B. subtilis* before contact with SWCNTs; (**c**, **e**, and **g**) *E. coli* after incubated with SWCNTs for 10, 60, and 120 min, respectively; and (**d**, **f**, and **h**) *B. subtilis* after incubated with SWCNTs for 10, 60, and 120 min, respectively. All scale bars are 1 mm (Reproduced with permission of the *Royal Society of Chemistry*)

(2) a rather steep increase of conductivity while fluctuations steadily declined (the breakdown of large CNT agglomerates, more homogeneous dispersion), and (3) the conductivity increase which was almost negligible after 650 kJ/L (no further sonication longer required). The optimum conditions were found to be as follows: sonication energy requirement is 650 kJ/L of the solution irrespective of the CNT concentration, 750 W sonicator, and 13 min of sonication.

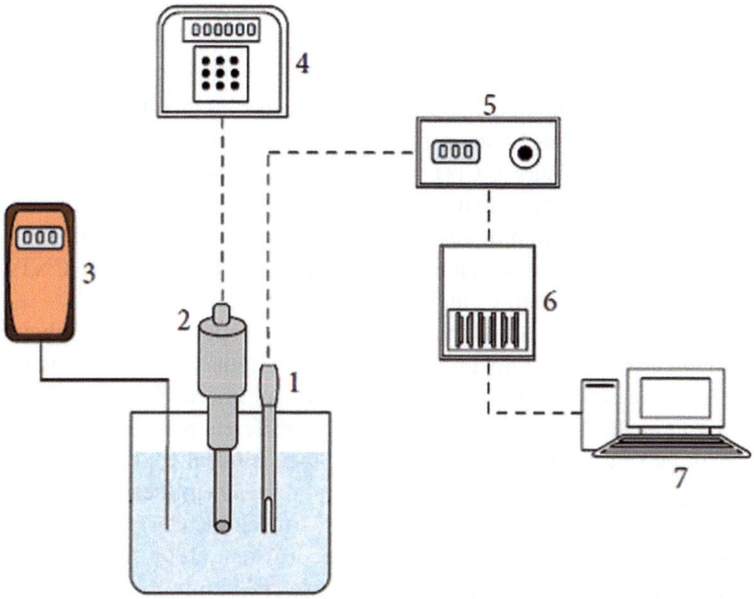

Fig. 5.23 Schematic of experimental setup: (*1*) conductivity probe, (*2*) sonicator probe, (*3*) thermometer, (*4*) sonicator, (*5*) conductivity meter, (*6*) data acquisition system, (*7*) computer (Reproduced with permission of *Hindawi Publishing Corporation*)

As it is seen from the selected material above, spectroscopy, microscopy, scattering, and other methods are applied for the characterization of CNT dispersions within solution, although not so frequently as for solid intermediates or final nanotube-based products. Since CNT dispersion and solubilization is frequently an important intermediate step for further applications, its profound study is needed for their correct choice.

5.5 Toxicity of CNT Dispersions

The CNTs are well known to be toxic, and the toxicity depends on their, for instance, structure (SWCNTs or MWCNTs), surface area, length, degree of aggregation, grade of oxidation, presence of functional groups, fabrication method, dose, and concentration. Biomedical use of dispersed CNTs or accidental inhalation of CNT aerosols could lead to undesirable effects, for example, prominent pulmonary inflammation, oxidative stress, induction of cytotoxic effects on lungs, or apoptosis in different cell types. CNTs contain many residual catalyst metals, such as Fe, Co, Mo, and Ni, remaining even after post-purification treatments, which affect biological cells causing harmful effects and inducing inflammatory and granulomatous reactions [146]. Both well-dispersed MWCNTs and SWCNTs have an inflammatory potential [147], in particular pulmonary toxicity [148–150].

There are several books [151, 152] and reviews [153] in this field. The handbook [154] describes, in particular, the following aspects and effects of CNT toxicity: interaction with dispersing medium and cytotoxicity assays, effect of impurities and dispersion, dermal exposure to CNTs, pulmonary response to CNTs, toxic response to CNTs in the intra-abdominal cavity, CNTs and immunity (allergic immune responses, etc.), recognition of CNTs by macrophages, CNT interactions with the cardiovascular homeostasis, genotoxicity and mutagenicity of CNTs, and biodistribution, biodegradation, and pharmacokinetics of CNTs.

CNT dispersions have a huge potential for clinical applications, but they can be used once their toxicity is fully understood. In vitro studies are useful in identifying the determinants of their toxicity; in vivo studies are informative in demonstrating that different methods of administration result in different pathologies. Overall, the CNTs may be safe for human use; however, their clinical application is possible after having good standardization and appropriate ways to measure nanotoxicity, when more than one measurement techniques to check CNT toxicity are used [155]. Their toxicity studies are sometimes inconsistent [156], but tendency of its dependence on CNT agglomeration is clearly observed.

Length and Agglomeration Effects Studying the toxicity of CNTs in dispersions, several effects have been established. Thus, CNT toxicity depends on their length [157]. Thus, the shorter MWCNTs are more toxic in zebrafish embryos after injection [158]. Another important point is related with CNT agglomeration due to van der Waals forces between CNTs: these effects should be taken into account. The reported results in this field are sometimes contradictory and depend on many other factors. Thus, comparing dispersed CNTs with solid commercial CNTs, it was proven for SWCNTs that well-dispersed individual SWCNTs are *more toxic* than agglomerates due to better physical puncturing of bacterial membranes and ruin the cell integrity [159]. Also, SWCNTs were well dispersed (diameter of bundle, 0.2 μm; length of bundle, 0.7 μm) by ultrasonication to conduct an inhalation study [160]. Wistar rats were exposed to them for 4 weeks, leading to the fact that these SWCNTs did not induce neutrophil inflammation in the lung under these conditions.

Talking more on the importance of CNT agglomeration, we note the following results on the toxicity of DWCNTs, investigated [161] to three aquatic organisms, namely, *Pseudokirchneriella subcapitata*, *Daphnia pulex*, and *Poecilia reticulata*. It was indicated that *ionic strength* enhanced DWCNT agglomeration, while humic acid stabilized the CNTs and in turn inhibited the formation of aggregates. In the presence of humic acid high DWCNTs, acute toxicity toward *D. pulex* and *P. reticulata* was observed, but ionic strength led to opposite effect irrespective of DWCNT form. The toxicity of DWCNTs in the aquatic systems (i) is dependent on media properties and (ii) is likely to proceed at different rates to organisms at different trophic levels.

Dispersant Nature Different results on agglomeration were described in other reports. Thus, well-dispersed SWCNTs were found to be *less toxic* than agglomerates, depending on the dispersant used [162]. In case of applying the surfactant Tween 80, the SWCNTs had fewer aggregates and were found to be less toxic; the acid-treated SWNTs were less toxic to cells using the neutral red cell viability assay. However, the authors noted that it was hard to differentiate whether this effect arose due to increased dispersion or increasing purity of the SWNTs. Dispersing SWCNTs in water using a series of natural (amylase, gum arabic, and Suwannee River natural organic matter) and synthetic (Triton X-100 and polyvinylpyrrolidone) dispersants (Fig. 5.24), whose molecules can be noncovalently attached to the SWCNT surface via different physisorption mechanisms, an absence of any toxicity to WB-F344 rat liver epithelial cells when the dispersant was nontoxic was noted [18]. So, a strong dependence of the toxicity of SWCNT suspensions on the *toxicity of the dispersant* was suggested. As a conclusion, *noncovalent functionalization with nontoxic dispersants* is a good method for the preparation of stable aqueous biocompatible SWCNTs. Indeed, the surfactant nature is extremely important for biomedical applications and directly related with possibility of toxic effects. Thus, the natural lung surfactant Survanta® was used to disperse SWCNTs in a biological medium [163]. At physiologically relevant concentrations, Survanta® produced well-dispersed SWCNTs without causing a cytotoxic or fibrogenic effect. Non-dispersed SWCNTs did not exhibit these effects, suggesting the importance of dispersion status of SWCNTs on bioactivities.

Disappearance of CNTs from Living Organisms The lung toxicity of nanoscale dispersed SWCNTs (mean diameter ~1 nm) was studied in mice [164]. The toxicity of SWCNTs in vivo was found to be attributable to aggregation of the nanomaterial rather than the large aspect ratio of the individual nanotubes. In addition, nanoscale dispersed SWCNTs disappear from the lung over time (Fig. 5.25). Biocompatible nanoscale dispersion provides a scalable method to generate purified preparations of SWCNTs with minimal toxicity, thus allowing them to be used safely in commercial and biomedical applications.

Preparation method of CNT dispersions is also important. Thus, for comparison, chronic toxicity of DWCNTs, dispersed using *sonication* (so-DWCNTs) and *stirring* (st-DWCNTs), was investigated with the *diatom Thalassiosira pseudonana*, *copepod Tigriopus japonicus*, and *medaka Oryzias melastigma* [165]. It was concluded that so-DWCNTs are consistently significantly more toxic than st-DWNTs, and dispersion method and size of aggregations should be considered in DWCNT toxicity testing. We note that, as it is seen throughout the book, the ultrasonication is a very frequent and main method for CNT dispersion, used in combination with chemical methods, so this fact above should be taken into account in biochemical experiments.

Exposure Time Acute and chronic immune responses over a short- (1, 3, 6, and 12 h) and long-time (from 1 to 5 weeks) exposure to MWCNTs by optical, electron, and immunohistochemical approaches were analyzed [166]. Immunocytochemical characterization using specific markers showed that in these inflammatory

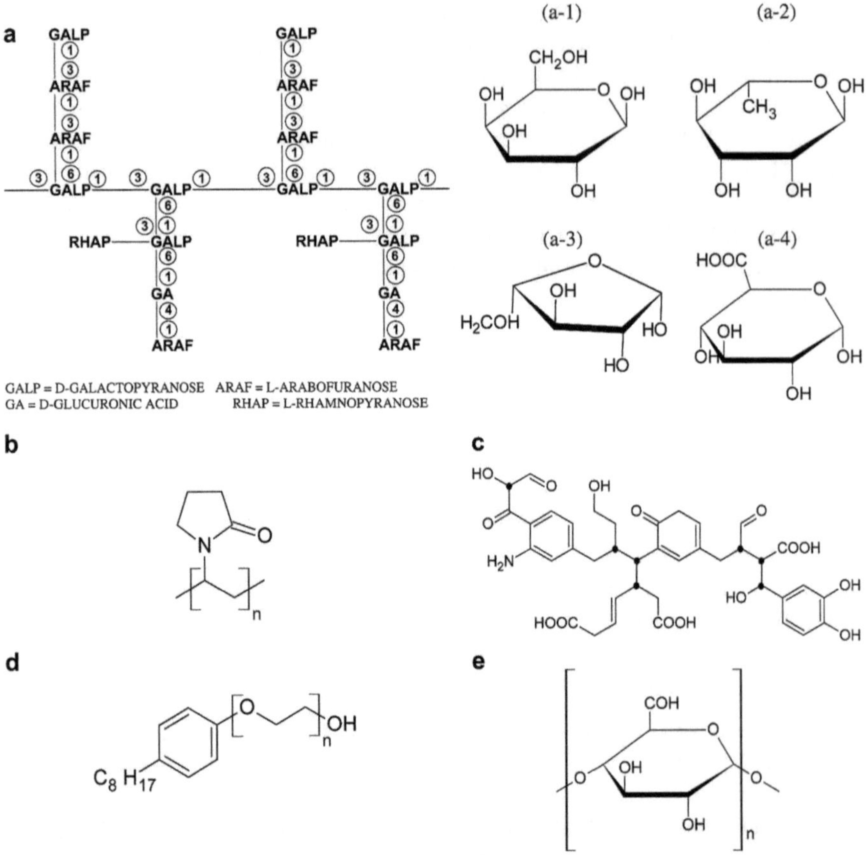

Fig. 5.24 Molecular structure of solubilizers: (**a**) gum arabic (*a-1*) galactose, (*a-2*) rhamnose, (*a-3*) arabinose, (*a-4*) glucuronic acid; (**b**) poly(vinylpyrrolidone); (**c**) a building block of humic acids, which has a compositional similarity to SRNOM; *dots* represent chiral centers; (**d**) Triton X-100; (**e**) amylose (Reproduced with permission of the *Elsevier Science*)

processes, the monocytes–macrophages (CD45+, CD68+) are the most involved cells. These immunocompetent cells are characterized by sequence of events starting from the expression of pro-inflammatory cytokines (in particular IL-18) and amyloidogenesis.

As a conclusion, the integrating the nano-toxicological profile of different kinds of CNTs and their dispersions with a life cycle perspective must be a prerequisite for the development of nano-technologically based applications in a safest possible manner [167]. It seems that environmental and health impacts have not been sufficiently studied up to this moment, and therefore, more deep research is needed.

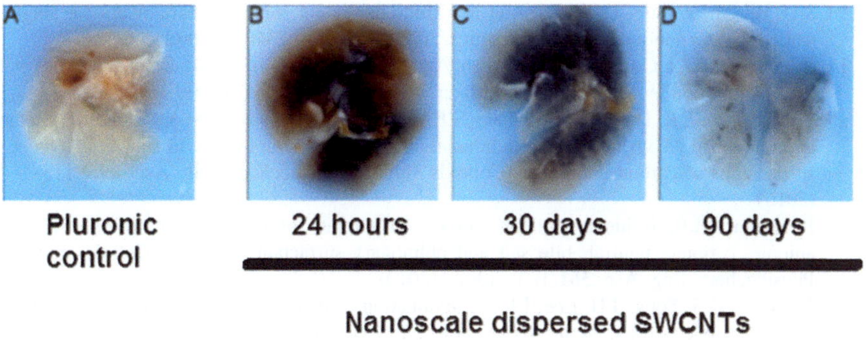

Fig. 5.25 Mice were treated with Pluronic alone or nanoscale dispersed SWCNTs in Pluronic. Paraffin blocks containing fixed embedded lungs that were removed (**a**, **b**) 24 h, (**c**) 30 days, and (**d**) 90 days after the instillation (Reproduced with permission of the *American Chemical Society*)

References

1. H. Li, J.C. Nie, S. Kunsági-Máte, Modified dispersion of functionalized multi-walled carbon nanotubes in acetonitrile. Chem. Phys. Lett. **492**(4–6), 258–262 (2010)
2. J. Wang, D. Früchtl, Z. Sun, J.N. Coleman, W.J. Blau, Control of optical limiting of carbon nanotube dispersions by changing solvent parameters. J. Phys. Chem. C **114**(13), 6148–6156 (2010)
3. G. Sun, Z. Liu, G. Chen, Dispersion of pristine multi-walled carbon nanotubes in common organic solvents. Nano **5**(2), 103–109 (2010)
4. M. Ghislandi, E. Tkalya, S. Schillinger, C.E. Koning, G. De With, High performance graphene- and MWCNTs-based PS/PPO composites obtained *via* organic solvent dispersion. Compos. Sci. Technol. **80**, 16–22 (2013)
5. S.D.F. Brandão, D. Andrada, A.F. Mesquita, A.P. Santos, H.F. Gorgulho, R. Paniago, M.A. Pimenta, C. Fantini, C.A. Furtado, The influence of oxygen-containing functional groups on the dispersion of single-walled carbon nanotubes in amide solvents. J. Phys. Condens. Matter **22**(33.), art. no), 334222 (2010)
6. M.W. Forney, J.C. Poler, Significantly enhanced single-walled carbon nanotube dispersion stability in mixed solvent systems. J. Phys. Chem. C **115**(21), 10531–10536 (2011)
7. P.-C. Ma, N.A. Siddiqui, E. Mäder, J.-K. Kim, Correlation between electrokinetic potential, dispersibility, surface chemistry and energy of carbon nanotubes. Compos. Sci. Technol. **71** (14), 1644–1651 (2011)
8. Q. Cheng, S. Debnath, L. O'Neill, T.G. Hedderman, E. Gregan, H.J. Byrne, Systematic study of the dispersion of SWCNTs in organic solvents. J. Phys. Chem. C **114**(11), 4857–4863 (2010)
9. D.H. Kim, Y.S. Yun, H.-J. Jin, Difference of dispersion behavior between graphene oxide and oxidized carbon nanotubes in polar organic solvents. Curr. Appl. Phys. **12**(3), 637–642 (2012)
10. F. Torrens, Calculations on solvents and co-solvents of single-wall carbon nanotubes: Cyclopyranoses. Theochem **757**(1), 183–191 (2005)
11. H.T. Ham, Y.S. Choi, I.J. Chung, An explanation of dispersion states of single-walled carbon nanotubes in solvents and aqueous surfactant solutions using solubility parameters. J. Colloid Interface Sci. **286**(1), 216–223 (2005)
12. H. Gui, H. Chen, C.Y. Khripin, et al., Facile and low-cost length sorting of single-wall carbon nanotubes by precipitation and applications for thin-film transistors. Nanoscale **8**, 3467–3473 (2016)

13. S.D. Bergin, Z. Sun, D. Rickard, et al., Multicomponent solubility parameters for single-walled carbon nanotube solvent mixtures. ACS Nano **3**(8), 2340–2350 (2009)
14. X. Peng, J. Jia, X. Gong, Z. Luan, B. Fan, Aqueous stability of oxidized carbon nanotubes and the precipitation by salts. J. Hazard. Mater. **165**(1–3), 1239–1242 (2009)
15. P. Alafogianni, K. Dassios, S. Farmaki, S.K. Antiohos, T.E. Matikas, N.-M. Barkoula, On the efficiency of UV–vis spectroscopy in assessing the dispersion quality in sonicated aqueous suspensions of carbon nanotubes. Colloids Surf. A Physicochem. Eng. Asp. **495**, 118–124 (2016)
16. P. Kumar, H.B. Bohidar, Aqueous dispersion stability of multi-carbon nanoparticles in anionic, cationic, neutral, bile salt and pulmonary surfactant solutions. Colloids Surf. A Physicochem. Eng. Asp. **361**(1–3), 13–24 (2010)
17. J.S. Kim, K.S. Song, J.H. Lee, I.J. Yu, Evaluation of biocompatible dispersants for carbon nanotube toxicity tests. Arch. Toxicol. **85**(12), 1499–1508 (2011)
18. A.L. Alpatova, W. Shan, P. Babica, B.L. Upham, A.R. Rogensues, S.J. Masten, E. Drown, A.K. Mohanty, E.C. Alocilja, V.V. Tarabara, Single-walled carbon nanotubes dispersed in aqueous media via non-covalent functionalization: Effect of dispersant on the stability, cytotoxicity, and epigenetic toxicity of nanotube suspensions. Water Res. **44**(2), 505–520 (2010)
19. V. Datsyuk, P. Landois, J. Fitremann, A. Peigney, A.M. Galibert, B. Soulaa, E. Flahaut, Double-walled carbon nanotube dispersion *via* surfactant substitution. J. Mater. Chem **19**, 2729–2736 (2009)
20. B. Sohrabi, N. Poorgholami-Bejarpasi, N. Nayeri, Dispersion of carbon nanotubes using mixed surfactants: Experimental and molecular dynamics simulation studies. J. Phys. Chem. B **118**, 3094–3103 (2014)
21. O. Deriabinaa, N. Lebovka, L. Bulavin, A. Goncharuk, Regulation of dispersion of carbon nanotubes in a mixture of good and bad solvents. Cornell University Library, arXiv:1304.5679, https://arxiv.org/abs/1304.5679. Accessed on 9 Feb 2017
22. C.-Y. Chen, C.T. Jafvert, Photoreactivity of carboxylated single-walled carbon nanotubes in sunlight: Reactive oxygen species production in water. Environ. Sci. Technol. **44**(17), 6674–6679 (2010)
23. L. Zhou, Y. Zhang, Q. Wang, C. Ferronato, X. Yang, Photochemical behavior of carbon nanotubes in natural waters: Reactive oxygen species production and effects on •OH generation by Suwannee River fulvic acid, nitrate, and Fe (III). Environ. Sci. Pollut. Res. **23**(19), 19520–19528 (2016)
24. M.N. Ahmad, J.-Y. Xie, Y.-H. Ma, W.-T. Yang, Surface functionalization of single-walled carbon nanotubes using photolysis for enhanced dispersion in an organic solvent. Xinxing Tan Cailiao/ New Carbon Mater **25**(2), 134–140 (2010)
25. M.A. Hamon, K.L. Stensaas, M.A. Sugar, K.C. Tumminello, A.K. Allred, Reacting soluble single-walled carbon nanotubes with singlet oxygen. Chem. Phys. Lett. **447**(1), 1–4 (2007)
26. M. Li, M. Boggs, T.P. Beebe, C.P. Huang, Oxidation of single-walled carbon nanotubes in dilute aqueous solutions by ozone as affected by ultrasound. Carbon **46**(3), 466–475 (2008)
27. K.J. Ziegler, J. Shaver, R.H. Hauge, R.E. Smalley; I.M. Marek, Ozonation of carbon nanotubes in fluorocarbons. US20060159612, 2006
28. R.E. Smalley, K.J. Ziegler, J. Shaver, R.H. Hauge, I.M. Marek, Ozonation of carbon nanotubes in fluorocarbons. WO07050096, 2007
29. C. Song, P.E. Pehrsson, W. Zhao, Recoverable solution reaction of HiPco carbon nanotubes with hydrogen peroxide. J. Phys. Chem. B **109**(46), 21634–21639 (2005)
30. C. Niu, D. Moy, J. Ma, A. Chishti Modification of nanotubes oxidation with peroxygen compounds. US6872681, 2005
31. S.A. Ntim, O. Sae-Khow, C. Desai, F.A. Witzmann, S. Mitra, Size dependent aqueous dispersibility of carboxylated multiwall carbon nanotubes. J. Environ. Monit. **14**(10), 2772–2779 (2012)

32. T. Coccini, E. Roda, D.A. Sarigiannis, P. Mustarelli, E. Quartarone, A. Profumo, L. Manzo, Effects of water-soluble functionalized multi-walled carbon nanotubes examined by different cytotoxicity methods in human astrocyte D384 and lung A549 cells. Toxicology **269**(1), 41–53 (2010)
33. B. Zhao, H. Hu, S. Niyogi, M.E. Itkis, M.A. Hamon, P. Bhowmik, M.S. Meier, R.C. Haddon, Chromatographic purification and properties of soluble single-walled carbon nanotubes. J. Am. Chem. Soc. **123**(47), 11673–11677 (2001)
34. S. Niyogi, H. Hu, M.A. Hamon, P. Bhowmik, B. Zhao, S.M. Rozenzhak, J. Chen, R.C. Haddon, Chromatographic purification of soluble single-walled carbon nanotubes (s-SWCNTS). J. Am. Chem. Soc. **123**(4), 733–734 (2001)
35. N. Tagmatarchis, A. Zattoni, P. Reschiglian, M. Prato, Separation and purification of functionalised water-soluble multi-walled carbon nanotubes by flow field-flow fractionation. Carbon **43**(9), 1984–1989 (2005)
36. H. Jia, Y. Lian, M.O. Ishitsuka, T. Nakahodo, Y. Maeda, T. Tsuchiya, T. Wakahara, T. Akasaka, Centrifugal purification of chemically modified single-walled carbon nanotubes. Sci. Technol. Adv. Mater. **6**(6), 571–581 (2005)
37. Y. Lian, Y. Maeda, T. Wakahara, T. Nakahodo, T. Akasaka, S. Kazaoui, N. Minami, H. Tokumoto, Spectroscopic study on the centrifugal fractionation of soluble single-walled carbon nanotubes. Carbon **43**(13), 2750–2759 (2005)
38. J.M. Hughes, D. Aherne, S.D. Bergin, A. Oneill, P.V. Streich, J.P. Hamilton, J.N. Coleman, Using solution thermodynamics to describe the dispersion of rod-like solutes: Application to dispersions of carbon nanotubes in organic solvents. Nanotechnology **23**(26.), art. no), 265604 (2012)
39. J. Gigault, B. Grassl, G. Lespes, A new analytical approach based on asymmetrical flow field-flow fractionation coupled to ultraviolet spectrometry and light scattering detection for SWCNT aqueous dispersion studies. Analyst **137**(4), 917–923 (2012)
40. I.D. Rosca, F. Watari, M. Uo, T. Akasaka, Oxidation of multiwalled carbon nanotubes by nitric acid. Carbon **43**(15), 3124–3131 (2005)
41. C.G. Salzmann, B.T.T. Chu, G. Tobias, S.A. Llewellyn, M.L.H. Green, Quantitative assessment of carbon nanotube dispersions by Raman spectroscopy. Carbon **45**(5), 907–912 (2007)
42. O. Kleinerman, A.N.G. Parra-Vazquez, et al., Cryogenic-temperature electron microscopy direct imaging of carbon nanotubes and graphene solutions in superacids. J. Microsc. **259**(1), 16–25 (2015)
43. Y. Chen, J. Chen, H. Hu, M.A. Hamon, M.E. Itkis, R.C. Haddon, Solution-phase EPR studies of single-walled carbon nanotubes. Chem. Phys. Lett. **299**(6), 532–535 (1999)
44. M. Gallo, A. Favila, D. Glossman-Mitnik, DFT studies of functionalized carbon nanotubes and fullerenes as nanovectors for drug delivery of antitubercular compounds. Chem. Phys. Lett. **447**(1), 105–109 (2007)
45. D. Paolucci, M. Marcaccio, C. Bruno, F. Paolucci, N. Tagmatarchis, M. Prato, Voltammetric quantum charging capacitance behaviour of functionalised carbon nanotubes in solution. Electrochim. Acta **53**(11), 4059–4064 (2008)
46. G.A. Rivas, M.D. Rubianes, M.C. Rodriguez, N.F. Ferreyra, G.L. Luque, M.L. Pedano, S.A. Miscoria, C. Parrado, Carbon nanotubes for electrochemical biosensing. Talanta **74** (3), 291–307 (2007)
47. M. Melle-Franco, M. Marcaccio, D. Paolucci, F. Paolucci, V. Georgakilas, D.G. Guldi, M. Prato, F. Zerbetto, Cyclic voltammetry and bulk electronic properties of soluble carbon nanotubes. J. Amer. Chem. Soc. **126**(6), 1646–1647 (2004)
48. J. Martí, M.C. Gordillo, Structure and dynamics of liquid water adsorbed on the external walls of carbon nanotubes. J. Chem. Phys. **119**(23), 12540–12546 (2003)
49. S.M. Fatemi, M. Foroutan, Recent developments concerning the dispersion of carbon nanotubes in surfactant/polymer systems by MD simulation. J. Nanostruct. Chem. **6**, 29–40 (2016)

50. J.-Y. Chang, A. Ghule, J.-J. Chang, S.-H. Tzing, Y.-C. Ling, Opening and thinning of multiwall carbon nanotubes in supercritical water. Chem. Phys. Lett. **363**(5), 583–590 (2002)
51. B. Benedict, P.E. Pehrsson, W. Zhao, Optically sensing additional sonication effects on dispersed HiPco nanotubes in aerated water. J. Phys. Chem. B **109**(16), 7778–7780 (2005)
52. B.D. Holt, V. Roginskaya, B. Van Houten, M.F. Islam, K.N. Dahl, Dispersed single wall carbon nanotubes do not impact mitochondria structure or function, but technical issues during analysis could yield incorrect results. J. Mater. Chem. B **5**, 369–374 (2017)
53. C. Kuroda, H. Haniu, K. Ajima, et al., The dispersion state of tangled multi-walled carbon nanotubes affects their cytotoxicity. Nanomaterials **6**, 219 (2016.) 10 pp
54. M.O. Lisunova, N.I. Lebovka, O.V. Melezhyk, Y.P. Boiko, Stability of the aqueous suspensions of nanotubes in the presence of nonionic surfactant. J. Colloid Interface Sci. **299**(2), 740–746 (2006)
55. A. Hirano, W. Gao, X. He, J. Kono, Destabilization of surfactant-dispersed carbon nanotubes by anions. Nanoscale Res. Lett. **12**, 81 (2017)
56. H. Dumortier, S. Lacotte, G. Pastorin, R. Marega, W. Wu, D. Bonifazi, J.-P. Briand, A. Bianco, Functionalized carbon nanotubes are non-cytotoxic and preserve the functionality of primary immune cells. Nano Lett. **6**(7), 1522–1528 (2006)
57. H. Wilson, S. Ripp, L. Prisbrey, et al., Electrical monitoring of sp^3 defect formation in individual carbon nanotubes. J. Phys. Chem. C **120**, 1971–1976 (2016)
58. M.D. Vo, D.V. Papavassiliou, D.V., Effect of sodium dodecyl sulfate adsorption on the behavior of water inside single walled carbon nanotubes with dissipative particle dynamics simulation. Molecules **21**, 500 (2016.) 16 pp
59. J. Njuguna, O.A. Vanli, R. Liang, Spectral methods for dispersion characterization of carbon nanotubes in aqueous suspensions. J. Spectrosc. **2015**., Article ID 463156, 11 (2015)
60. Y. Zhang, S. Tan, H. Lopez. Determination of carbon nanotube concentration in a solution by fluorescence measurement. US20060141634 (2006)
61. A.O. Borovskaya, B.Z. Idiatullin, O.S. Zueva, Carbon nanotubes in the surfactants dispersion: Formation of the microenvironment. J. Phys. Conf. Ser. **690**, 012030 (2016.) 6 pp
62. M.H. Bocanegra-Bernal, J. Echeberria, J. Ollo, A. Garcia-Reyes, C. Domínguez-Riosa, A. Reyes-Rojas, A. Aguilar-Elguezabal, A comparison of the effects of multi-wall and single-wall carbon nanotube additions on the properties of zirconia toughened alumina composites. Carbon **49**(5), 1599–1607 (2011)
63. M. Asgari, E. Lohrasbi, Comparison of single-walled and multiwalled carbon nanotubes durability as Pt support in gas diffusion electrodes. ISRN Electrochemistry **2013**., ID 564784, 7 (2013)
64. L. Ji, W. Chen, J. Bi, S. Zheng, Z. Xu, D. Zhu, P. Alvarez, Adsorption of tetracycline on single-walled and multi-walled carbon nanotubes as affected by aqueous solution chemistry. Environ. Toxicol. Chem. **29**(12), 2713–2719 (2010)
65. K.M. Liew, C.H. Wong, X.Q. He, M.J. Tan, Thermal stability of single and multi-walled carbon nanotubes. Phys. Rev. B **71**, 075424 (2005)
66. S. Ferguson, N. McGara, B. Cavness, D. Gonzales, S. Williams, Spectra of radiation emitted by single-walled and multi-walled carbon nanotubes during multiple microwave irradiation and cooling cycles. Int. J. Nano Sci. Nanotechnol **4**(1), 71–79 (2013)
67. E.E. Urena-Benavides, M.J. Kayatin, V.A. Davis, Dispersion and rheology of multiwalled carbon nanotubes in unsaturated polyester resin. Macromolecules **46**(4), 1642–1650 (2013)
68. E.E. Urena-Benavides, M.J. Kayatin, V.A. Davis, dispersion of single-walled and multiwalled carbon nanotubes in unsaturated polyester resin. *ANTEC*, 2012, http://www.4spe.org/Resources/resource.aspx?ItemNumber=6471
69. A.N. Parra-Vasquez, N. Behabtu, M.J. Green, C.L. Pint, C.C. Young, J. Schmidt, E. Kesselman, A. Goyal, P.M. Ajayan, Y. Cohen, Y. Talmon, R.H. Hauge, M. Pasquali, Spontaneous dissolution of ultralong single- and multiwalled carbon nanotubes. ACS Nano **4** (7), 3969–3978 (2010)

70. V.A. Davis, A.N. Parra-Vasquez, et al., True solutions of single-walled carbon nanotubes for assembly into macroscopic materials. Nat. Nanotechnol. **4**, 830–834 (2009)
71. P. Zhao, E. Einarsson, R. Xiang, Y. Murakami, S. Maruyama, Controllable expansion of single-walled carbon nanotube dispersions using density gradient ultracentrifugation. J. Phys. Chem. C **114**(11), 4831–4834 (2010)
72. K. Moshammer, F. Hennrich, M.M. Kappes, Selective suspension in aqueous sodium dodecyl sulfate according to electronic structure type allows simple separation of metallic from semiconducting single-walled carbon nanotubes. Nano Res. **2**, 599–606 (2009)
73. R. Krupke, F. Hennrich, H. von Lohneysen, M.M. Kappes, Separation of metallic from semiconducting single-walled carbon nanotubes. Science **301**, 344–347 (2003)
74. B.R. Burg, J. Schneider, M. Muoth, L. Durrer, T. Helbling, N.C. Schirmer, T. Schwamb, C. Hierold, D. Poulikakos, Aqueous dispersion and dielectrophoretic assembly of individual surface-synthesized single-walled carbon nanotubes. Langmuir **25**(14), 7778–7782 (2009)
75. N. Kumar Mehra, V. Mishra, N.K. Jain, A review of ligand tethered surface engineered carbon nanotubes. Biomaterials **35**, 1267–1283 (2014)
76. D.W. Chang, I.-Y. Jeon, J.-B. Baek, L. Dai, Efficient dispersion of singlewalled carbon nanotubes by novel amphiphilic dendrimers in water and substitution of the pre-adsorbed dendrimers with conventional surfactants and lipids. Chem. Commun. **46**(42), 7924–7926 (2010)
77. K. Imasaka, U. Khaled, S. Wei, J. Suehiro, pH dependence of water-solubility of single-walled carbon nanotubes treated by microplasma in aqueous solution. http://www.ispc-conference.org/ispcproc/papers/367.pdf. Accessed on 25 June 2016.
78. O. Matarredona, H. Rhoads, Z. Li, J.H. Harwell, L. Balzano, D.E. Resasco, Dispersion of single-walled carbon nanotubes in aqueous solutions of the anionic surfactant NaDDBS. J. Phys. Chem. B **107**, 13357–13367 (2003)
79. N. Hadidi, F. Kobarfard, N. Nafissi-Varcheh, R. Aboofazeli, Optimization of single-walled carbon nanotube solubility by noncovalent PEGylation using experimental design methods. Int. J. Nanomedicine **6**, 737–746 (2011)
80. H. Yilmaz, B. Rasulev, J. Leszczynski, Modeling the dispersibility of single walled carbon nanotubes in organic solvents by quantitative structure-activity relationship approach. Nanomaterials **5**, 778–791 (2015)
81. A. Ansón-Casaos, J.M. González-Domínguez, I. Lafragüeta, J.A. Carrodeguas, M. Teresa Martíneza, Optical absorption response of chemically modified single-walled carbon nanotubes upon ultracentrifugation in various dispersants. Carbon **66**, 105–118 (2014)
82. N. Ould-Moussa, C. Blanca, C. Zamora-Ledezma, O.D. Lavrentovich, I.I. Smalyukhd, et al., Dispersion and orientation of single-walled carbon nanotubes in a chromonic liquid crystal. Liq. Cryst. **40**(12), 1628–1635 (2013)
83. L. Shao, G. Tobias, C.G. Salzmann, B. Ballesteros, S.Y. Hong, A. Crossley, B.G. Davis, M.L.H. Green, Removal of amorphous carbon for the efficient sidewall functionalisation of single-walled carbon nanotubes. Chem. Commun. **21**, 5090–5092 (2007)
84. L. Rodríguez-Pérez, R. García, M.A. Herranz, N. Martín, Modified SWCNTs with amphoteric redox and solubilizing properties. Chem. Eur. J. **20**, 7177–7286 (2014)
85. Y.Y. Huang, E.M. Terentjev, Dispersion of carbon nanotubes: Mixing, sonication, stabilization, and composite properties. Polymer **4**, 275–295 (2012)
86. J. Campo, Y. Piao, S. Lam, et al., Enhancing single-wall carbon nanotube properties through controlled endohedral filling. Nanoscale Horiz. **1**, 317–324 (2016)
87. M. Mananghaya, E. Rodulfo, G.N. Santos, A.R. Villagracia, Theoretical investigation on the Solubilization in water of functionalized single-wall carbon nanotubes. J. Nanotechnol **2012**, 6 (2012., Article ID 780815)
88. A. Kitaygorodskiy, W. Wang, S.-Y. Xie, Y. Lin, K.A.S. Fernando, X. Wang, L. Qu, Y.-P. Sun, NMR detection of single-walled carbon nanotubes in solution. J. Am. Chem. Soc. **127** (20), 7517–7520 (2005)
89. H. Yoon, M. Yamashita, S. Ata, D.N. Futaba, T. Yamada, K. Hata, Controlling exfoliation in order to minimize damage during dispersion of long SWCNTs for advanced composites. Sci. Rep **4**, 3907 (2014)

90. E. Heister, C. Lamprecht, V. Neves, et al., Higher dispersion efficacy of functionalized carbon nanotubes in chemical and biological environments. ACS Nano **4**(5), 2615–2626 (2010)
91. S. Endoh, J. Maru, A. Miyauchi, Characterization of shape of carbon nanotubes, CNTs dispersed in water by sonication. J. Soc. Powder Technol Japan **49**(6.), Special Issue of 47th Particle Technology Meeting), 483–488 (2012)
92. M. Campo, A. Jiménez-Suárez, A. Ureña, Effect of type, percentage and dispersion method of multi-walled carbon nanotubes on tribological properties of epoxy composites. Wear **324–325**, 100–108 (2015)
93. M.J. Green, Analysis and measurement of carbon nanotube dispersions: Nanodispersion versus macrodispersion. Polym. Int. **59**, 1319–1322 (2010)
94. T. Liu, Z. Xiao, S. Luo, In-situ structural characterization of single-walled carbon nanotubes in dispersion, in *Carbon Nanotubes – Synthesis, Characterization, Applications*, ed. by S. Yellampalli (Ed), (InTech, Rijeka, 2011), pp. 241–264. doi:10.5772/16465. Available from: https://www.intechopen.com/books/carbon-nanotubes-synthesis-characterization-applications/in-situ-structural-characterization-of-single-walled-carbon-nanotubes-in-dispersion
95. X. Chang, M. Henderson, D. Bouchard, Multi-instrumental characterization of carbon nanotubes dispersed in aqueous solutions. Presented at 248th ACS National Meeting and Exposition, San Francisco, 10–14 August 2014
96. J. Jidraph Njuguna, O. Arda Vanli, R. Liang, A review of spectral methods for dispersion characterization of carbon nanotubes in aqueous suspensions. J. Spectrosc. **2015**., Article ID 463156, 11 (2015)
97. Z.F. Li, G.H. Luo, W.P. Zhou, F. Wei, R. Xiang, Y.P. Liu, The quantitative characterization of the concentration and dispersion of multi-walled carbon nanotubes in suspension by spectrophotometry. Nanotechnology **17**(15), 3692
98. B. Yang, L. Ren, L. Li, X. Tao, Y. Shi, Y. Zheng, The characterization of the concentration of the single-walled carbon nanotubes in aqueous dispersion by UV-vis-NIR absorption spectroscopy. Analyst **138**, 6671–6676 (2013)
99. S. Attal, R. Thiruvengadathan, O. Regev, Determination of the concentration of single-walled carbon nanotubes in aqueous dispersions using UV−visible absorption spectroscopy. Anal. Chem. **78**(23), 8098–8104 (2006)
100. Y. Shi, L. Ren, D. Li, H. Gao, B. Yang, Optimization conditions for single-walled carbon nanotubes dispersion. J. Surf. Eng. Mater. Adv. Technol **3**, 6–12 (2013)
101. J.A. Fagan, M.L. Becker, J. Chun, P. Nie, B.J. Bauer, J.R. Simpson, A. Hight-Walker, E.K. Hobbie, Centrifugal length separation of carbon nanotubes. Langmuir **24**, 13880–13889 (2008)
102. S. Shahnawaz, B. Sohrabi, M. Najafi, The investigation of functionalization role in multi-walled carbon nanotubes dispersion by surfactants. doi:10.3390/ecsoc-18-f002. http://www.usc.es/congresos/ecsoc/18/files/F/f002/index.pdf
103. M.-c. Yang, M.-y. Li, S. Luo, R. Liang, Real-time monitoring of carbon nanotube dispersion using dynamic light scattering and UV-vis spectroscopy. Int. J. Adv. Manuf. Technol. **82**(1), 361–367 (2016)
104. L. Reinert, M. Zeiger, S. Suarez, V. Presser, F. Mucklicha, Dispersion analysis of carbon nanotubes, carbon onions, and nanodiamonds for their application as reinforcement phase in nickel metal matrix composites. RSC Adv. **5**, 95149–95159 (2015)
105. J.H. Lehman, M. Terrones, E. Mansfield, K.E. Hurst, V. Meunier, Evaluating the characteristics of multiwall carbon nanotubes. Carbon **49**, 2581–2602 (2011)
106. C.S.S.R. Kumar, *Raman Spectroscopy for Nanomaterials Characterization* (Springer, Berlin, 2012), p. 645
107. Y. Wang, D. Vasileva, S.P. Zustiak, I. Kuljanishvili, Raman spectroscopy enabled investigation of carbon nanotubes quality upon dispersion in aqueous environments. Biointerphases **12**(1), 011004 (2017)

108. M.S. Dresselhaus, A. Jorio, M. Hofmann, G. Dresselhaus, R. Saito, Perspectives on carbon nanotubes and graphene raman spectroscopy. Nano Lett. **10**(3), 751–758 (2010)
109. V.M. Irurzun, M. Pilar Ruiz, D.E. Resasco, Raman intensity measurements of single-walled carbon nanotube suspensions as a quantitative technique to assess purity. Carbon **48**, 2873–2881 (2010)
110. B.R. Priya, H.J. Byrne, Quantitative analysis of dispersion and doping of individual carbon nanotubes in water based solutions using absorption and Raman spectroscopy. Phys. Stat. Sol. (b) **245**(10), 1964–1966 (2008)
111. M.-M. Li, Z.-P. Wu, M. Zhao, W.-B. Zhang, Y.-Y. Hu, Y.-H. Yin, Preparation of homogeneously dispersed and highly concentrated double-walled carbon nanotubes as catalyst support. Rare Metals **35**(4), 337–343 (2016)
112. N. Izard, A. Pénicaud, E. Anglaret, Raman studies of suspensions and solutions of singlewall carbon nanotubes. Funct. Carbon Nanotubes (2005), number HH11.12005, arXiv:cond-mat/0507079. 6pp. https://arxiv.org/ftp/cond-mat/papers/0507/0507079.pdf; https://arxiv.org/abs/cond-mat/0507079v1
113. T. Thuy Nguyen, S. Uan Nguyen, D. Tam Phuong, D. Chien Nguyen, A. Tuan Mai, Dispersion of denatured carbon nanotubes by using a dimethylformamide solution. Adv. Nat. Sci. Nanosci. Nanotechnol. **2**, 035015 (2011.) 4 pp
114. D. Douroumis, D.G. Fatouros, N. Bouropoulos, K. Papagelis, D. Tasis, Int. J. Nanomedicine **2**(4), 761–766 (2007)
115. J. Hee Kim, M. Kataoka, D. Shimamoto, et al., Raman and fluorescence spectroscopic studies of a DNA-dispersed double-walled carbon nanotube solution. ACS Nano **4**(2), 1060–1066 (2010)
116. F. Villalpando-Paez, H. Son, D. Nezich, et al., Raman spectroscopy study of isolated double-walled carbon nanotubes with different metallic and semiconducting configurations. Nano Lett. **8**(11), 3879–3886 (2008)
117. J. Zhao, Morphology and dispersion of pristine and modified carbon Nanofibers in water, in *Nanofibers*, ed. by A. Kumar (Ed), (INTECH, Rijeka, Croatia, 2010), pp. 269–294
118. Studying the Internal Structure of Carbon Nanotubes with SAXS. http://www.anton-paar.com. Accessed on 8 May 2017
119. J.J. Hernández, M.C. García-Gutiérrez, A. Nogales, D.R. Rueda, T.A. Ezquerra, Small-angle X-ray scattering of single-wall carbon nanotubes dispersed in molten poly(ethylene terephthalate). Compos. Sci. Technol. **66**, 2629–2632 (2006)
120. S. Pujari, Shear-induced anisotropy of concentrated multiwalled carbon nanotube suspensions using x-ray scattering. J. Rheol. **55**, 1033 (2011)
121. E. Ramos, W.A. Pardo, M. Mir, J. Samitier, Dependence of carbon nanotubes dispersion kinetics on surfactants. Nanotechnology **28**(13), 135702 (2017)
122. C. Doe, T.-H. Kim, S.R. Kline, S.-M. Choi Small angle neutron scattering study of isolated single wall carbon nanotubes in water. Transactions of the Korean Nuclear Society Spring Meeting, Jeju, Korea, May 10–11, 2007
123. H. Wang, W. Zhou, D.L. Ho, et al., Dispersing single-walled carbon nanotubes with surfactants: A small angle neutron scattering study. Nano Lett. **4**(9), 1789–1793 (2004)
124. E.S. Kastrisianaki-Guyton, L. Chen, S.E. Rogers, T. Cosgrove, J.S. van Duijneveldt, Adsorption of sodium dodecylsulfate on single-walled carbon nanotubes characterised using small-angle neutron scattering. J. Colloid Interface Sci. **472**, 1–7 (2016)
125. K. Yurekli, C.A. Mitchell, R. Krishnamoorti, Small-angle neutron scattering from surfactant-assisted aqueous dispersions of carbon nanotubes. J. Am. Chem. Soc. **126**, 9902–9903 (2004)
126. B.J. Bauer, E.K. Hobbie, M.L. Becker, Small-angle neutron scattering from labeled single-wall carbon nanotubes. Macromolecules **39**, 2637–2642 (2006)
127. W. Zhou, M.F. Islam, H. Wang, et al., Small angle neutron scattering from single-wall carbon nanotube suspensions: Evidence for isolated rigid rods and rod networks. Chem. Phys. Lett. **384**, 185–189 (2004)
128. L.A. Hough, M.F. Islam, B. Hammouda, A.G. Yodh, P.A. Heiney, Structure of semidilute single-wall carbon nanotube suspensions and gels. Nano Lett. **6**(2), 313–317 (2006)

129. M. Granite, A. Radulescu, Y. Cohen, Small-angle neutron scattering from aqueous dispersions of single-walled carbon nanotubes with pluronic F127 and poly(vinylpyrrolidone). Langmuir **28**, 11025–11031 (2012)
130. K. Nasouri, A. Mousavi Shoushtari, M. Reza Mohaddes Mojtahedi, Synthesis and characterization of highly dispersed multi-walled carbon nanotubes/polyvinylpyrrolidone composite nanofibers for EMI shielding application. Polymer. Compos. 1548-0569 (2016). doi:10.1002/pc.23774. http://dx.doi.org/10.1002/pc.23774
131. X. Fu, J. Wang, J. Ding, H. Wu, Y. Dong, Y. Fu, Quantitative evaluation of carbon nanotube dispersion through scanning electron microscopy images. Compos. Sci. Technol. **87**, 170–173 (2013)
132. R. Herrera-Basurto, A.I. López-Lorente, M. Valcárcel, Scanning electron microscopy of carbon nanotubes dispersed in ionic liquid: Solvent influence study. Microchem. J. **122**, 137–143 (2015)
133. B. Baykal, V. Ibrahimova, G. Er, E. Bengu, D. Tuncel, Dispersion of multi-walled carbon nanotubes in an aqueous medium by water-dispersible conjugated polymer nanoparticles. Chem. Commun. **46**, 6762–6764 (2010)
134. J. Mast, E. Verleysen, P.-J. De Temmerman, Physical characterization of nanomaterials in dispersion by transmission electron microscopy in a regulatory framework. in *Advanced Transmission Electron Microscopy*, ed. by F. L. Deepak, A. Mayoral, R. Arenal (Eds), (Springer, Cham, 2015)
135. Q. Quanxiang Li, J.S. Church, A. Kafi, M. Naebe, B.L. Fox, An improved understanding of the dispersion of multi-walled carbon nanotubes in non-aqueous solvents. J. Nanopart. Res. **16**, 2513, 12 pp (2014)
136. M. Bottini, A. Magrini, N. Rosato, A. Bergamaschi, T. Mustelin, Dispersion of pristine single-walled carbon nanotubes in water by a thiolated organosilane: Application in supramolecular nanoassemblies. J. Phys. Chem. B **110**(28), 13685–13688 (2006)
137. A. Prakash Periasamy, Y.-H. Ho, S.-M. Chen, Multiwalled carbon nanotubes dispersed in carminic acid for the development of catalase based biosensor for selective amperometric determination of H_2O_2 and iodate. Biosens. Bioelectron. **29**, 151–158 (2011)
138. A. Jagusiak, B. Piekarska, T. Pańczyk, et al., Dispersion of single-wall carbon nanotubes with supramolecular Congo red – Properties of the complexes and mechanism of the interaction. Beilstein J. Nanotechnol. **8**, 636–648 (2017)
139. S. Hirayama, T. Hayashida, K. Umemura, Atomic force microscopy imaging of dialyzed single-walled carbon nanotubes dispersed with sodium dodecyl sulfate. Int. J. Smart Nano Mater. **4**(2), 119–127 (2013)
140. S. Kumar, I. Kaur, N. Kumari, et al., Atomic force microscope manipulation of multiwalled and single walled carbon nanotubes with reflux and ultrasonic treatments. Appl. Nanosci. **4**, 19–26 (2014)
141. Z. Dong, U.C. Wejinya, Electrical properties measurement of Carbon Nanotubes using Atomic Force Microscope for nano sensor applications. 2010, Xiamen, China, doi: 10.1109/NEMS.2010.5592261. Published in: 5th IEEE International Conference on Nano/Micro Engineered and Molecular Systems (NEMS)
142. D. Yoon, J.-B. Choi, C.-S. Han, Y.-J. Kim, S. Baik, The quantitative characterization of the dispersion state of single-walled carbon nanotubes using Raman spectroscopy and atomic force microscopy. Carbon **46**, 1530–1534 (2008)
143. Z. Zhang, Y. Che, R.A. Smaldone, M. Xu, B.R. Bunes, J.S. Moore, L. Zang, Reversible dispersion and release of carbon nanotubes using foldable oligomers. J. Am. Chem. Soc. **132** (40), 14113–14117 (2010)
144. S. Liu, A.K. Ng, R. Xu, et al., Antibacterial action of dispersed single-walled carbon nanotubes on Escherichia coli and Bacillus Subtilis investigated by atomic force microscopy. Nanoscale **2**, 2744–2750 (2010)
145. S. Sadiq Ali, M. Shahabuddin, M. Asif, In situ monitoring of dispersion dynamics of carbon nanotubes during sonication using electrical conductivity measurements. J. Nanomater. **2015**., Article ID 479053, 8 (2015)

146. J. Kolosnjaj, H. Szwarc, F. Moussa, Toxicity studies on carbon nanotubes: State of the art. ECS Trans. **35**(25), 121–134 (2011)
147. Y. Morimoto, M. Hirohashi, M. Horie, et al., Pulmonary toxicity of well-dispersed single-wall carbon nanotubes following intratracheal instillation. J. Nano Res. **18-19**, 9–25 (2012)
148. M. Horie, M. Stowe, T. Kambara, et al., Pulmonary inflammation of well-dispersed multi-wall carbon nanotubes following intratracheal instillation: Toxicity by fiber of 1–5 μm in length. Materials **5**, 2833–2849 (2012)
149. J. Kayat, V. Gajbhiye, R. Kumar Tekade, et al., Pulmonary toxicity of carbon nanotubes: A systematic report. Nanomedicine **7**(1), 40–49 (2011)
150. G.M. Mutlu, M.C. Hersam, D. Urich, et al., Highly dispersed and purified preparations of single-walled carbon nanotubes do not cause detectable pulmonary toxicity. Am. J. Respir. Crit. Care Med. **179**, A5251 (2009)
151. K. Donaldson, C. Poland, R. Duffin (eds.), *The Toxicology of Carbon Nanotubes* (Cambridge University Press, Cambridge, 2012), p. 264
152. P. Gustavsson, M. Hedmer, J. Rissler, *Carbon Nanotubes: Exposure, Toxicology and Protective Measures in the Work Environment* (Swedish Work Environment Authority, Stockholm, Sweden, 2011)
153. C.P. Firme III, P.R. Bandaru, Toxicity issues in the application of carbon nanotubes to biological systems. Nanomed. Nanotechnol. Biol. Med. **6**, 245–256 (2010)
154. Movia, D.; Giordani, S. Toxicity of Carbon Nanotubes. Handbook of Green Chemistry. Weinheim: Wiley-VCH 2012, 8(7), 175–216.
155. A. Jafar, Y. Alshatti, A. Ahmad, Carbon nanotube toxicity: The smallest biggest debate in medical care. Cogent Med **3**, 1217970 (2016)
156. S. Yazdan Madani, A. Mandel, A.M. Seifalian, A concise review of carbon nanotube's toxicology. Nano Rev. **4**, 1–14 (2013). doi:10.3402/nano.v4i0.21521
157. K. Kostarelos, The long and short of carbon nanotube toxicity. Nat. Biotechnol. **26**, 774–776 (2008)
158. J. Cheng, S. Han Cheng, Influence of carbon nanotube length on toxicity to zebrafish embryos. Int. J. Nanomedicine **7**, 3731–3739 (2012)
159. S. Liu, L. Wei, L. Hao, et al., Sharper and faster nano darts kill more bacteria: A study of antibacterial activity of individually dispersed pristine single-walled carbon nanotube. ACS Nano **3**(12), 3891–3902 (2009)
160. Y. Morimoto, M. Hirohashi, N. Kobayashi, et al., Pulmonary toxicity of well-dispersed single-wall carbon nanotubes after inhalation. Nanotoxicology **6**(7), 766–775 (2012)
161. L.P. Lukhele, B.B. Mamba, N. Musee, V. Wepener, Acute toxicity of double-walled carbon nanotubes to three aquatic organisms. J. Nanomater. **2015**., Article ID 219074, 19 (2015)
162. H.C. Nerl, C. Cheng, A.E. Goode, S.D. Bergin, B. Lich, M. Gass, A.E. Porter, Imaging methods for determining uptake and toxicity of carbon nanotubes *in vitro* and *in vivo*. Nanomedicine **6**(5), 849–865 (2011)
163. L. Wang, V. Castranova, A. Mishra, et al., Dispersion of single-walled carbon nanotubes by a natural lung surfactant for pulmonary in vitro and in vivo toxicity studies. Part. Fibre Toxicol. **7**, 31 (2010)
164. G.M. Mutlu, G.R. Budinger, A.A. Green, et al., Biocompatible nanoscale dispersion of single-walled carbon nanotubes minimizes in vivo pulmonary toxicity. Nano Lett. **10**(5), 1664–1670 (2010)
165. K.W. Kwok, K.M. Leung, E. Flahaut, J. Cheng, S.H. Cheng, Chronic toxicity of double-walled carbon nanotubes to three marine organisms: Influence of different dispersion methods. Nanomedicine **5**(6), 951–961 (2010)
166. R. Girardello, S. Tasselli, N. Baranzini, R. Valvassori, M. de Eguileor, A. Grimaldi, Effects of carbon nanotube environmental dispersion on an aquatic invertebrate, Hirudo medicinalis. PLoS One **10**(12), e0144361 (2015)
167. G. Tejral, N. Reddy Panyala, J. Havel, Carbon nanotubes: Toxicological impact on human health and environment. J. Appl. Biomed. **7**, 1–13 (2009)

Chapter 6
Cost and Main Applications of Soluble CNTs

Main Applications In addition to such wide applications of soluble CNTs as biosensors [1–5], electrodes in biofuel cells [6], composites [7–11], advanced polymers [12–15], cements [16], agriculture [17], radar-absorbing materials [18], etc., mentioned throughout the text, we would like to underline the following uses. Soluble CNTs are applied mainly in nanomedicine [19] for different purposes, the most important of which is the *drug delivery* [20]. Within the family of nanomaterials, CNTs have emerged as a new alternative and efficient tool for transporting and translocating therapeutic molecules. It has become possible after the recent discovery of their capacity to penetrate into the cells. CNT can be loaded with active molecules by forming stable covalent bonds or supramolecular assemblies based on noncovalent interactions.

As it was described above, CNT can be functionalized with bioactive peptides, proteins, nucleic acids, and drugs and used to deliver their cargos to cells and organs. Once the cargos are carried into various cells, tissues, and organs, they are able to express their biological function [21]. Because functionalized CNTs display low toxicity and are not immunogenic, such systems hold great potential in the field of nanobiotechnology and nanomedicine [22]. Thus, SWCNTs and MWCNTs were solubilized via the esterification of nanotube-bound carboxylic acids by oligomeric polyethylene glycol compounds [23]. The obtained water-soluble samples were used as starting materials in reactions with bovine serum albumin (BSA) protein in ambient aqueous solutions, yielding SWCNT-BSA and MWCNT-BSA conjugates. Related information on drug delivery and other related applications of CNTs is given in [24–29].

CNT-Alg gel (alginate hydrogel) was reported as useful scaffold material in *tissue engineering* with the sidewalls of CNTs acting as active sites for chemical functionalization [30]. Soluble CNTs were dispersed in sodium alginate solution as a cross-linker. As a result, the CNT-Alg gel showed faster gelling and higher mechanical strength than the conventional alginate gel. *Elimination of cancer cells* using SWCNTs is reported in [31]. The behavior of the functionalized,

O.V. Kharissova, B.I. Kharisov, *Solubilization and Dispersion of Carbon Nanotubes*, DOI 10.1007/978-3-319-62950-6_6

water-soluble SWCNTs under exposition to a noninvasive, 13.56 MHz RF field was studied. Then, human cancer cell lines were incubated with various concentrations of SWCNTs and then treated in the RF field. As a result, SWCNTs targeted to cancer cells may allow noninvasive RF field treatments to produce lethal thermal injury to the malignant cells. As it was shown in another "anticancer" report, the platinum(IV) complex *c,c,t*-[$Pt(NH_3)_2Cl_2(OEt)(O_2CCH_2CH_2CO_2H)$] (**6.1**), which is nearly nontoxic to testicular cancer cells, displays a significantly enhanced cytotoxicity profile when attached to the surface of amine-functionalized soluble SWCNTs (**6.2**) [32]. The authors noted that by linking additional groups, such as cancer cell-targeting moieties, to the platinated SWCNTs as longboat passengers, it may be possible to achieve highly selective constructs for use in clinical trials. Other related biomedical applications for CNTs are generalized in recent comprehensive reviews [33–39].

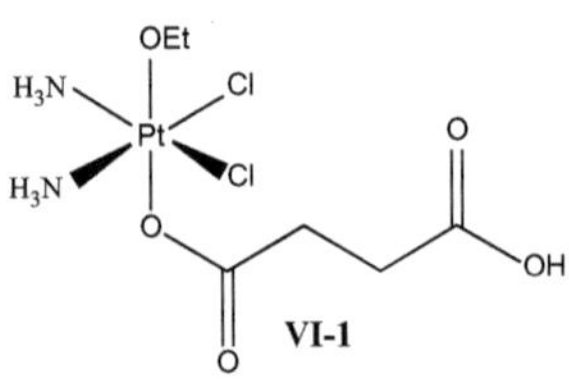

VI-1

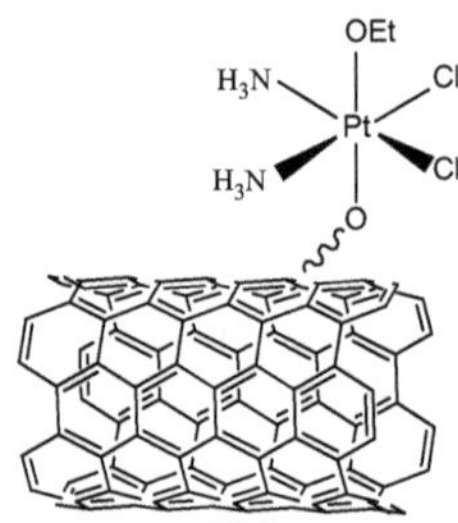

VI-2

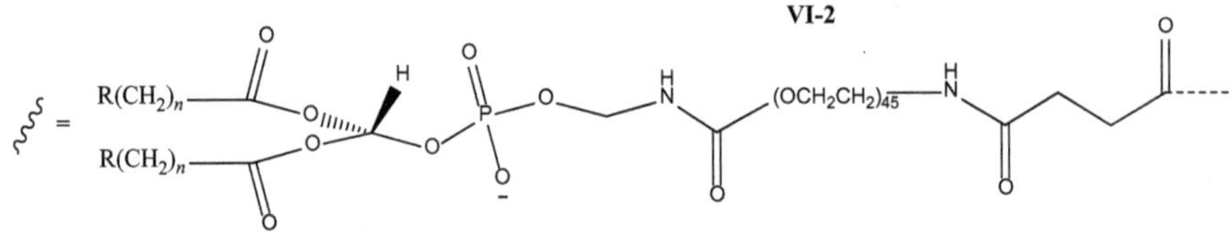

Considerable efforts have been dedicated to preparation, characterization, and search of applications for the *films* on the basis (or starting from) CNTs [40–45]. Thus, free-standing film made of a SWCNT–polyvinyl alcohol (PVA) composite material was fabricated by pretreatment of SWCNTs with ultrasonication in water with the aid of a surfactant that promotes unbundling of aggregated SWNCTs [46]. Large-scaled and homogeneous MWCNT films were fabricated using the soluble octadecylamine-modified MWCNT at r.t. by spin-coating method [47]. Composite film containing titania electrostatically linked to oxidized MWCNTs (TiO_2-s-MWCNTs) was prepared from a suspension of TiO_2 nanoparticles in soluble carbon nanotubes. Photoinduced charge transfer between the MWCNT and TiO_2 was proposed.

A representative application for such films is the modification of electrode surfaces. Thus, SEM and electrochemical studies of water-soluble SWCNTs, prepared via noncovalent functionalization by *Congo red* through a physical grinding treatment and immobilized on the surface of a glassy carbon electrode, showed that

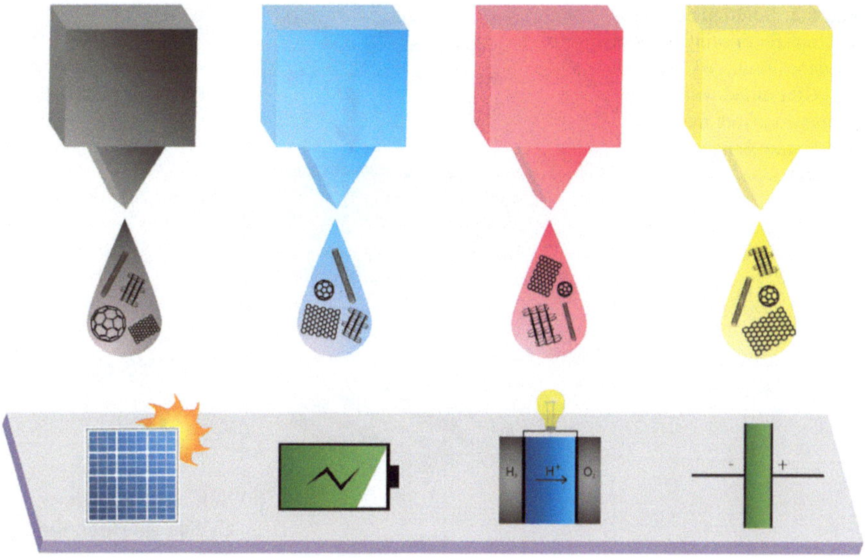

Fig. 6.1 Schematic representing the printing process of nanostructured carbon materials for solar cells, batteries, fuel cells, and supercapacitors [51] (Reproduced with permission of *Elsevier Science*)

SWCNTs formed uniform films with porous network structures of nanosizes on the electrode surface, which were stable in neutral and acidic solutions but were unstable in basic media [48]. Examination of potential applications of the films demonstrated that the water-soluble SWCNTs were the ideal materials for constructing SWCNT-based electrochemical sensing films, as well as for *flexible and printed electronics* [49, 50] (Figs. 6.1, 6.2, and 6.3).

The directed assembly of SWCNTs at lithographically defined positions on gate oxide surfaces was reported [53], allowing for the high yield (~90%) and parallel fabrication of SWCNT *device arrays.* The method is based on SWCNT functionalization through diazonium chemistry, further assembling with HfO_2 surface and heating to 600 °C. The precision, ease, and resultant high yield of this method provide a promising route to the parallel fabrication of large-scale carbon nanotube electronics. Conducting textiles (*dyes*) can be simply prepared by incorporating carbon nanotubes through a dyeing approach, immersing textiles in an aqueous sulfonated polyaniline–carbon nanotube dispersion acting as a dye [54]. In comparison with textiles dyed with sulfonated polyaniline, the conductivity and capacitance considerably increased. The patent [55] is dedicated to hair coloring using modified CNTs. Other important applications are related with the use of CNTs in *solar cells* [56–58].

Cost Analysis Despite of a relatively high number of experimental reports and reviews on the methods of CNTs solubilization and dispersion, the information on scalability of these processes is practically absent. Standard "low-cost process"

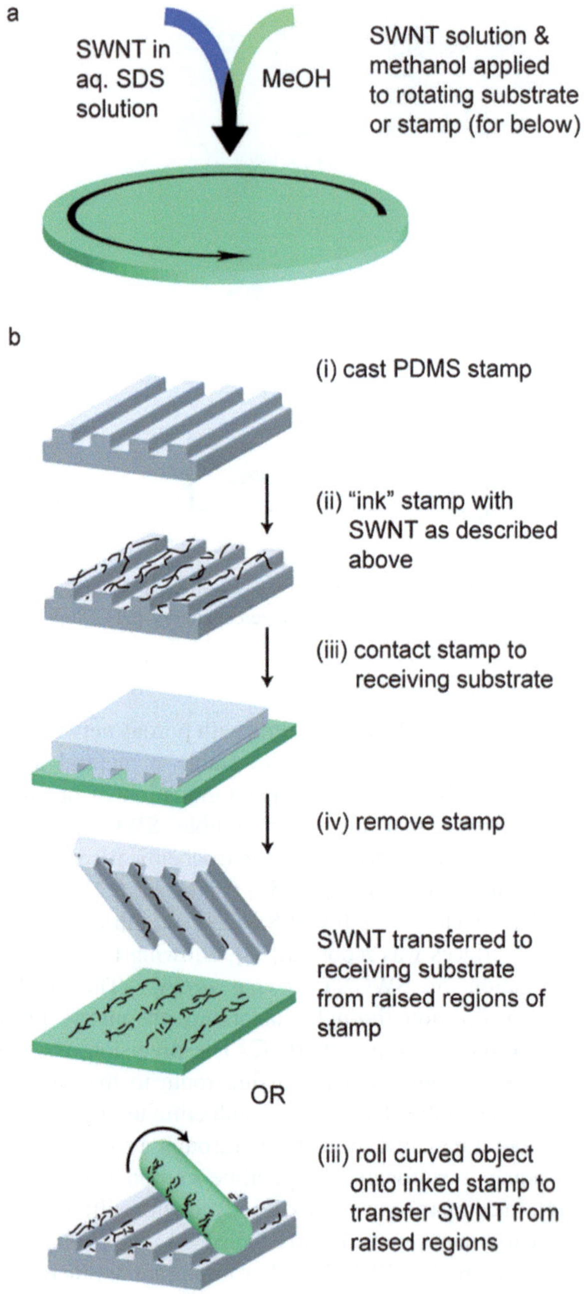

Fig. 6.2 Solution casting and transfer printing process for depositing SWCNTs [52] (Reprinted with permission from the *American Chemical Society*)

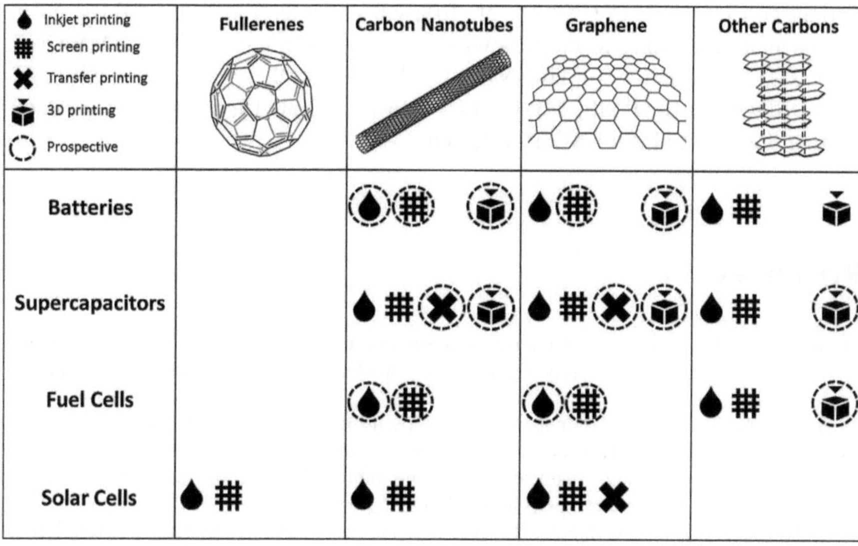

Fig. 6.3 Summary of the current state of printed nanomaterials for energy storage and conversion applications. Each symbol represents a printing technique that has been used and reported for the given application and nanomaterial. A dashed circle indicates that printing technique has not yet been used for the corresponding application and nanomaterial, but is a good candidate for future research [51] (Reproduced with permission of *Elsevier Science*)

expressions are frequent for a considerable percentage of reports on CNTs dispersion, without any supporting quantitative data and costs. Obviously, a hypothetical possibility of scaling up a CNT solubilization method should depend on a series of such factors as the cost of raw materials (solvents and surfactants), cost of CNTs (which depends on their type, purity, quantity, functionalization, and demand), cost of use of dispersion techniques (for instance, ultrasonication or hydrothermal equipment), and workforce varying according to location, scale of company, taxes, and other factors.

In case of chemical-based solubilization, three classes of molecules are mainly used for CNT dispersion, among others [59]. Surfactants are used because they are easily available and their cost is not high. Polymers and biopolymers (nucleic acids and peptides) are also very efficient in the dispersion process. The cost of CNTs has been changing considerably in the past few years, being dependent on their purity and type as well as the supplier [60] [61]. The MWCNT cost is usually between US $ 0.5 and 100/g, while DWCNTs can be obtained for around US$ 10/g. The cost of SWCNTs also varies a lot, usually between US$ 20 and 2000/g.

Several companies, suppliers of nanomaterials, give direct information on the cost of CNTs dispersions. For example, the *NanoLab* company [62] provides functionalized carboxylated MWCNTs–COOH suspended in water, ethanol, acetone, or dimethylformamide, which are stable for months and can be centrifuged to 3000 rpm for 10 min without evidence of settling. The prices vary from 600 to 800 USD/L at concentration 1 g/L. The *Nanoamorphous Materials* [63] supplies distinct concentration of CNTs (with different diameters) in water, DMF, and *n*-butyl acetate according to the following examples of rates: water, 1% MWCNTs (200 USD/kg); water, 3% MWCNTs (195 USD/500 mL); water, 0.1% SWCNTs (300 USD/200 mL); DMF, MWCNTs 7–8% (469 USD/L); metallic or semiconducting SWCNTs (1500 USD/10 mL); etc.

References

1. M. Eguílaz, C.J. Venegas, A. Gutiérrez, Carbon nanotubes non-covalently functionalized with cytochrome c: A new bioanalytical platform for building bienzymatic biosensors. Microchem. J. **128**, 161–165 (2016)
2. S. Hou, A. Zhang, M. Su, Nanomaterials for biosensing applications. Nano **6**, 58, 4 pp (2016)
3. M. Durga Prakash, S.R. Krishna Vanjari, C. Shekhar Sharma, S. Govind Singh, Ultrasensitive, label free, chemiresistive nanobiosensor using multiwalled carbon nanotubes embedded electrospun SU-8 nanofibers. Sensors **16**, 1354, 15 pp (2016)
4. G. Hughes, K. Westmacott, K.C. Honeychurch, A. Crew, R.M. Pemberton, J.P. Hart, Recent advances in the fabrication and application of screen-printed electrochemical (bio)sensors based on carbon materials for biomedical, agri-food and environmental analyses. Biosensors **6**, 50, 39 pp (2016)
5. X. Sun, Z. Gong, Y. Cao, X. Wang, Acetylcholinesterase biosensor based on poly (diallyldimethylammonium chloride)-multi-walled carbon nanotubes-graphene hybrid film. Nano Micro Lett **5**(1), 47–56 (2013)
6. B. Chan Kim, I. Lee, S.-J. Kwon, et al., Fabrication of enzyme-based coatings on intact multi-walled carbon nanotubes as highly effective electrodes in biofuel cells. Sci Rep **7**, 40202 (2017)
7. Z. Jiang, D. Chen, Y. Yu, J. Miao, Y. Liu, L. Zhang, Composite fibers prepared from multi-walled carbon nanotubes/cellulose dispersed/dissolved in ammonium/dimethyl sulfoxide mixed solvent. RSC Adv. **7**, 2186–2192 (2017)
8. J. Foldyna, V. Foldyna, M. Zelenák, Dispersion of carbon nanotubes for application in cement composites. Procedia Eng **149**, 94–99 (2016)
9. T. Jarolim, M. Labaj, R. Hela, K. Michnova, Carbon Nanotubes in Cementitious Composites: Dispersion, Implementation, and Influence on Mechanical Characteristics. Adv. Mater. Sci. Eng. **2016**, Article ID 7508904, 6 pp (2016)
10. M.G. Raucci, M. Alvarez-Perez, D. Giugliano, S. Zeppetelli, L. Ambrosio, Properties of carbon nanotube-dispersed Sr-hydroxyapatite injectable material for bone defects. Regen Biomater. **3**(1), 13–23 (2016)
11. Y. Dror, W. Salalha, W. Pyckhout-Hintzen, et al., From carbon nanotube dispersion to composite nanofibers. Progr. Colloid Polym. Sci. **130**, 64–69 (2005)
12. J.-S. Kim, G.-W. Kim, Hysteresis compensation of piezoresistive carbon nanotube/polydimethylsiloxane composite-based force sensors. Sensors **17**, 229, 12 pp (2017)
13. S.-H. Park, J. Bae. Polymer composite containing carbon nanotubes and their applications. Rec. Patents Nanotechn. 11(2), pp. 109–115 (2017), http://benthamscience.com/journals/recent-patents-on-nanotechnology; https://www.ncbi.nlm.nih.gov/pubmed/27978788; http://benthamscience.com/journals/recent-patents-on-nanotechnology/article/146781/
14. S. Boukheir, A. Len, J. Füzi, V. Kenderesi, Structural characterization and electrical properties of carbon nanotubes/epoxy polymer composites. J. Appl. Polym. Sci. **134**(8), 44514 (2017)
15. M. Shigeta, K. Kamiya, M. Uejima, S. Okada, Dispersion of carbon nanotubes in organic solvent by commercial polymers with ethylene chains: Experimental and theoretical studies. Jpn. J. Appl. Phys. **54**, 035101 (2015)

16. S.-H. Jang, S. Kawashima, H. Yin, Influence of carbon nanotube clustering on mechanical and electrical properties of cement pastes. Materials **9**, 220, 11 pp (2016)
17. A. Mukherjee, S. Majumdar, A.D. Servin, L. Pagano, O.P. Dhankher, J.C. White, Carbon nanomaterials in agriculture: A critical review. Front. Plant Sci. **7**, article 172, 16 pp (2016)
18. B. Dong Che, L.-T.T. Nguyen, B. Quoc Nguyen, et al., Effects of carbon nanotube dispersion methods on the radar absorbing properties of MWCNT/epoxy nanocomposites. Macromol. Res. **22**(11), 1221–1228 (2014)
19. V.S.W. Chan, Nanomedicine: An unresolved regulatory issue. Regul. Toxicol. Pharmacol. **46** (3), 218–224 (2006)
20. Z. Liu, X. Sun, N. Nakayama-Ratchford, H. Dai, Supramolecular chemistry on water- soluble carbon nanotubes for drug loading and delivery. ACS Nano **1**(1), 50–56 (2007)
21. C. Klumpp, K. Kostarelos, M. Prato, A. Bianco, Functionalized carbon nanotubes as emerging nanovectors for the delivery of therapeutics. BBA-Biomembranes **1758**(3), 404–412 (2006)
22. A. Bianco, K. Kostarelos, M. Prato, Applications of carbon nanotubes in drug delivery. Curr. Opin. Chem. Biol. **9**(6), 674–679 (2005)
23. K. Fu, W. Huang, Y. Lin, D. Zhang, T.W. Hanks, A.M. Rao, Y.-P. Sun, Functionalization of carbon nanotubes with bovine serum albumin in homogeneous aqueous solution. J. Nanosci. Nanotechn. **2**(5), 457–461 (2002)
24. L.W. Zhang, L. Zeng, A.R. Barron, N.A. Monteiro-Riviere, Biological interactions of functionalized single-wall carbon nanotubes in human epidermal keratinocytes. Int. J. Toxicology **26**(2), 103–113 (2007)
25. X. Dong, Z. Sun, X. Wang, D. Zhu, L. Liu, X. Leng, Simultaneous monitoring of the drug release and antitumor effect of a novel drug delivery system-MWCNTs/DOX/TC. Drug Deliv. **24**(1), 143–151 (2017)
26. S. Kumar, R. Rani, N. Dilbaghi, K. Tankeshwar, K.-H. Kim, Carbon nanotubes: A novel material for multifaceted applications in human healthcare. Chem. Soc. Rev. **46**, 158–196 (2017)
27. S. Sharma, N. Kumar Mehra, K. Jain, N. Kumar Jain, Effect of functionalization on drug delivery potential of carbon nanotubes. Art. Cells, Nanomed., Biotechn. **44**(8), 1851–1860 (2016)
28. P. Shantaram Uttekar, A. Malhari Kulkarni, P. Namdeorao Sable, P. Digambar Chaudhari, Surface modification of carbon nano tubes with nystatin for drug delivery applications. Indian J. Pharm. Edu. Res. **50**(3), 385–390 (2016)
29. T. Ohta, Y. Hashida, F. Yamashita, M. Hashida, Development of novel drug and Gene delivery carriers composed of single-walled carbon nanotubes and designed peptides with PEGylation. J. Pharm. Sci. **105**(9), 2815–2824 (2016)
30. M. Kawaguchi, T. Fukushima, T. Hayakawa, N. Nakashima, Y. Inoue, S. Takeda, K. Okamura, K. Taniguchi, Preparation of carbon nanotube-alginate nanocomposite gel for tissue engineering. Dent. Mater. J. **25**(4), 719–725 (2006)
31. C.J. Gannon, P. Cherukuri, B.I. Yakobson, L. Cognet, J.S. Kanzius, C. Kittrell, B.R. Weisman, S.A. Curley, Carbon nanotube-enhanced thermal destruction of cancer cells in a noninvasive radiofrequency field. Cancer **110**(12), 2654–2665 (2007)
32. R.P. Feazell, N. Nakayama-Ratchford, H. Dai, S.J. Lippard, Soluble single-walled carbon nanotubes as longboat delivery Systems for Platinum(IV) anticancer drug design. J. Am. Chem. Soc. **129**, 8438–8439 (2007)
33. Y. Hwang, S.-H. Park, J. Woo Lee, Applications of functionalized carbon nanotubes for the therapy and diagnosis of cancer. Polymers **9**, 13, 26 pp (2017)
34. N.M. Bardhan, 30 years of advances in functionalization of carbon nanomaterials for biomedical applications: A practical review. (Annual Issue: Early career scholars in materials Science). J. Mater. Res. **32**(1), 107–127 (2017)
35. E. Heister, E.W. Brunner, G.R. Dieckmann, I. Jurewicz, A.B. Dalton, Are carbon nanotubes a natural solution? Applications in biology and medicine. ACS Appl. Mater. Interfaces **5**(6), 1870–1891 (2013)

36. I. Jesion, M. Skibniewski, E. Skibniewska, et al., Graphene and carbon nanocompounds: Biofunctionalization and applications in tissue engineering. Biotechnol. Biotechnol. Equip. **29**(3), 415–422 (2015)
37. R. Amezcua, A. Shirolkar, C. Fraze, D.A. Stout, Nanomaterials for cardiac myocyte tissue engineering. Nano **6**, 133, 15 pp (2016)
38. J. Venkatesan, R. Ramjee Pallela, S.-K. Kim, Applications of carbon nanomaterials in bone tissue engineering. J. Biomed. Nanotechnol. **10**, 3105–3123 (2014)
39. N. Burblies, J. Schulze, H.-C. Schwarz, Coatings of different carbon nanotubes on platinum electrodes for neuronal devices: Preparation, cytocompatibility and interaction with spiral ganglion cells. PLoS One **11**(7), e0158571 (2016)
40. J.L. Hernandez-Lopez, E.R. Alvizo-Paez, S.E. Moya, J. Ruiz-Garcia, Ordered carbon nanotube thin films produced by the trapping of water-soluble single-wall carbon nanotubes at the air/water interface. Carbon **45**(12), 2448–2450 (2007)
41. J. Li, Y. Zhang, Large-scale aligned carbon nanotubes films. Physica E **33**(1), 235–239 (2006)
42. M.A. Hayat Nawaz, S. Rauf, et al., One step assembly of thin films of carbon nanotubes on screen printed interface for electrochemical aptasensing of breast cancer biomarker. Sensors **16**, 1651, 15 pp (2016)
43. F. Li, B. Tang, J. Xiu, S. Shufen Zhang, Hydrophilic modification of multi-walled carbon nanotube for building photonic crystals with enhanced color visibility and mechanical strength. Molecules **21**, 547, 9 pp (2016)
44. Meng, X., Liu, Y., Huang, M., Cao, J.-P. Flexible perfluoroalkoxy films filled with carbon nanotubes and their electric heating property. J. Appl. Polym. Sci., 2017, **134**(18), 44782), 6 pp (2017)
45. A. Almowarai, Y. Ueno, Y. Show, Fabrication of CNT dispersion fluid by wet-jet milling method for coating on bipolar plate of fuel cell. J. Nanomater. **2015**, Article ID 315017, 7 pp (2015)
46. A.G. Rozhin, Y. Sakakibara, M. Tokumoto, H. Kataura, Y. Achiba, Near-infrared nonlinear optical properties of single-wall carbon nanotubes embedded in polymer film. Thin Solid Films **464**, 368–372 (2004)
47. K. Yu, Z. Zhu, M. Xu, Q. Li, W. Lu, Q. Chen, Soluble carbon nanotube films treated using a hydrogen plasma for uniform electron field emission. Surf. Coat. Technol. **179**(1), 63–69 (2004)
48. C. Hu, X. Chen, S. Hu, Water-soluble single-walled carbon nanotubes films: Preparation, characterization and applications as electrochemical sensing films. J. Electroanal. Chem. **586** (1), 77–85 (2006)
49. J. Zaumseil, Single-walled carbon nanotube networks for flexible and printed electronics. Semicond. Sci. Technol. **30**, 074001, 20 pp (2015)
50. Kumar, S., Cola, B.A., Jackson, R., Graham, S. A review of carbon nanotube ensembles as flexible electronics and advanced packaging materials. J. Electron. Packag., 2011, 133, 020906, 12 pp.
51. S. Lawes, A. Riese, Q. Sun, N. Cheng, X. Sun, Printing nanostructured carbon for energy storage and conversion applications. Carbon **92**, 150–176 (2015)
52. M.A. Meitl, Y. Zhou, A. Gaur, S. Jeon, M.L. Usrey, M.S. Strano, et al., Solution casting and transfer printing single-walled carbon nanotube films. Nano Lett. **4**(9), 1643–1647 (2004)
53. G.S. Tulevski, J. Hannon, A. Afzali, Z. Chen, P. Avouris, C.R. Kagan, Chemically assisted directed assembly of carbon nanotubes for the fabrication of large-scale device arrays. J. Am. Chem. Soc. **129**(39), 11964–11968 (2007)
54. M.I.H. Panhuis, J. Wu, S.A. Ashraf, G.G. Wallace, Conducting textiles from single-walled carbon nanotubes. Synth. Met. **157**(8), 358–362 (2007)
55. Huang, X., Kobos, R. K., Xu, G. Hair coloring and cosmetic compositions comprising carbon nanotubes. US7276088, 2007
56. A.J. Miller, R.A. Hatton, S.R.P. Silva, Interpenetrating multiwall carbon nanotube electrodes for organic solar cells. Appl. Phys. Lett. **89**, 133117 (2006)

57. H.A. Alturaif, Z.A. ALOthman, J.G. Shapter, S.M. Wabaidur, Use of carbon nanotubes (CNTs) with polymers in solar cells. Molecules **19**, 17329–17344 (2014)
58. T. Grace, L.P. Yu, C. Gibson, et al., Investigating the effect of carbon nanotube diameter and wall number in carbon nanotube/silicon heterojunction solar cells. Nano **6**, 52, 13 pp (2016)
59. C. Klumpp, K. Kostarelos, M. Prato, A. Bianco, Functionalized carbon nanotubes as emerging nanovectors for the delivery of therapeutics. Biochim. Biophys. Acta Biomembr. **1758**(3), 404–412 (2006)
60. https://sites.google.com/site/cntcomposites/cost-and-production. Accessed on May 8, 2017
61. M. Bierdel, S. Buchholz, V. Michele, L. Mleczko, R. Rudolf, M. Voetz, A. Wolf, Phys. Status Solidi **244**, 3939–3943 (2007)
62. http://www.nano-lab.com/nanotubesuspensions.html. Accessed on May 8, 2017
63. http://www.nanoamor.com/carbon_nanotube_dispersions. Accessed on May 8, 2017

Chapter 7
Training on CNT Solubilization

In this section, we present several selected experimental procedures, directly borrowed from original articles. We hope this material will be very useful as educational material for M.Sc. and Ph.D. students, working in the areas of biochemistry and drug delivery, fabrication of CNT thin films from dispersion, application of CNT composites with polymers, etc.

7.1 Ultrasonication

SWCNTs synthesized by the HiPco process, both as-produced (batch CNI 002) and in purified form (batch CNI 26-0036B-2), were purchased from *Carbon Nanotechnologies*. In the experiments, the SWCNTs were dispersed in water by means of the surfactant sodium dodecylbenzene sulfonate (SDBS), which has recently been shown to allow the dispersion of the tubes as individuals in aqueous solution. Nanotube dispersions were prepared by a combined tip and bath ultrasonication approach as follows. A small piece of HiPco SWCNT mat was added to an aqueous solution (5 mg/mL) of the surfactant and subjected to ultrasonic treatment in a tip sonicator (Dr. Hielscher UP 200 s). A small number of pulses (usually five), with 0.5-s-on/0.5-s-off pulse cycles at ~40 W/cm^2, were applied to a volume of 1.5 mL. Subsequently, the obtained suspension was centrifuged (Eppendorf 5417C centrifuge) at 20,200 g for 30 min. The upper ~80% of the resulting supernatant was then carefully decanted and subjected to ultrasonication in a bath sonicator (Branson 1510, 80 kHz) for time periods ranging from a few minutes to several hours [1]. Reproduced with permission of the *American Chemical Society*.

O.V. Kharissova, B.I. Kharisov, *Solubilization and Dispersion of Carbon Nanotubes*, DOI 10.1007/978-3-319-62950-6_7

7.2 Plasma Treatment

A schematic setup of the plasma reactor and a flowchart of the functionalization process are shown in Fig. 2.8: 100 mg of MWCNT powder (Sigma-Aldrich, outer diameter = 10–30 nm; inner diameter = 3–10 nm; length = 1–10 μm; purity >90%) is added to 10–20 mL of pure ethanol (Wako Pure Chemicals Co., purity >95%) and sonicated at room temperature using a supersonic homogenizer (Sonics Vibra-Cell, VC 130, Sonics & Materials Inc., $f = 20$ kHz, 6.0 mm ϕ probe) at an input power of 10–20 W for 15–120 min. The suspension is dried under reduced pressure and soaked in 0.0–0.30 mole (5 mL) of citric acid (Wako Pure Chemicals Co., assay >98%) solution for 0–120 h. The MWCNTs in the citric acid solution are then placed on the lower electrode (SUS, 50 mm ϕ) of the reactor, which is evacuated to ca. 400 Pa using a rotary pump at a very slow rate. When the wet phase starts to disappear, oxygen gas is introduced into the reactor at a rate of 0–10 sccm, and the background chamber pressure is kept at about 400 Pa. Then the plasma reaction is carried out for 10–30 min by an RF input power of $P_{rf} = 100$–300-W, $f = 13.56$ MHz. The reflected RF power is minimized (<20 W) by controlling the matching current during the plasma reaction. It is noted that when the samples become fully dried before starting the plasma reaction, they are less reactive for the oxygen plasma. However, when the plasma is started in the wet phase, the water molecules and part of the citric acid molecules evaporate with the processing time but are considered to remain inside the chamber and in the gas container connected to the chamber, contributing to the functionalization process. After the treatment, the MWCNTs are washed at least three times using pure water (Wako Pure Chemicals Co., distilled water) and dried under reduced pressure at room temperature [2]. Reproduced with permission from *Hindawi Publishing Corporation.*

7.3 Shear Force Mixing

For polymer wrapping with SFM, 0.5 g/L PFO-BPy (American Dye Source, MW ¼ 34 kg/mol) was dissolved in 140 mL toluene before adding 0.38 g/L CoMoCAT raw material. Hence, a maximum of 17 mg [5, 6] SWCNTs was in the initial dispersion. SFM using a Silverson L5M-A mixer (Fig. 2.14) was then applied at maximum speed (10,230 rpm) for a given time. The temperature was kept constant at 20 °C with a cooling bath. The dispersion step was followed by centrifugation at 60,000 g (Beckman Coulter Avanti J26XP centrifuge) for 45 min with an intermediate supernatant extraction and centrifuge tube exchange after 15 min. For SWCNT recycling the pellet obtained after centrifugation and supernatant extraction was reused instead of raw CoMoCAT material. Fresh toluene (140 mL) and polymer (0.5 g/L) were added to the pelletized nanotubes and SFM was repeated [3]. Reproduced with permission of *Elsevier Science.*

7.4 Hydrothermal Method

The experiments were carried out by using MWCNTs mixed with different organic solvents in the concentration of 80 mg/L. The different organic solvents used for dispersion were dichloromethane, ethanol, isopropyl alcohol, and hexane. The individual solutions along with MWCNTs were placed in ultrasonicator (Sidilu, 20 kHz, 300 kV) and sonicated for 15 min. Later, 20 mL of the sonicated solutions was placed in Teflon liner with Morey-type autoclaves. The autoclaves were heated constantly using a hot air oven at critical temperature of individual solvents for 5 h. Subsequently, after 5 h, the autoclaves were removed from the oven and cooled down to room temperature. The resultant materials in liner with the solvent were collected carefully in a beaker. They were later sonicated for further dispersion and centrifuged (KEMI, 159 W/HP, 230 V, 50 Hz) at 10,000 rpm for 15 min to check the stability, because during centrifugation the bundled MWCNTs get settled down and supernatant contains finely dispersed MWCNTs (at lower concentration). These resultant supernatant solutions containing finely dispersed MWCNTS were used for further characterization [4]. Reproduced with permission of *Hindawi Publishing Corporation.*

7.5 Ozonation

SWCNTs were treated in an all-glass-quartz system illustrated in Fig. 3.1. Vacuum annealing was carried out in a quartz tube under a 10^{-2}–10^{-3} Torr vacuum. An 11 mg sample of purified nanotubes was weighed by difference (accuracy $\pm$ 0.0001 g) into a small quartz thimble equipped with small hooks used for mechanical transfer from the O_3 treatment chamber, through air, into a modified quartz BET measurement tube which is also used for annealing. The 270 cm^3 O_3 treatment chamber was of wide bore, and this geometry was found to be necessary to avoid large O_3 concentration gradients, retarding the reaction kinetics and leading to irreproducible results. Ozone (93–97%) treatment was carried out at 22.5 Torr and 300 K for 10 min. The thimble was then transferred to a 12 mm quartz measurement tube and vacuum-annealed for 1 h at 523 K. An isotherm at 77.3 K was then measured using N_2. Then the sample was vacuum-annealed at 1073 K for 1 h, and a second N2 isotherm was measured. To increase accuracy, the volume of the standard measurement tube was decreased by inserting a 10-mm-diameter glass rod. The O_3-etching cycle was repeated six times, causing considerable loss of nanotube mass due to oxidation. The zeroth cycle was carried out as above, omitting the O_3 treatment [5]. Reproduced with permission of the *American Chemical Society.*

7.6 Preparation of DWCNT Suspensions of Varying Ionic Strengths (Ca^{2+})

Moderately hard reconstituted waters with varying ionic Ca^{2+} strengths were prepared by varying the amount of calcium carbonate. Reconstituted waters with ionic strengths of 0.00449 M Ca^{2+}, 0.0226 M Ca^{2+}, and 0.0339 M Ca^{2+} were prepared. The salts used to prepare reconstituted water were kept constant when Ca $^{2+}$ was varied. The salts were 0.2400 g of $NaHCO_3$, 0.0200 g of KCL, and 0.1500 g of $MgSO_4$, and these were added to 5 L of Milli-Q water. The DWCNT suspensions were prepared by adding 50 mg of DWCNTs to 100 mL of the prepared waters with varying ionic strengths. The suspensions were sonicated for 2 h and were shaken at 30 min intervals. The suspensions were then refrigerated at 4 °C for 24 h [6]. Reproduced with permission of the *Hindawi Publishing Corporation.*

7.7 Use of MnO_2

Suspensions of MnO_2 and MWCNT in water, containing dispersants, were ultrasonically agitated during 30 min before the analysis, and fabrication of MnO_2–MWCNT composites allowed the formation of stable aqueous suspensions of positively charged MWCNT and negatively charged MnO_2 [7]. Reproduced with permission of the *Royal Society of Chemistry.*

7.8 Use of Surfactant Dodecyl Flavin FC12

A mixture of 1 mg of SWNT and 1 mg of FC12 (N-dodecyl isoalloxazine) was added to 4 mL of p-Xylene. The resulting mixture was exposed to brief (5 min) bath sonication (Branson 1510 sonicator, 70 W) and further subjected to 1-h tip sonication [40% power (18.8 W/mL); probe tip diameter, 13 mm; VCX 750, Sonics & Materials], cooled with a water circulator maintained at 10 °C. After varying centrifugation steps (i.e., 5000, 10,000, 20,000, and 30,000 g, no. 1 fixed-angle rotor, Supra 22K, Hanil or Wisespin CF-10, DAIHAN Scientific Co., Ltd.) for 2 h, 80% supernatant was collected [8]. Reproduced with permission of the *American Chemical Society.*

7.9 Use of a Coordination Compound (Ruthenium Polypyridyl Complex)

Different amounts of purified HiPco (Batch # 195.1) and CoMoCAT (Sigma-Aldrich) SWCNTs were weighted in separate vials, followed by the addition of 1 mg/ml RuPy1 aqueous solution to give different initial concentration of the nanotube in solution. The resultant mixture was sonicated with a probe sonicator (Sartorius, LABSONIC® M) for 30 min and then centrifuged (VWR centrifugator) for 45 min at 12,000 *g*. The supernatant was collected carefully and diluted for further analysis [9]. Reproduced with permission of the *Royal Society of Chemistry.*

7.10 Use of 1,10-Binaphthyl-Incorporated Conjugated Polymers

A 1.0 mg portion of SWCNTs was added into 1 mL of N-methyl-2-pyrrolidone (NMP) in which a 1.0 mg portion of a desired polymer was dissolved. The resulting mixture was then sonicated for 1 h in an ice bath with a probe tip sonicator (8–10 W). After centrifugation of the solution at 12,200 g for 30 min, the supernatant was obtained to get a SWCNT/polymer dispersion [10]. Reproduced with permission of the *Elsevier Science.*

7.11 Use of DNA

A 1 mg of HiPco SWNTs was mixed with 1 mL $(GT)_{15}$ (single-stranded DNA) solution (1 mg/ml in 10 mM Tris, 1 mM EDTA, pH 7.6) and sonicated with a cup-horn sonicator at 3 W for 1 h in an ice/water bath. The mixture was then centrifuged to remove any insoluble materials, and the supernatant was filtered through a disk membrane (with pore size of 100 nm). The pellet of DNA-SWNTs on the membrane was washed with copious amounts of water to remove buffer components and unbound DNA [11]. Reproduced with permission of the *American Chemical Society.*

7.12 Use of Proteins

Aqueous dispersions of SWCNTs were prepared by dispersing the SWCNTs in dimethylformamide followed by washing with Milli-Q water. A known concentration of SWNTs (50 μg/mL) was dispersed in a protein solution (10 mg/mL). Ultrasonication of the dispersion was carried out for predetermined times using a bath sonicator (Model 50 T, VWR International, West Chester, PA) with rated power of 45 W. The sonication efficiency and, hence, the quality of the dispersion varied with the volume of the solution sonicated. For best dispersions, the final sonication volume was kept at 4 mL. The solutions thus obtained were subjected to ultracentrifugation at 123,000 g for 30 min. The upper 60% of the supernatant after ultracentrifugation was carefully collected and subjected to another round of ultracentrifugation at 185,000 *g* for 30 min. The upper 75% of this supernatant was collected and used for further analysis [12]. Reproduced with permission of the *American Chemical Society*.

References

1. J.I. Paredes, M. Burghard, Dispersions of individual single-walled carbon nanotubes of high length. Langmuir **20**, 5149–5152 (2004)
2. M. Jellur Rahman, T. Mieno, Water-dispersible multiwalled carbon nanotubes obtained from citric-acid-assisted oxygen plasma functionalization. J. Nanomater. **2014**, Article ID 508192, 9 pp (2014)
3. A. Graf, Y. Zakharko, S.P. Schießl, C. Backes, M. Pfohl, B.S. Flavel, J. Zaumseil, Large scale, selective dispersion of long single-walled carbon nanotubes with high photoluminescence quantum yield by shear force mixing. Carbon **105**, 593–599 (2016)
4. K. Jagadish, S. Srikantaswamy, K. Byrappa, L. Shruthi, M. Ramesh Abhilash, Dispersion of multiwall carbon nanotubes in organic solvents through hydrothermal supercritical condition. J. Nanomater. **2015**, Article ID 381275, 6 pp (2015)
5. O. Byl, J. Jie Liu, J.T. Yates Jr., Etching of carbon nanotubes by ozone. A surface area study. Langmuir **21**, 4200–4204 (2005)
6. L.P. Lukhele, B.B. Mamba, N. Musee, V. Wepener, Acute toxicity of double-walled carbon nanotubes to three aquatic organisms. J. Nanomater. **2015**, Article ID 219074, 19 pp (2015)
7. Y. Liu, I. Zhitomirsky, Aqueous electrostatic dispersion and heterocoagulation of multiwalled carbon nanotubes and manganese dioxide for the fabrication of supercapacitor electrodes and devices. RSC Adv. **4**, 45481–45489 (2014)
8. M. Park, S. Kim, H. Kwon, et al., Selective dispersion of highly pure large-diameter semiconducting carbon nanotubes by a flavin for thin-film transistors. ACS Appl. Mater. Interfaces **8**, 23270–23280 (2016)
9. K. Huang, A. Saha, K. Dirian, C. Jiang, P.-L. Chu, J.M. Tour, D.M. Guldi, A.A. Martí, Carbon nanotubes dispersed in aqueous solution by ruthenium(II) polypyridyl complexes. Nanoscale **8**, 13488–13497 (2016)
10. S.H. Min, H.-I. Kim, K.-S. Kim, et al., Selective dispersion of single-walled carbon nanotubes by binaphthyl based conjugated polymers: Integrated experimental and simulation approach. Polymer **96**, 63–69 (2016)
11. C. Hu, Y. Zhang, G. Bao, et al., DNA functionalized single-walled carbon nanotubes for electrochemical detection. J. Phys. Chem. B Lett. **109**, 20072–20076 (2005)
12. S.S. Karajanagi, H. Yang, P. Asuri, et al., Protein-assisted solubilization of single-walled carbon nanotubes. Langmuir **22**(4), 1392–1395 (2006)

Chapter 8
Conclusions and Further Outlook

8.1 Set of Available Solubilization Methods

The field of carbon nanotubes continues to be one of the hottest topics in the nanoscience and nanotechnology, so the development of their functionalization and dispersion methods is and will be an important research area, at least for the next 10 years. A series of contemporary techniques are being used for CNT solubilization, from *physical* (classic ultrasound, plasma treatment, UV light, dielectrophoresis [1, 2], gel electrophoresis [3, 4], density-gradient ultracentrifugation [5, 6], irradiation techniques, and chromatography [7], among others) to *chemical* and *biological*, applying *inorganic* (other carbon allotropes, iodine, bromine, metallic sodium in liquid ammonia, CO_2, peroxides, ozone, metal salts, and mineral acids) and *organic* (acids, salts, polymers, dyes, natural products, in particular diazonium salts [8]; functionalization with porphyrins, amines, pyrene, polymers, sugars, etc.) compounds and *biomolecules* (DNA, peptides, amino acids, etc.), as well as micelles on their basis and some metal complexes. CNTs can be functionalized by *oxidation* (peroxy acids; metal oxidants, such as osmium tetroxide, potassium permanganate, chromates; ozone; oxygen; superoxides) or *reduction* reactions (interactions with thiols, carbenes, dienes, etc.).

Frequently, physical action (more frequently ultrasound, more rarely hydrothermal technique) is combined with chemical/biological treatment. In some cases, successive steps can be applied, for instance, use of low- and high-weight surfactants, mineral acid treatment for creation of –OH and –COOH groups, and their further interaction with organic molecules. Carbon nanotube dispersion in nematic liquid crystals is also known. Many of these chemical agents above are *amphiphiles*, known to be suitable molecules to disperse CNTs in water by shielding their highly hydrophobic surfaces. Tailored anionic, nonionic, and cationic photopolymerizable amphiphiles (objects of permanent attention [9, 10] as good dispersants for CNTs) were designed to achieve programmed pH-dependent

O.V. Kharissova, B.I. Kharisov, *Solubilization and Dispersion of Carbon Nanotubes*, DOI 10.1007/978-3-319-62950-6_8

dispersions of CNTs [11]. All of these methods suffer from problems with scalability, effectiveness, and often dissolution sensitivity.

Each discussed method leads to an improvement of CNT *solubility*, frequently considerable. The formed dispersions could be stable for long periods of time, from several weeks to some months, even sometimes remained stable after centrifugation. Sometimes, unpredictable results could occur, for example, in case of action of humic acid to effectively disperse MWCNTs, but not SWCNTs, into stable suspensions under the studied conditions or when structures of amorphous carbon and carbon particles of MWCNTs were completely eliminated and the tips of nanotubes opened applying planetary ball mill.

8.2 Important Observations and Considerations

It has been suggested that van der Waals interaction, π–π stacking interactions between aromatic rings in organic compounds and CNTs, and hydrophobic interaction are major factors that are responsible for the CNT dispersion. One of the main approaches to disperse and exfoliate carbon nanotubes and graphene is based on the use of *surfactants* [12]. Choosing them to be able to stabilize carbon nanotubes in water, it is necessary to employ such dispersing agents that (a) strongly adsorb on the nanotube surface; (b) present hydrophilic groups, better if rigid, that extend toward the aqueous phase; (c) are not very mobile on the nanotube surface; and (d) show aggregates with structure dependent on nanotube diameter and chirality. Using surfactants, tremendous progress has been made to *stabilize CNT dispersions*, although many questions await answer to design surfactant formulations that selectively stabilize nanotubes monodispersed in diameter and chirality. Although the efficiency of anionic, cationic, and nonionic surfactants has been demonstrated to different extents, the exact mechanism by which carbon nanotubes and the different surfactants interact is still uncertain. As it is seen, considerable research efforts have already been devoted to optimize and develop processes for their effective separation. Currently, techniques for separating bundles of carbon nanotubes into homogeneous dispersion are still under development, although a few methods have been successful at the laboratory scale.

For appropriate *solvent selection*, it was established for SWCNT case that (1) heavier solvents (and small in size) most probably are better solvents for SWCNTs and (2) higher polarizability of the solvent molecule increases the dispersibility. Among solvents, water is the most preferable due to much more existing and potential CNT applications, in particular more for medical and biological purposes. N-Methyl-2-pyrrolidone (NMP), N-dodecyl-pyrrolidone (N12P), acetone, tetrahydrofuran (THF), N,N-dimethylformamide (DMF), N, N-dimethylacetamide (DMA), and cyclohexyl-pyrrolidone (CHP) have been considered as suitable solvents for CNT dispersion; meanwhile, in contrast, much precipitation can be obviously observed for systems of the CNTs in water, ethanol, and toluene. Non-functionalized CNTs can be solubilized in suitably chosen organic solvents, and, furthermore, their solubility could be understood in terms of the *Hansen solubility parameters* (HSPs) [13].

Some curious but naturally determined and explained observations were emphasized by researchers, in particular on better dispersibility of the CNTs with bigger diameters (treatment with plasma or graphene oxide) or more favorable suspension of CNTs using negatively charged nanodiamonds as compared to positively charged particles, possibly caused by electrostatic interactions. The presence of *π-conjugated systems* that can potentially interact with CNTs to induce its dispersion is considered as a contributing factor, as well as both hydrophilic and hydrophobic moieties in amphiphilic bifunctional molecules. Additionally, a huge number of *polyaromatic compounds* (especially pyrene derivatives) and a variety of polymers and biomolecules are noted to be used as surfactants or functionalizing agents. As an example, bifunctional molecules with pyrene groups exhibited high CNT solubility in common organic solvents with very different polarities such as tetrahydrofuran, toluene, and *n*-hexane. Several special studies have been carried out in the areas of influence of solvent and light on CNT dispersibility, combinations and abilities of surfactants, CNT cytotoxicity, etc.

8.3 Comparison, Advantages, and Disadvantages of Methods

As it is seen above, a lot of very distinct techniques have been applied for CNT solubilization. The most common method for fast creation of CNT dispersion is the ultrasonication; every laboratory has at least ultrasonic baths, which in majority of cases are sufficient for needed grade of dispersion without damage of CNTs. As a disadvantage, the CNTs could be partially destroyed under long high-power ultrasonic treatment. In addition, ultrasonication can be easily combined with addition of surfactants and other organic compounds to increase dispersion effect. UV and visible light and thermal annealing of CNTs are also available, but not so efficient in comparison with other methods, so their use is limited. Plasma treatment of CNTs, irradiation techniques and labeling, mechanical methods, pulsed streamer discharge, microwave, and hydrothermal method are effective, but they require special high-cost equipment and conditions and are not always available, so their use is also limited. The cryogenic crushing, pulsed laser light, and oscillating magnetic fields belong to exotic techniques.

Biological methods are mainly used for biomedical purposes requiring aqueous media. That is why the chemical methods, on the basis of a grand variety of solubilizing agents of distinct nature, are the most popular, common, and reported elsewhere. Using them, especially in combination with ultrasonication, one can get CNT dispersions on different stability, in a variety of hydrophilic and hydrophobic solvents, with and without CNT functionalization, at distinct costs including those satisfactory for commercialization. Several selected methods and strategies shown

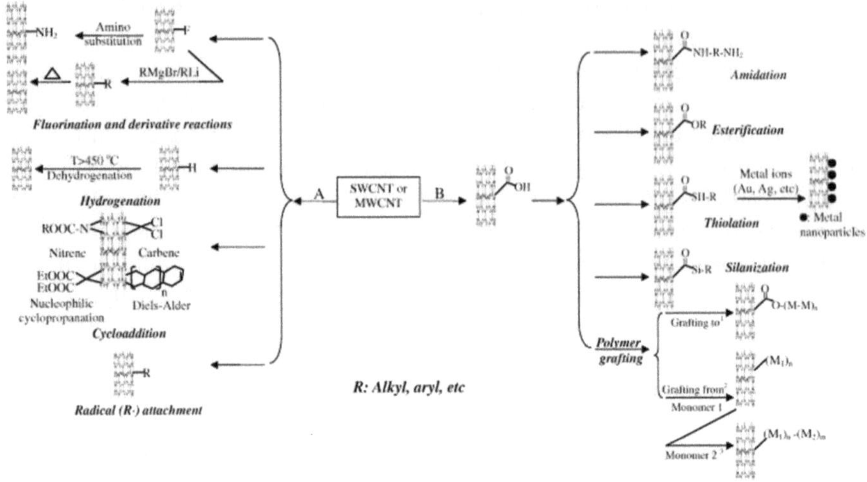

Fig. 8.1 Strategies for covalent functionalization of CNTs ((**a**) direct sidewall functionalization, (**b**) defect functionalization). [1]Carboxylic CNTs react with reactive polymers; [2]Functionalized CNTs act as initiators to initiate polymerization; [3]Living polymerization, CNT-copolymer can be obtained; n, m: Degree of polymerization (Reproduced with permission from *Elsevier Science*)

in Fig. 8.1 [14], based on organic reactions and leading to CNT functionalization [15], emphasize their universality for CNT solubilization. At the same time, not all methods are sometimes desirable for use, since some of them, for example, can heavily damage SWCNTs (Fig. 5.4) and MWCNTs. Scalability is a separate problem, which needs a profound investigation, taking into account the (1) possibility to scale up a laboratory-developed process, (2) availability and cost of equipment and chemicals, (3) possible economical effect, and (4) environmental aspects based on toxicity of CNTs and chemicals used.

8.4 Topics of Special Attention

We believe that a special attention should be given to the following topics in a greater scale. *First*, the development of CNTs-based soluble systems possessing magnetic properties for magnetically driven drug delivery could be of an interest, since classic functionalized core-shell Fe-Au or Fe_xO_y-Au nanoparticles could be substituted by more economic equivalent nanomaterials. *Second*, CNT-supported magnetic nanomaterials containing ultrasmall metal atoms, oxides, or mixed oxides (i.e., ferrites) could be useful for the purposes of catalysis. *Third*, water-soluble CNTs or their composites, which could be immediately transformed to a sponge in the interface "petroleum–water," could be potentially used for petroleum removal in oil spill sites, under the conditions of efficiency and low cost. The last could be reached using, in particular, natural materials contributing to CNT solubilization.

Other desirable ideas for the development of CNT dispersion area are shown below. Indeed, the topics of special interest can vary in a very broad range depending on research interests of researchers and continuously changing demands.

8.5 Recommendations and Future Work

Despite many efforts, realized for the last two decades in the field of CNT solubilization, several available niches are clearly seen. In our opinion, *first of all*, use of inorganic chemical compounds and especially inorganic natural products is extremely insufficiently represented. Inside the inorganic compounds, we would pay attention on inorganic peroxides, not only H_2O_2 but also its derivatives such as BaO_2 and CaO_2, among others, which are a permanent source of hydrogen peroxide due to slow interaction with water. As representative inorganic natural products, only sodium montmorillonite and Laponite are known to be used for CNT dispersion, i.e., this area is completely out of development, although it could be attractive for large-scale applications of carbon nanotube dispersions, since natural hydrophilic minerals have generally low cost and are compatible with the environment. *Second*, organic peroxides are not widely applied for CNT solubilization. They are generally well soluble in organic solvents and may be used more extensively for CNT oxidation and further interchange of formed –COOH groups by organic radicals. It is recommended to use them together with substances, leading to their decomposition in mild conditions, for instance, in Fenton systems, for creation of free radicals, leading to CNT oxidation and functionalization. *Third*, the use of biomolecules is also far from sufficient number of reports (about 100 available articles only). Microbes, fungi, bacteria, and other living cells are completely excluded; their soluble composites are almost unknown, although decorated and functionalized CNTs are applied for creation of antimicrobial media, drug delivery, and tumor treatment.

8.6 Expectations and Final Remark

The ability to manipulate the dispersion state of CNTs leads to a variety of useful applications for functionalized CNT dispersions both in water and organic solvents, including fabrication of devices, creation of membranes and optical sensors, drug delivery and other medical uses, and possible regulation of environmental spread and organic pollutants (for instance, dyes) adsorbed onto carbon nanotubes. Taking into account the facts that carbon nanotube area is currently an enormous field of research and applications and CNT consumption is being increased annually, the demand for their obtaining in solubilized/dispersed forms in liquid and solid phases will be obviously dramatically enhanced in the next decade.

References

1. R. Krupke, F. Hennrich, H. von Lohneysen, M.M. Kappes, Separation of metallic from semiconducting single-walled carbon nanotubes. Science **301**, 344–347 (2003)
2. B.R. Burg, J. Schneider, M. Muoth, L. Durrer, T. Helbling, N.C. Schirmer, T. Schwamb, C. Hierold, D. Poulikakos, Aqueous dispersion and dielectrophoretic assembly of individual surface-synthesized single-walled carbon nanotubes. Langmuir **25**(14), 7778–7782 (2009)
3. K. Moshammer, F. Hennrich, M.M. Kappes, Selective suspension in aqueous sodium dodecyl sulfate according to electronic structure type allows simple separation of metallic from semiconducting Single-walled carbon nanotubes. Nano Res. **2**, 599–606 (2009)
4. A.A. Vetcher, J.-H. Fan, I.A. Vetcher, T. Lin, S.M. Abramov, R. Draper, M.E. Kozlov, R.H. Baughman, S.D. Levene, Electrophoretic fractionation of carbón nanotube dispersion on agarose gels. Int. J. Nanosci. **06**, 1–7 (2007). doi:10.1142/S0219581X07004304
5. M.S. Arnold, A.A. Green, J.F. Hulvat, S.I. Stupp, M.C. Hersam, Sorting carbon nanotubes by electronic structure using density differentiation. Nat. Nanotechnol. **1**, 60–65 (2006)
6. P. Zhao, E. Einarsson, R. Xiang, Y. Murakami, S. Maruyama, Controllable expansion of single-walled carbon nanotube dispersions using density gradient ultracentrifugation. J. Phys. Chem. C **114**(11), 4831–4834 (2010)
7. M. Zheng, A. Jagota, E.D. Semke, B.A. Diner, R.S. McLean, S.R. Lustig, R.E. Richardson, N.G. Tassi, DNA-assisted dispersion and separation of carbon nanotubes. Nat. Mater. **2**(5), 338–342 (2003)
8. B.I. Kharisov, O.V. Kharissova, H. Leija Gitierrez, U. Ortiz Méndez, Recent advances on the soluble carbon nanotubes. Ind. Eng. Chem. Res. **48**(2), 572–590 (2009)
9. I. Schwyzer, R. Kaegi, L. Sigg, A. Magrez, B. Nowack, Influence of the initial state of carbon nanotubes on their colloidal stability under natural conditions. Environ. Pollut. **159**(6), 1641–1648 (2011)
10. D.W. Chang, I.-Y. Jeon, J.-B. Baek, L. Dai, Efficient dispersion of single walled carbon nanotubes by novel amphiphilic dendrimers in water and substitution of the pre-adsorbed dendrimers with conventional surfactants and lipids. Chem. Commun. **46**(42), 7924–7926 (2010)
11. C. Thauvin, A. Perino, E. Contal, E. Morin, P. Schultz, S. Meunier, A. Wagner, Programmed dispersions of MWNTs in aqueous media by coating with photopolymerizable synthetic amphiphiles. J. Phys. Chem. C **115**(15), 7319–7322 (2011)
12. E.E. Tkalya, M. Ghislandi, G. de With, C.E. Koning, The use of surfactants for dispersing carbon nanotubes and graphene to make conductive nanocomposites. Curr. Opin. Colloid Interf. Sci. **17**(4), 225–232 (2012)
13. S. Detriche, J.B. Nagy, Z. Mekhalif, J. Delhalle, Surface state of carbon nanotubes and Hansen solubility parameters. J. Nanosci. Nanotech. **9**(10), 6015–6025 (2009)
14. P.-C. Ma, N.A. Siddiqui, G. Marom, J.-K. Kim, Dispersion and functionalization of carbon nanotubes for polymer-based nanocomposites: A review. Compos. Part A **41**, 1345–1367 (2010)
15. M. Khazaee, W. Xia, G. Lackner, et al., Dispersibility of vapor phase oxygen and nitrogen functionalized multi-walled carbon nanotubes in various organic solvents. Sci Rep **6**, 26208, 10 pp (2016)

Index

O.V. Kharissova, B.I. Kharisov, *Solubilization and Dispersion of Carbon Nanotubes*, DOI 10.1007/978-3-319-62950-6

MIX
Papier aus verantwortungsvollen Quellen
Paper from responsible sources
FSC® C105338

If you have any concerns about our products,
you can contact us on
ProductSafety@springernature.com

In case Publisher is established outside the EU,
the EU authorized representative is:
Springer Nature Customer Service Center GmbH
Europaplatz 3, 69115 Heidelberg, Germany

Printed by Libri Plureos GmbH
in Hamburg, Germany